Springer Series on Polymer and Composite Materials

The "Springer Series on Polymer and Composite Materials" publishes monographs and edited works in the areas of Polymer Science and Composite Materials. These compound classes form the basis for the development of many new materials for various applications. The series covers biomaterials, nanomaterials, polymeric nanofibers, and electrospun materials, polymer hybrids, composite materials from macro- to nano-scale, and many more; from fundamentals, over the synthesis and development of the new materials, to their applications. The authored or edited books in this series address researchers and professionals, academic and industrial chemists involved in the areas of Polymer Science and the development of new Materials. They cover aspects such as the chemistry, physics, characterization, and material science of Polymers, and Polymer and Composite Materials. The books in this series can serve a growing demand for concise and comprehensive treatments of specific topics in this rapidly growing field. The series will be interesting for researchers working in this field and cover the latest advances in polymers and composite materials. Potential topics include, but are not limited to:

Fibers and Polymers:

- Lignocellulosic biomass and natural fibers
- Polymer nanofibers
- Polysaccharides and their derivatives
- Conducting polymers
- Surface functionalization of polymers
- Bio-inspired and stimuli-responsive polymers
- Shape-memory and self-healing polymers
- Hydrogels
- Rubber
- Polymeric foams
- Biodegradation and recycling of polymers

Bio- and Nano- Composites:

- Fiber-reinforced composites including both long and short fibers
- Wood-based composites
- Polymer blends
- Hybrid materials (organic-inorganic)
- Nanocomposite hydrogels
- Mechanical behavior of composites
- The Interface and Interphase in polymer composites
- Biodegradation and recycling of polymer composites
- Applications of composite materials

San-Lang Wang • Masakazu Anpo
Editors

Chitin and Chitosan

Springer

Editors
San-Lang Wang
Tamkang University
New Taipei, Taiwan

Masakazu Anpo
Osaka Metropolitan University
Sakai, Japan

ISSN 2364-1878 ISSN 2364-1886 (electronic)
Springer Series on Polymer and Composite Materials
ISBN 978-981-95-7456-8 ISBN 978-981-95-7457-5 (eBook)
https://doi.org/10.1007/978-981-95-7457-5

This Springer imprint is published by the registered company Springer Nature Singapore Pte Ltd.
The registered company address is: 152 Beach Road, #21-01/04 Gateway East, Singapore 189721, Singapore

Preface

Chitin and its derivative chitosan, which are versatile materials mainly obtained from shrimp or crab shells, squid pens, and the cell walls of fungi, ranked second only to cellulose in abundance among biopolymers. They offer a unique combination of biocompatibility, biodegradability, and numerous functional properties that make them invaluable across diverse applications, from agricultural, food, environmental, and biomedical fields to pharmaceuticals.

Chitin and chitosan have attracted considerable attention due to their adaptability and multifaceted properties. Research societies for chitin and chitosan have also been established, such as the Japanese Society for Chitin and Chitosan, the Taiwan Society for Chitin and Chitosan, and the European Chitin Society (EUCHIS). Many international symposia on chitin and chitosan are also held regularly, such as the International Conference on Chitin and Chitosan (ICCC), Asia Pacific Chitin and Chitosan Symposium (APCCS), and EUCHIS conferences. We have, thus, had the privilege of observing the phenomenal development of chitin and chitosan research over many years. What began as a niche area of academic interest has blossomed into a highly productive, multidisciplinary field with considerable industrial potential. This development has been driven by the focused efforts of devoted researchers worldwide, thus further expanding the unlimited usages for these natural polymers.

Written by fifteen experts, as introduced below, this book is a major contribution to the biopolymer field, serving as a guide to microbial reuse and regeneration of chitin-containing fishery by-products (Chaps. 1 and 2), the modification of chitin and chitosan derivatives as well as their properties (Chaps. 3, 4, and 5), and their various applications (Chaps. 6, 7, 8, 9, 10, 11, 12, 13, 14, and 15). Each chapter not only covers the fundamental research carried out but also presents the various topics and perspectives related to these biomasses. A highly academic as well as practical endeavor, we believe this book will be a definitive reference for chitin and chitosan research, stimulating interest in this field among future generations.

New Taipei Taiwan
Sakai Japan
November 20th, 2025

San-Lang Wang
Masakazu Anpo

Contents

Utilization of Chitinous Fishery Wastes for Enzyme Production and Their Application 1
Chien Thang Doan, Thi Ngoc Tran, Anh Dzung Nguyen, and San-Lang Wang

Production and Application of Bioactive Compounds from Chitinous Materials by Microbial Fermentation 21
Thi Hanh Nguyen, San-Lang Wang, Anh Dzung Nguyen, and Van Bon Nguyen

Molecular Mechanisms of Chitosan-Based Hydrogel Formation: From Carbamate Chemistry to Polyelectrolyte Complexation and Nanocomposite Design 41
Hiroshi Tamura and Tetsuya Furuike

Chemical Synthesis of Chito-Hetero-Oligosaccharides 63
Yuko Yoneda and Toshinari Kawada

Utilization of Azide–Alkyne Cycloaddition, "Click Chemistry" for the Synthesis of Chitosan Derivatives and Conjugates 85
Sankar Rathinam, Aathira Pradeep, Maonian Xu, Rangasamy Jayakumar, and Már Másson

Veterinary Clinical Applications of Chitin and Chitosan 103
Yoshiharu Okamoto

Chitosan Nanoparticle Systems and Biocomposites for Nutraceutical Delivery 133
Jayachandran Venkatesan, Neema Kumari, Pandurang Appana Dalavi, Kranti Kiran Reddy Ealla, and Se-Kwon Kim

Chitosan-Based Biomedical Hydrogels 149
Lihua Li, Lu Lu, Changren Zhou, Yifan Zhang, Zexiang Zheng, Xinshuo Liu, and Junxiao Xiang

Chitosan Hydrogels in Tissue Engineering and Regenerative Medicine 171
Sona Sunil and Jayakumar Rangasamy

Chitosan in Periodontal Healthcare 195
Wanpen Tachaboonyakiat

Nanomaterials Based on Chitosan as Potential Drug Delivery Carriers .. 211
Anh Dzung Nguyen, Van Bon Nguyen, Vu Hoa Tran, Chien Thang Doan, and San-Lang Wang

Thiolated Chitosan: A Versatile Mucoadhesive Polymer for Enhanced Mucosal Drug Delivery 237
Manasa N, Swornahuti Panda, Jerifa Sonowal, Sudarshan Naidu Chilamakuri, and Subramanian Natesan

Application of Chitosan in Textile Science and Engineering 259
Ying Zhang, Mei-Chen Lin, Ching-Wen Lou, Ting-Ting Li, and Jia-Horng Lin

Application of Chitosan in Membranes 281
Wei-Song Hung

Treatment of Chitin Production Wastewater Containing High Salinity from Shrimp Scraps with By-Product Chitosan, Chemical Coagulants and Biological Processes 293
Tan Phong Nguyen and Nguyen Van Nhi Tran

Utilization of Chitinous Fishery Wastes for Enzyme Production and Their Application

Chien Thang Doan, Thi Ngoc Tran, Anh Dzung Nguyen, and San-Lang Wang

Abstract Chitinous fishery wastes, including shrimp shells, crab shells, and squid pens, are abundant by-products rich in protein and chitin. Therefore, this waste source is an ideal substrate for microbial fermentation to produce high-value products such as enzymes. As highly specific biocatalysts operating under mild conditions, enzymes have become indispensable tools across diverse industrial sectors. This chapter discusses microbial enzymes derived from chitinous fishery wastes, including chitinases, chitosanases, and proteases, highlighting their important applications.

Keywords Chitinase · Chitinous fishery waste · Chitosanase · Crab shells · Enzyme · Protease · Shrimp shells · Squid pens

C. T. Doan · T. N. Tran
Faculty of Natural Sciences and Technology, Tay Nguyen University, Buôn Ma Thuột, Dak Lak, Vietnam
e-mail: dcthang@ttn.edu.vn; ttngoc@ttn.edu.vn

A. D. Nguyen
Institute of Biotechnology and Environment, Tay Nguyen University, Buôn Ma Thuột, Dak Lak, Vietnam
e-mail: nadzung@ttn.edu.vn

S.-L. Wang (✉)
Department of Chemistry, Tamkang University, New Taipei City, Taiwan
e-mail: sabulo@mail.tku.edu.tw

S.-L. Wang, M. Anpo (eds.), *Chitin and Chitosan*, Springer Series on Polymer and Composite Materials, https://doi.org/10.1007/978-981-95-7457-5_1

1 Introduction

According to Food and Agriculture Organization (2024), the total production of aquatic animals has surged over the decades, rising from 19 million tons in 1950 to an all-time high of over 185 million tons in 2022, with an average annual growth rate of 3.2 percent [1]. Consequently, vast amounts of waste are generated while processing these aquatic resources. It is estimated that fishery by-products can account for up to 60–80% of the initial live weight [2], leading to a global waste stream of millions of tons annually. This issue highlights a significant global challenge regarding effective waste management for the seafood industry. Traditional disposal methods, such as landfilling, consume valuable land and pose a considerable groundwater and soil pollution risk. At the same time, conversion into low-value animal feed fails to capture the immense biochemical potential locked within this resource.

Chitinous fishery wastes, including squid pens, shrimp, and crab shells (Fig. 1), are particularly interesting among the diverse types of fishery wastes. Globally, it is estimated that around 3.8 million tons of shrimp shells and 6–8 million tons of crab shells are released per year [3, 4]. From a scientific perspective, this huge waste mass is not a liability but an untapped reservoir of valuable biopolymers. A typical chemical analysis of shrimp shell waste reveals a composition rich in valuable organic compounds, such as protein and chitin [3]. Chitin, a structural polysaccharide, is the second most abundant biopolymer on Earth after cellulose and serves as the primary raw material for producing high-value derivatives, such as chitosan and chito-oligosaccharides [5]. Therefore, the conventional disposal of crustacean shells represents an environmental burden and a significant economic loss of a valuable and renewable resource.

In light of this situation, microbial fermentation has emerged as a promising and sustainable approach to valorize chitinous fishery wastes. Utilizing chitinous fishery by-products as the substrate for enzyme production offers a dual advantage that

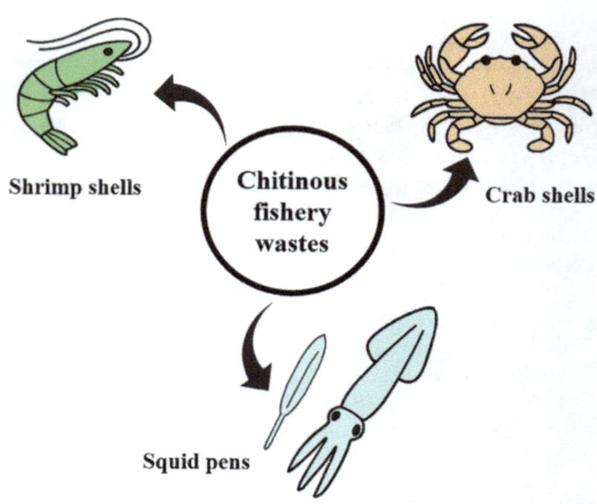

Fig. 1 Types of chitinous fishery wastes

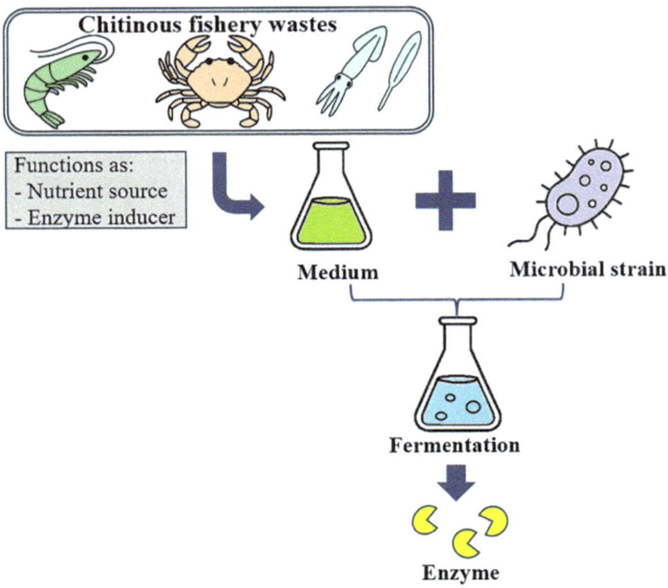

Fig. 2 Bioconversion of chitinous fishery wastes for enzyme production

Table 1 Benefits of utilizing fishery wastes for enzyme production

Aspect	Description
Economic benefits	Businesses acquire free raw material (fishery waste). Turn a negative cost (waste disposal) into positive profits. Reduces the cost of supplemental nutrients, as chitinous fishery waste alone can supply essential nutrients for microbial growth.
Environmental benefits	Reduces pollution from millions of tons of fishery waste generated annually. Avoids harsh acid treatment of chitin, thereby reducing toxic chemical wastewater.

makes it exceptionally efficient. First, the waste is a nearly complete nutrient medium, providing a carbon/nitrogen source (from chitin and protein) and essential trace minerals required for microbial growth. Second, the polymers within chitinous fishery waste function as potent and natural inducers for enzyme synthesis (Fig. 2). Chitin strongly stimulates microorganisms to secrete chitinases and chitosanases, while the abundant proteins induce the biosynthesis of proteases. This in situ induction mechanism eliminates the need for expensive, fine inducers and supplementary nutrients (such as fermentable sugars, peptone, yeast extract, or beef extract) in the fermentation medium. Notably, 30–40% of the enzyme production cost comes from culture medium [6]. Thus, using chitinous fishery wastes to produce microbial enzymes can achieve dual economic and environmental benefits (Table 1 and Fig. 3).

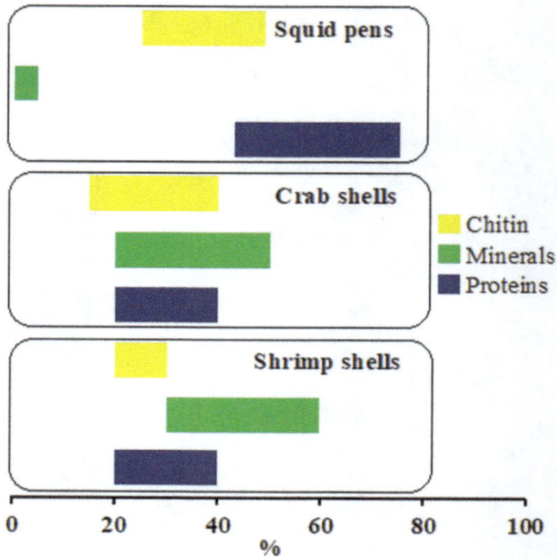

Fig. 3 Average composition (chitin, minerals, and proteins) of shrimp shells, crab shells, and squid pens

2 Chitinous Fishery Wastes: Types and Potential as Substrates for Microbial Enzyme Production

2.1 Shrimp Shells

Shrimp is one of the most widely consumed seafood commodities in the world. Global shrimp production continues to grow, reaching 5.6 million tons in 2023 and projected to reach 7.28 million tons by 2025. The strong growth of the shrimp processing industry also means generating a massive volume of by-products. This industry is estimated to discard approximately 3.8 million tons of waste annually. Depending on the species, size, and processing method, by-products can account for 30% to 70% of the total raw shrimp weight, with the head and shell alone constituting about 50% [3]. This presents both a significant environmental challenge and an abundant source of biological resources.

The chemical composition of dried shrimp shells is the deciding factor in their potential for reuse, comprising three main components:

– Protein (20–40%): Shrimp shells contain a remarkably high protein content, which is characterized by a predominant proportion of essential amino acids compared to non-essential ones [3]. This composition offers a superior advantage when utilizing shrimp waste as a fermentation substrate. The protein serves two critical functions. First, it is a rich and complex nitrogen source, providing the essential amino acids for microbial growth. Second, the protein acts as a natural inducer for the biosynthesis of proteolytic enzymes. Indeed, several studies have identified shrimp waste as the most efficient and potential carbon and nitrogen source for microbial protease production, outperforming other media. These fermentation processes use only the basal medium without adding other carbon or nitrogen sources, which reduces production costs [7–9].

- Minerals (30–60%): The mineral component is mainly calcium carbonate with a small amount of calcium phosphate. The mineral salts in shrimp shells play a dual role in fermentation. On one hand, these mineral salts can serve as a source of essential ions (such as Ca^{2+}) for microbial growth. This inherent mineral content often eliminates the need for additional Ca^{2+} supplementation in the culture medium. On the other hand, the high mineral content is primarily viewed as a significant challenge. It provides a complex, rigid structure to the shrimp shell, creating a physical barrier that severely limits the ability of microorganisms to access the valuable chitin and protein locked within the matrix. To overcome this issue, a demineralization step is often considered. This pretreatment can be performed chemically using acids (such as HCl) or through a greener, biological approach using fermentation with bacteria. For instance, *Pseudomonas aeruginosa* K-187 exhibits better chitinase biosynthesis on a medium containing shrimp and crab shell powder pre-treated with HCl and NaOH [10].
- Chitin (20–30%): Chitin is the crustacean shells' most valuable target component. It exists as α-chitin, the most common and structurally stable allomorph [3]. In microbial fermentation, chitin plays a dual role: it serves as the primary carbon source for microbial growth and, crucially, acts as an indispensable inducer for the biosynthesis of chitinolytic enzymes. Shrimp shells are also one of the primary sources used to prepare chitin. However, isolating chitin from its natural matrix is a significant challenge. The conventional chemical extraction process is notoriously wasteful, costly, and environmentally damaging. Large volumes of harsh chemicals, typically hydrochloric acid (HCl) for demineralization and sodium hydroxide (NaOH) for deproteinization, are required to remove the minerals and protein [2]. This chemical-intensive process generates a massive pollution load, which is exceptionally difficult and expensive to treat. New eco-friendly technologies have been explored in response to these drawbacks, especially biotechnological routes such as fermentation and enzyme catalysis. While promising, these processes can also have limitations, such as being expensive, time-consuming, and challenging to operate continuously on a large scale, sometimes requiring supplementary chemical steps [8]. This reality highlights the potential of an integrated approach, which is a central theme of this chapter. Two key objectives can be achieved simultaneously by combining the fermentation for protease production directly with the protein removal step. The protease produced in situ by the microorganisms carries out the deproteinization of the shells, replacing the harsh NaOH treatment and creating a synergistic process economically and environmentally superior to traditional methods.

2.2 Crab Shells

The global processing of crabs generates a massive waste stream, with estimates placing the annual volume of discarded crab shells between 6 and 8 million tons [4]. This high volume is mainly due to the fact that a significant portion of the crab's body is inedible, with the edible meat accounting for only about 40% of the total body weight.

When considered a substrate for fermentation, the composition of crab shells—approximately 20–40% protein, 20–50% mineral, and 15–40% chitin—presents a unique set of challenges and opportunities [11]. The high mineral content of crab shells is a critical issue affecting their viability in fermentation processes, as it creates a formidable physical barrier that hinders the ability of microorganisms to access and utilize the nutrients within the shell. Consequently, a demineralization pre-treatment step is often considered, and some studies indicate that fermentation efficiency for enzyme production is indeed higher when using demineralized crab shells [10, 12]. Despite these challenges, crab shells contain significant amounts of chitin and protein, making them a viable substrate for producing valuable enzymes, namely chitinases/chitosanases and proteases. Several microorganisms have been identified as effective candidates for fermenting crab shell waste to produce enzymes, such as *Bacillus*, *Paenibacillus*, and *Pseudomonas* [10, 12–14].

However, a more innovative approach involves integrating the production of enzymes directly with the process of chitin extraction. This concept is highly logical, as protease-producing bacteria can efficiently hydrolyze the shell's protein matrix, accomplishing a biological deproteinization step. In one study, crab shell waste was fermented using six different protease-producing *Bacillus* species (*B. subtilis* A26, *B. mojavensis* A21, *B. pumilus* A1, *B. amyloliquefaciens* An6, *B. licheniformis* NH1, and *B. cereus* BG1) to produce both chitin and a nutrient-rich fermented-crab supernatant. Remarkably, when cultured in a medium containing only crab shells as the substrate, fermentation achieved a demineralization rate of 80–90% and a deproteinization rate of 80–95% [14]. This demonstrates a highly efficient, multi-objective process where deproteinization, demineralization, and the production of a valuable enzyme-rich supernatant can be achieved simultaneously.

2.3 Squid Pens

In 2020, global seafood production reached 214 million tons. Within this massive industry, the squid fishery is a significant contributor, representing approximately 4.3% of all marine catches by volume and a notable 7% by value worldwide. However, a substantial portion of the catch becomes waste during processing. It is estimated that more than 40% of a squid's total body weight is discarded as by-products, which include the viscera, skin, and the pen. The squid pen, or the gladius, is an internalized shell that accounts for about 0.2% of the body's total weight. It serves a crucial structural role, as an attachment point for major muscle groups and a protective barrier for the visceral organs [15].

By-products from the squid processing industry present a unique and interesting case. It is essential to distinguish between two different parts often referred to as "cuttlebone": the cuttlebone proper from cuttlefish, and the gladius (or "squid pen") from squid. Cuttlebone is primarily composed of aragonite-form calcium carbonate with only a minor amount of chitin, making it an unsuitable raw material for enzyme production. However, cuttlebone is found in several food products and traditional Eastern medicine. In contrast, the squid pen is an extremely valuable and unique source of chitin. Its composition is approximately 25–49% chitin and 43–75% protein, with only trace amounts of mineral salts (0.03–1.9%) [16]. The high content of both protein and chitin makes the squid pen an ideal substrate for the fermentative production of proteases, chitinases, and chitosanases.

The most significant feature of the squid pen is that its chitin exists in the β-chitin form. This significantly contrasts with the α-chitin found in shrimp and crab shells. The structural difference between α-chitin and β-chitin has important implications for their use as fermentation substrates. In α-chitin, the polymer chains are arranged in an antiparallel fashion, which allows for extensive and powerful intermolecular hydrogen bonding. This creates a highly compact, crystalline, and stable structure that is highly resistant to chemical and enzymatic degradation. Conversely, β-chitin features a parallel arrangement of its polymer chains. This structural organization results in significantly weaker intermolecular hydrogen bonds, leading to a looser, more hydrated, and more accessible structure. This higher reactivity allows solvents and microbial enzymes to penetrate the polymer matrix with much greater ease, thereby dramatically increasing its biodegradability. Consequently, this enhanced substrate bioavailability often translates to superior microbial performance. It is a well-established principle that enzyme yield highly depends on the specific substrate's composition and form. Therefore, it is frequently observed that many microbes exhibit significantly higher chitinolytic activity when cultured on a medium containing β-chitin from squid pens compared to a medium containing the more recalcitrant α-chitin from crustacean shells. Furthermore, the negligible mineral content is a significant advantage. The costly, time-consuming, and polluting demineralization step, mandatory for crustacean shells, can be greatly simplified or eliminated before the squid pens are used in fermentation. Indeed, squid pens are effective substrates for the production of both chitinase, chitosanase, and protease [2, 5, 6, 17].

Despite these superior chemical properties, the widespread use of squid pens in industrial fermentation is limited by practical challenges. The supply is less abundant or common than shrimp and crab processing waste. Additionally, collecting and separating the pens during processing can be more complex. These logistical issues currently constrain the large-scale application of this otherwise ideal substrate.

3 Production and Application of Target Enzymes

3.1 Chitinase/Chitosanase

3.1.1 Production

Chitinase (EC 3.2.1.14) is a group of hydrolytic enzymes that break the β-1,4-glycosidic bonds within the chitin polymer chain to release low-molecular-weight products. These enzymes are categorized into two main groups based on their mechanism of action: exochitinases and endochitinases. Endochitinases randomly cleave chitin at internal sites, while exochitinases act at the terminal ends of chitin oligosaccharides. Exochitinases are further divided into chitobiosidases, which liberate chitobiose ($(GlcNAc)_2$), and end N-acetyl-β-glucosaminidases, which release N-acetyl-glucosamine (GlcNAc). A broad spectrum of organisms, including viruses, bacteria, insects, fungi, plants, and mammals, can produce chitinase [5]. Among these, microbial chitinases hold significant industrial value, driving considerable interest in optimizing their production processes. Notably, bacteria belonging to the genera *Paenibacillus, Bacillus, Streptomyces, Serratia*, and *Pseudomonas* have demonstrated the ability to synthesize chitinase on media containing chitinous fishery wastes such as shrimp shells, crab shells, and squid pens [2].

Chitosanase (EC 3.2.1.132), a glycosidic hydrolase, can endo-hydrolyze the β-1,4-glycosidic bonds of partially acetylated chitosan from the reducing end. Chitosanases are found in diverse sources, including bacteria, fungi, viruses, and some plants [6]. Similar to chitinase, many bacterial strains have been found to synthesize chitosanase on media containing chitinous fishery wastes [2]. The discovery of microbial strains capable of synthesizing chitinase and chitosanase using chitinous fishery wastes is a crucial prerequisite for applying these waste materials in the industrial production of microbial chitinase and chitosanase.

In the enzyme production process, the choice of substrate is a decisive factor for both technical efficiency and economic viability. For chitinase production, the most natural and effective inducer is chitin itself [13]. However, using purified chitin versus utilizing chitinous fishery wastes represents two vastly different strategies, each with distinct advantages and disadvantages. In theory, purified chitin is an ideal substrate as it directly provides the inducer for the microorganism. In practice, however, its use on an industrial scale faces several significant hurdles. Chitin is a polymer with a highly stable crystalline structure (α-chitin form), rendering it insoluble in water and very difficult to degrade. For microorganisms to access and utilize it, the chitin must often be processed into an amorphous, suspended form known as "colloidal chitin" typically using strong acids like concentrated HCl. This energy-intensive process generates a large volume of acidic wastewater from the washing stage, requiring costly treatment and posing a potential environmental hazard.

Furthermore, purified chitin is a commercial chemical with a considerable price tag. Using it as the primary substrate would drive the production cost of chitinase to a very high level, thereby reducing the product's market competitiveness. This is a common issue in enzyme production, where replacing expensive pure substrates with low-cost chitinous fishery waste is a key strategy for economic viability. Additionally, purified chitin primarily supplies a carbon source and an inducer. For microorganisms to grow and thrive, the culture medium must still be supplemented with other organic nutrients, especially a nitrogen source (such as peptone and yeast extract), which are expensive components.

Utilizing chitinous fishery waste effectively overcomes most disadvantages, transforming a costly process into an economical and sustainable solution. Fishery waste can be considered a nearly complete fermentation medium. It simultaneously provides an inducer and carbon source from the chitin within the substrate, a nitrogen source from the abundant protein content and trace minerals. This approach helps eliminate or minimize the need for expensive purified nutrients, simplifying the medium's composition and significantly reducing production costs. Moreover, by providing comprehensive nutrition and inducers, microorganisms' chitinase/chitosanase biosynthesis capacity on chitinous fishery waste media has been reported to be higher than purified chitin, for instance, *Paenibacillus* sp. TKU052 demonstrated 1.53 times higher chitinase biosynthesis capacity on crab shell medium than on purified chitin medium [13]. Using chitinous fishery waste as a substrate not only offers greater nutritional abundance but also leads to a shorter time to achieve maximum enzyme yield. *P. elgii* TKU051 could produce its highest chitosanase level in a medium containing squid pen powder on day three (0.779 U/mL), whereas it took up to 6 days when using a colloidal chitin medium (0.718 U/mL). In many cases, the richness of chitinous fishery waste is such that bacterial strains require only the waste material supplemented with a basic mineral salt medium (such as K_2HPO_4 and $MgSO_4$) for robust growth and enzyme production [5, 6, 10, 13].

The optimal type of chitinous fishery waste for chitinase/chitosanase fermentation varies depending on the microorganism. Generally, all three forms of chitinous fishery wastes (shrimp shells, crab shells, and squid pens) can be effectively utilized as substrates for chitinase/chitosanase biosynthesis. Some microorganisms, such as *Serratia marcescens*, effectively use shrimp shells for chitinase production [18]. For crab shells, *Paenibacillus* sp. TKU052 is an example [13]. Some *Paenibacillus* strains also efficiently synthesize chitosanase on media containing squid pens [6]. *S. thermocarboxydus* TKU045 could exhibit the highest chitinase activity on medium containing squid pen powder [5]. This highlights that identifying the appropriate type of chitinous fishery waste is necessary in the enzyme production process.

The pre-treatment of raw materials before the fermentation process is also worth considering. Generally, chitinous fishery waste is typically processed by drying and grinding into powder. Grinding the chitinous fishery waste increases the accessibility of microorganisms to the nutrients within the substrate, thereby enhancing microbial growth and enzyme biosynthesis. For squid pens as a substrate, no additional processing steps beyond drying and grinding are usually required. However, for shrimp and crab shells, due to their high mineral content, many chitinase/chitosanase production processes require an additional chemical pre-treatment step, specifically demineralization with acid (HCl) or deproteinization with alkali (NaOH).

Some studies also include a boiling step before drying and grinding, as seen in the fermentation of shrimp shells for chitinase production using *S. marcescens* WF [18].

Achieving maximum enzyme yield requires careful optimization of the fermentation conditions. The concentration of the chitinous fishery waste itself is a critical parameter that varies depending on the microorganism. To optimize the biosynthesis of chitinase/chitosanase, both classical and modern statistical methods have been applied. The classical approach, One-Factor-At-a-Time, involves changing one variable while keeping others constant. While simple, this method often fails to identify the interactive effects between parameters. Modern statistical techniques like Response Surface Methodology (RSM) are employed for more comprehensive optimization. RSM allows for the simultaneous study of multiple variables (e.g., substrate concentration, pH, temperature) and, crucially, captures the interactive effects between them to find optimal conditions. Thanks to RSM, optimal conditions for chitinase/chitosanase biosynthesis have been found for several microorganisms. For instance, using RSM, *P. elgii* TKU051 optimally synthesized chitosanase with a squid pen powder amount of 5.278% (w/v), an initial pH value of 8.93, an incubation temperature of 38 °C, and an incubation duration of 5.73 days [6]. Similarly, *S. marcescens* WF's optimal conditions for chitinase biosynthesis were dry shrimp shell powder at 6%, pH 6.5, and a temperature of 28 °C during fermentation for up to 72 h [18].

3.1.2 Application

The chitinolytic enzymes, primarily composed of chitinases and chitosanases, offer a broad spectrum of industrial applications, ranging from bulk biochemical production to high-value specialty products. These applications mainly revolve around the controlled enzymatic hydrolysis of chitin and chitosan. In this section, we focus on the applications of chitinases/chitosanases originating from the bioconversion of chitinous fishery wastes.

- Production of N-acetyl-D-glucosamine: N-acetyl-D-glucosamine, the monomeric building block of the chitin polymer, is a key target for complete hydrolysis. This process is typically achieved through the synergistic action of a chitinase enzyme cocktail. Initially, endochitinase randomly cleaves chitin chains into soluble chitin oligosaccharides. This crucial depolymerization significantly increases the number of chain ends available for subsequent enzymatic attack. Following this, exochitinases, particularly N-acetyl-β-glucosaminidases, act on these oligomers and their terminal ends to release the final GlcNAc. This two-step approach is considerably more efficient than using a single enzyme type, for instance, a combination of *Paenibacillus* sp. TKU052 chitinase and *S. speibonae* TKU048 N-acetyl-β-D-glucosaminidase demonstrated a synergistic effect, achieving over 90% efficiency in GlcNAc production from chitin within 12–14 h [13]. The resulting GlcNAc is a valuable chemical in its own right and can also serve as a fermentable sugar for producing bioethanol. Thus, to unlock the

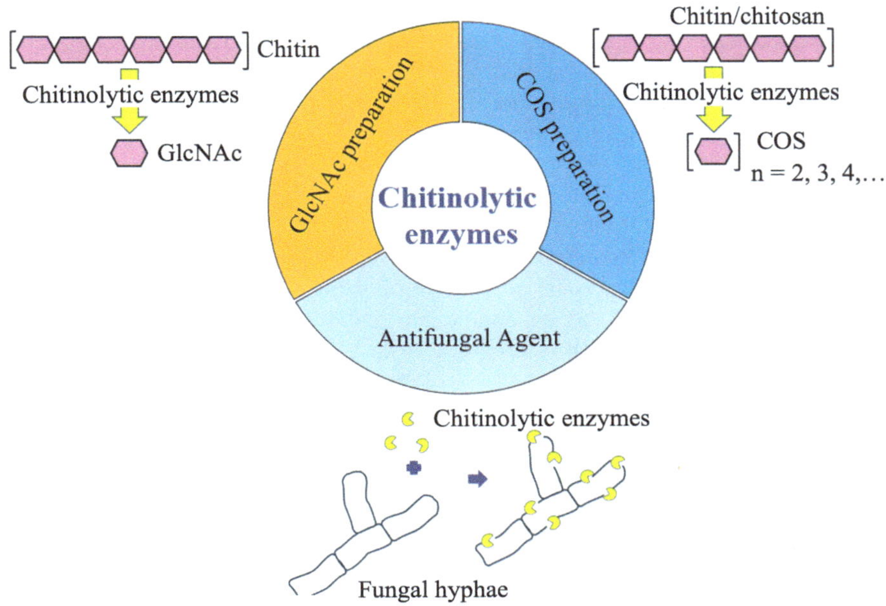

Fig. 4 Applications of microbial chitinolytic enzymes obtained from the conversion of chitinous fishery waste

potential applications of chitinases in bioethanol production, there is no other way to reduce the cost of enzymes hydrolyzing chitin to N-acetyl-D-glucosamine to a reasonable level. Microbial chitinases produced from chitinous fishery wastes are suitable in this context (Fig. 4).

– Production of bioactive chito-oligosaccharide (COS): One of the most valuable applications of chitinolytic enzymes is the production of bioactive COS. This process involves the controlled, partial hydrolysis of chitin or chitosan into short-chain oligomers (typically with a degree of polymerization (DD) from 2 to 20), rather than complete breakdown into monomers. While chemical acid hydrolysis is a standard method for producing commercial COS, it often yields a complex mixture of products that are difficult to purify and generate environmental pollutants. In contrast, enzymatic hydrolysis is a superior method as it is non-toxic, peculiar, and allows precise control over the reaction, eliminating the need for subsequent salt removal. The biological activities of COS, such as antioxidant and prebiotic effects, are critically dependent on their molecular structure, particularly their degree of polymerization and deacetylation. Therefore, selecting the suitable enzyme, substrate, and hydrolysis conditions makes it possible to produce tailored COS with desired functionalities. Microbial chitinolytic enzymes derived from fermenting fishery wastes have proven highly effective. For example, chitosanase from *P. elgii* TKU051, cultured on medium containing squid pen powder, demonstrated impressive effectiveness in hydrolyzing chitosan, yielding low-DP COS with a DP of 2 to 3 as the major product. The process is achieved under mild conditions—at a pH of 5.5, an incubation temperature of 58.3 °C, and a period of 6.8 h—resulting in a remarkable product yield of 690.587 mg/g (product/substrate). Notably, the antioxidant efficiency of the resulting COS surpasses that of commercially available options (food-grade COSs) and many other plant-derived antioxidants, making it an excellent choice for health and food applications [19]. In another study, chitinase from *S. thermocarboxydus* TKU045, cultured from a medium containing squid pen powder, showed the ability to hydrolyze chitin to produce COS with DP from 1 to 7 under optimal pH and temperature conditions of 4 and 60 °C (respectively). The obtained COS showed higher antioxidant activity than higher DP COS and simultaneously demonstrated prebiotic effects by stimulating the growth of *Lactobacillus lactis* [5].

– Antifungal agent for biocontrol: The antifungal properties of chitinases/chitosanases represent a significant application in agriculture. However, it should be noted that not all chitinases/chitosanases exhibit antifungal activity; thus, their antifungal properties must be thoroughly evaluated. The main reason chitinases/chitosanases can act as an antifungal agent is that the cell walls of most pathogenic fungi are high in chitin. Chitinases/chitosanases function as a mycolytic enzyme, breaking down this structural component. This process reduces cell wall integrity and osmotic imbalance, ultimately leading to the lysis of fungal hyphae. This mechanism allows chitinolytic enzymes to be used directly as an environmentally friendly bio-fungicide for plant protection, offering an alternative to synthetic chemical fungicides. In this respect, cost is a significant barrier to using chitinase as an antifungal agent. Therefore, a low-cost manufacturing process is imperative. Of course, using chitinous fishery waste as a cheap substrate for fermentation has received intense attention. Some bacterial strains, such as *Bacillus* and *Pseudomonas,* showed high antifungal chitinase biosynthesis ability in media with shrimp and crab shell powder as a carbon source [20]. Antifungal activity of chitosanase from *P. elgii* has also been reported. More specifically, chitosanase biosynthesis from the *P. elgii* TKU051 strain has been optimized using squid pens as the sole carbon/nitrogen source [6]. Thus, this demonstrates that producing antifungal preparations with chitinolytic enzymes from microbial fermentation of chitinous fishery wastes is entirely feasible.

3.2 *Protease*

3.2.1 Production

Proteases are among the most important and diverse groups of enzymes, playing a pivotal role in hydrolyzing peptide bonds (-CO-NH-) within protein molecules. This process breaks down protein chains into shorter peptides and, eventually, free amino acids. The importance of proteases is not limited to laboratory research. In fact, proteases are the most important group of industrial enzymes, accounting for more than 60% of the global enzyme market [17]. This dominance comes from their wide range of applications in key industries: detergents, food, leather, pharmaceuticals, and medical. With such huge market demand, developing efficient, low-cost, and sustainable protease production processes is a strategic goal. In this context, chitinous fishery wastes have emerged as a promising substrate source because of their low cost and unique chemical properties. This is a "negative cost" resource, as fishery processing factories often have to spend money to process them. Therefore, utilizing chitinous fishery wastes as fermentation substrates saves raw material costs and solves an environmental problem.

The most outstanding feature of fishery waste is that it is an almost complete nutrient source, providing both carbon and nitrogen sources for microorganisms. The nitrogen source is the most expensive component of microbial growth media [21]. Due to their rich carbon and nitrogen content, shrimp shells, squid pens, and crab shells can be used as the sole carbon and nitrogen source in the fermentation medium, eliminating the need to supplement other expensive nutrients such as peptone, yeast extract, or beef extract. This helps to significantly reduce production costs, which is one of the most significant cost drivers in industrial fermentation processes. The high protein content in the substrate is not only a nutrient source but also a potent inducer, stimulating microorganisms to increase extracellular protease biosynthesis. Some microbial strains include *B. parabrevis* TKU046, *P. elgii* TKU051, *Paenibacillus* sp. TKU052, *B. licheniformis* TKU004 can produce protease efficiently in medium containing only carbon and nitrogen sources from chitinous fishery wastes [7, 8, 12, 17]. Thus, the medium for cultivating these bacterial strains is relatively simple, including carbon/nitrogen sources from chitinous fishery waste and some mineral salts, such as K_2HPO_4 and $MgSO_4$. In another study, *B. parabrevis* TKU046 was cultivated in a medium containing only shrimp shells [8].

Microorganisms have demonstrated the ability to biosynthesize protease on various chitinous fishery wastes. Protease activity was affected by the nature of the strain and the type of chitinous fishery waste, for example, *Paenibacillus* sp. TKU047 biosynthesized protease best on shrimp heads, while *P. elgii* TKU051 biosynthesized protease best on squid pens [9, 17]. A pretreatment step is required for microorganisms to access and utilize this nutrient source efficiently. The most common pre-processing step is mechanical grinding of materials to form a fine powder. This is intended to break down the rigid physical structure of the shell and, most importantly, increase the exposed surface area, allowing microbial enzymes to attack and degrade the substrate more easily. However, some further processing is often applied for shrimp and crab shells, the most important of which is demineralization. Due to the high calcium carbonate content, a demineralization step with dilute acid (such as HCl) or biological fermentation is necessary to remove the mineral layer of shrimp shells or crab shells—*Paenibacillus* sp. TKU052 is an example of demineralized crab shells with the highest protease biosynthesis result (4.41 U/mL), higher than other chitinous fishery waste sources [12].

Optimizing protease biosynthesis conditions from fishery waste substrate sources is necessary to improve efficiency and reduce production costs. In one study, the strain *Paenibacillus* sp. TKU047 gave the highest protease activity (2.98 U/mL) with only 0.5% shrimp heads powder after 2 days of culture [9]. Notably, the combination of the Taguchi method and RSM in optimizing protease biosynthesis in shrimp by-product medium, simultaneously using three bacterial strains *Aeribacillus pallidus* VP3, *Lysinibacillus fusiformis* C250R, and *Anoxybacillus kamchatkensis* M1V, helped to obtain a total protease activity of up to 8182 U/mL [22].

3.2.2 Application

The application values of proteases are indisputable and have been commercialized globally. However, a notable advantage of proteases derived from the fermentation of chitinous fishery wastes is the ability to be directly integrated into treatment processes, creating a closed and sustainable production cycle. In this section, we focused on the applications of proteases derived from this specific source that have been studied.

- Biological deproteinization in the chitin preparation process: This is a strategic application and best demonstrates the circular economy philosophy. Chitin, a valuable polymer, is extracted mainly from the shells of shrimp and crabs. As analyzed, these materials contain a significant amount of protein. Hence, the deproteinization stage is mandatory and unavoidable to obtain chitin with high purity, meeting the requirements for biomedical and food applications. The traditional chemical method uses strong alkaline solutions, such as sodium hydroxide (NaOH), at high concentrations and temperatures. However, this method has inherent disadvantages: (1) it generates a large amount of alkaline wastewater that is polluting, requiring expensive neutralization costs; (2) Harsh processing conditions can alter the structure and reduce the molecular weight of chitin, reducing the quality of the final product. Besides, the chemical deproteinization generates wastewater with valuable compounds like chitooligosaccharides, peptides, and proteins. However, the high alkalinity makes recovery difficult, making these by-products often overlooked [3].

To overcome this problem, scientists have turned to a "greener method" of deproteinizing biologically. The most straightforward and practical approach is to ferment selected microbial strains (such as bacteria from the genera *Paenibacillus, Brevibacillus,* and *Bacillus*) directly on shrimp and crab shells [8, 14, 17]. Microorganisms secrete extracellular protease enzymes during the growth process to hydrolyze and "eat" the protein source in the shell. Thus, a single process can achieve two goals at the same time:

1. Raw material processing: Shrimp and crab shells are gently cleaned of protein, preserving the structure of chitin.
2. Product recovery: The fermentation broth becomes a source of crude protease enzyme, which can be recovered, concentrated, and used for other applications (Fig. 5).

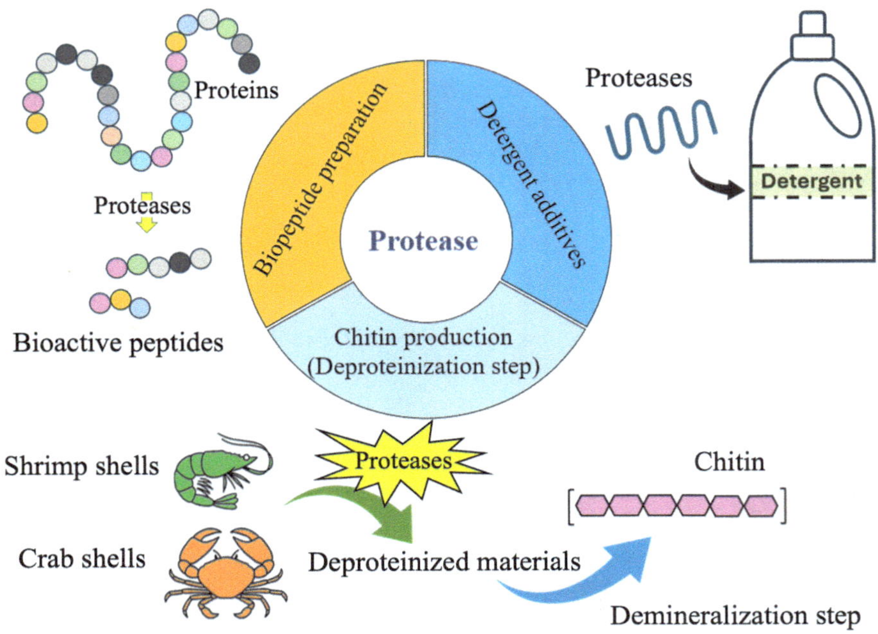

Fig. 5 Applications of microbial proteases obtained from the conversion of chitinous fishery waste

Several protease-producing microbial strains have been exploited for efficient deproteinization of chitinous fishery wastes. *B. parabrevis* TKU046, a strain that produces alkaline protease but not chitinolytic enzymes, has shown superior performance in liquid fermentation of shrimp shells, achieving a protein removal rate of 95% after 4 days. In particular, when processing cooked tiger shrimp shells, this strain removed up to 96.44% of the protein, leaving only 1.89% residue, while the recovered chitin had IR spectra similar to high-quality commercial chitin [8]. Similarly, *P. elgii* TKU051 demonstrated over 94% deproteinization of a variety of shrimp wastes (heads and shells, both fresh and cooked) after 7 days of fermentation, with only 1.51–2.68% protein residue [17]. This approach turns a costly processing step into a value-added manufacturing process, a perfect example of efficiency and sustainability.

- Preparation of bioactive peptides: Protease hydrolysis of proteins is the most efficient and well-controlled route to produce bioactive peptides. These low molecular weight amino acid fragments are easily absorbed, providing various health benefits. The widely studied bioactivities include antioxidant, prebiotic, Angiotensin-converting enzyme (ACE-inhibitory) inhibition, anti-bacterial, and anti-inflammatory properties [12]. The activity and efficacy of these peptides are highly dependent on the source of the protein, the specificity of the protease enzyme used, and the conditions of the hydrolysis process. In the context of the utilization of chitinous fishery waste, there are two main routes for the production of these peptides:

(1) Indirect route: Enzyme production first, hydrolysis later. In this approach, protease produced from fishery waste fermentation will be obtained (in crude or purified form), then used as a biological tool to hydrolyze other protein sources. The target protein source can be the protein fraction of fishery waste itself or other protein sources. For example, proteases from *B. licheniformis* TKU004, *P. macerans* TKU029, *P. mucilaginosus* TKU032, *Paenibacillus* sp. TKU042, *S. thermocarboxydus* TKU045, and *S. speibonae* TKU048 have been used to successfully hydrolyze by-products from fish, shrimp, crab, and squid, producing peptide-rich hydrolysates with higher antioxidant activity than the original raw materials [23]. Similarly, proteases from *Paenibacillus* sp. TKU052 has also been used to hydrolyze gelatin, producing antioxidant, ACE-inhibitory, and prebiotic activity peptides [12]. This route allows for reasonable control of the hydrolysis process but requires intermediate enzyme isolation and purification steps.
(2) Direct route: Co-production of peptides during fermentation. This is a more innovative, efficient, and integrated approach. Instead of isolating enzymes, the protein hydrolysis and peptide production occur in the fermenter. A typical study used *B. parabrevis* to ferment shrimp waste directly [8]. This process achieved three impressive simultaneous goals: (1) deproteinization of shrimp shells, facilitating the collection of chitin of high purity; (2) microorganisms synthesize and secrete protease into the medium; and (3) the fermented culture showed lactic acid bacteria growth-promoting activity (a form of prebiotic-like activity). Although this study did not analyze in detail which components of the fermentation broth contributed to the prebiotic activity, it is strongly hypothesized that the protease enzymes secreted by *B. parabrevis* hydrolyzed the abundant protein source in the shrimp shell, creating a "mixture" of peptides and amino acids. This mixture acted as a preferential, easily absorbed nutrient source, which allowed the lactic acid bacteria to thrive. In another study, PVQ9 (AVFPSIVGRPR), a peptide produced from the fermentation of shrimp waste using the protease-producing bacterium, *B. subtilis*, exhibited potent antibacterial activity against common foodborne Gram-positive bacteria, such as *Staphylococcus aureus*, *Bacillus cereus*, *Mammaliicoccus sciuri*, and *Kurthia gibsonii* [24].

This discovery has opened up a new and up-and-coming research and application direction: producing bioactive peptides directly from the fermentation process of waste without the need for intermediate enzyme purification steps. This "one-pot" approach will achieve multiple goals: simultaneously treating waste, purifying chitin, producing enzymes, and creating a high-value secondary product, a hydrolysate rich in active peptides. This significantly reduces operating costs and simplifies the process and effectively demonstrates the circular economy model in practice.

- Additives in the detergent industry: One of the most critical applications of proteases is in the detergent industry. Incorporating proteases into laundry powders, liquid detergents, dishwashing liquids, and contact lens cleaning solutions significantly enhances cleaning efficiency by degrading protein-based stains (such as blood, food residues, and bodily fluids). A detergent protease must meet three key criteria: (1) it should be alkaline to remain stable in high-pH conditions, (2) it must be thermostable to withstand high washing temperatures, and (3) it should resist surfactants and other harsh chemicals found in detergents. Interestingly, a protease was produced from the fermentation of shrimp head waste using *Paenibacillus* sp. TKU047 has demonstrated desirable properties. The protease was optimally active at pH 9, stable across pH 6–11, and thermotolerant with maximum activity at 70–80 °C. Remarkably, it retained 161% activity in the presence of sodium dodecyl sulfate and showed good compatibility with commercial detergents, maintaining over 70% activity with most formulations [9]. These features highlight TKU047 protease as a strong candidate for sustainable large-scale applications in the detergent industry.

In addition to the essential applications previously discussed, this document also examines other potential aspects of proteases obtained from the microbial fermentation of chitinous fishery waste:

- Anti-alpha glucosidase activity: During the screening of anti-α-glucosidase compounds, Doan et al. (2019) discovered a protease from *B. licheniformis* TKU004 that has a powerful ability to degrade yeast α-glucosidase (hydrolyzing more than 50% of α-glucosidase activity after 30 min and almost eliminating α-glucosidase activity after 3 h) [7]. The culture process also showed that the bacteria can produce this protease best in a medium containing shrimp heads powder (4.61 U/mL, 2 days). Due to its capacity to degrade yeast α-glucosidase, this protease may be used in biochemical and medicinal fields.
- Dye adsorbents from chitinous fishery wastes: Chitin is a material that can adsorb dyes. However, its cost makes chitin not an acceptable choice for treating dyes in wastewater. On the other hand, the chitin needed for dye treatment does not necessarily need to be high-purity chitin. In this case, chitin from the fermentation of chitinous fishery wastes with protease-producing bacteria strains is a viable alternative. Although this chitin may not be of high purity, it is cheaper. At the same time, the microbial activity during the fermentation process can give the chitin a suitable structure for the adsorption of dyes. In practice, the combination of protease and chitin production has been carried out with *P. elgii* TKU051 using shrimp wastes as substrates. The results showed that the remaining solid fraction produced chitin, and the protease-containing supernatant could isolate and purify enzymes. The obtained chitin was tested for its ability to absorb dyes. Notably, the results indicated that the effect of the obtained chitin was similar to that of commercial chitin in adsorbing Congo Red, with values of 90.48% and 90.91%, respectively [17].

4 Conclusion

Chitinous fishery wastes are inevitable products arising from the exploitation and processing of seafood, especially from shrimp, crab, and squid species. Instead of becoming an environmental burden, this protein- and chitin-rich waste source is being considered a potential substrate for creating high-value-added products. The production of microbial enzymes from chitinous fishery wastes has proven to be a multi-purpose solution, contributing to solving environmental problems and reducing treatment costs, while bringing economic benefits and opening up new technological opportunities, in line with the orientation of cleaner, greener production and moving towards the "zero waste" model. Extracellular enzymes such as chitinases, chitosanases, and proteases are produced effectively from these waste sources and demonstrate many outstanding applications. However, most of the current research is still limited to laboratory scale. Therefore, in the future, more in-depth research is needed to evaluate the efficiency on a larger scale, optimize costs, and build a circular and sustainable value chain from chitinous fishery wastes. Moreover, the ongoing screening and utilization of microbial strains that can produce enzymes from fishery waste containing chitin, with properties suitable for industrial-scale applications, should be continuously carried out.

References

1. FAO (2024) The state of world fisheries and aquaculture 2024: towards blue transformation. FAO, Rome
2. Wang C-H, Doan CT, Nguyen VB, Nguyen AD, Wang S-L (2019) Reclamation of fishery processing waste: a mini-review. Molecules 24:2234
3. Rossi N, Grosso C, Delerue-Matos C (2024) Shrimp waste upcycling: unveiling the potential of polysaccharides, proteins, carotenoids, and fatty acids with emphasis on extraction techniques and bioactive properties. Mar Drugs 22:153
4. Song X, Wei H, Zhou Y, Song W, Shi C, Mu C, Wang C, Wang X (2025) Utilization of crab shell waste for value-added bioplastics by *Pseudomonas*-based microbial cell factories. Int J Mol Sci 26:2543
5. Tran TN, Doan CT, Nguyen VB, Nguyen AD, Wang S-L (2019) The isolation of chitinase from *Streptomyces thermocarboxydus* and its application in the preparation of chitin oligomers. Res Chem Intermed 45:727–742
6. Doan CT, Tran TN, Tran TPH, Nguyen TT, Nguyen HK, Tran TKT, Vu BT, Trinh THT, Nguyen AD, Wang SL (2023) Chitosanase production from the liquid fermentation of squid pens waste by *Paenibacillus elgii*. Polymers 15:3724
7. Doan CT, Tran TN, Nguyen MT, Nguyen VB, Nguyen AD, Wang SL (2019) Anti-α-glucosidase activity by a protease from *Bacillus licheniformis*. Molecules 24:691
8. Doan CT, Tran TN, Nguyen VB, Nguyen AD, Wang SL (2019) Chitin extraction from shrimp waste by liquid fermentation using an alkaline protease-producing strain, *Brevibacillus parabrevis*. Int J Biol Macromol 131:706–715
9. Doan CT, Tran TN, Wen IH, Nguyen VB, Nguyen AD, Wang SL (2019) Conversion of shrimp head waste for production of a thermotolerant, detergent-stable, alkaline protease by *Paenibacillus* sp. Catalysts 9:798

10. Oh YS, Shieh IL, Tzeng YM, Wang SL (2000) Protease produced by *Pseudomonas aeruginosa* K-187 and its application in the deproteinization of shrimp and crab shell wastes. Enzym Microb Technol 27:3–10
11. Su W, Yu S, Wu D, Xia M, Wen Z, Yao Z, Tang J, Wu W (2019) A critical review of cast-off crab shell recycling from the perspective of functional and versatile biomaterials. Environ Sci Pollut Res 26:31581–31591
12. Doan CT, Tran TN, Nguyen VB, Nguyen AD, Wang SL (2020) Utilization of seafood processing by-products for production of proteases by *Paenibacillus* sp. TKU052 and their application in biopeptides' preparation. Mar Drugs 18:574
13. Doan CT, Tran TN, Wang SL (2021) Production of thermophilic chitinase by *Paenibacillus* sp. TKU052 by bioprocessing of chitinous fishery wastes and its application in N-acetyl-D-glucosamine production. Polymers 13:3048
14. Hajji S, Ghorbel-Bellaaj O, Younes I, Jellouli K, Nasri M (2015) Chitin extraction from crab shells by *Bacillus* bacteria: biological activities of fermented crab supernatants. Int J Biol Macromol 79:167–173
15. Nakasu PYS, Piccoli V, Ovejero-Pérez A, Kumar P, Al Ghatta A, Melanie S, Polesca C, Martínez L, Hallett JP (2025) Fractionation of squid pens with ionic liquids—an upgraded β-chitin and shellfish protein production. ACS Sustain Chem Eng 13:2649–2660
16. Messerli MA, Raihan MJ, Kobylkevich BM, Benson AC, Bruening KS, Shribak M, Rosenthal JJC, Sohn JJ (2019) Construction and composition of the squid pen from *Doryteuthis pealeii*. Biol Bull 237:1–15
17. Lee DH, Doan CT, Tran TN, Nguyen VB, Nguyen AD, Wang CL, Wang SL (2021) Proteases production and chitin preparation from the liquid fermentation of chitinous fishery by-products by *Paenibacillus elgii*. Mar Drugs 19:477
18. Mejía-Saulés JE, Waliszewski KN, García MA, Cruz-Camarillo R (2006) The use of crude shrimp shell powder for chitinase production by *Serratia marcescens* WF. Food Technol Biotechnol 44:95–100
19. Doan CT, Tran TN, Nguyen AD, Wang SL (2024) Enzymatic production of chitooligosaccharide using a GH Family 46 chitosanase from *Paenibacillus elgii* and its antioxidant activity. Catalysts 14:761
20. Chang WT, Chen CS, Wang SL (2003) An antifungal chitinase produced by *Bacillus cereus* with shrimp and crab shell powder as a carbon source. Curr Microbiol 47:102–108
21. Ye Y, Cai Y, Wang F, He Y, Yang Y, Guo Z, Liu M, Ren H, Wang S, Liu D et al (2025) Industrial microbial technologies for feed protein production from non-protein nitrogen. Microorganisms 13:742
22. Jabeur F, Mechri S, Kriaa M, Gharbi I, Bejaoui N, Sadok S, Jaouadi B (2020) Statistical experimental design optimization of microbial proteases production under co-culture conditions for chitin recovery from speckled shrimp *Metapenaeus monoceros* by-product. Biomed Res Int 2020:3707804
23. Tran TN, Doan CT, Nguyen VB, Nguyen AD, Wang SL (2022) Conversion of fishery waste to proteases by *Streptomyces speibonae* and their application in antioxidant preparation. Fishes 7:140
24. Tao W, Li W, Aweya JJ, Lin R, Jin R, Liang D, Ren Z, Shen Y (2025) *Bacillus subtilis*-fermented shrimp waste isolated peptide, PVQ9, and its antimicrobial mechanism on four Gram-positive foodborne bacteria. Food Microbiol 25:104654

Chien Thang Doan

Thi Ngoc Tran

Anh Dzung Nguyen

San-Lang Wang

Production and Application of Bioactive Compounds from Chitinous Materials by Microbial Fermentation

Thi Hanh Nguyen, San-Lang Wang, Anh Dzung Nguyen, and Van Bon Nguyen

Abstract Seafood processing generates millions of tons of chitin-rich byproducts annually, including shrimp heads and shells, crab shells, and squid pens. Traditionally underutilized as low-value animal feed or fertilizer, these materials are now recognized as abundant and sustainable resources for microbial fermentation. Rich in nutrients, chitinous marine byproducts provide an excellent substrate for microbial bioconversion, enabling the production of diverse bioactive compounds. Fermentation using strains such as *Pseudomonas aeruginosa*, *Serratia marcescens*, and *Paenibacillus* spp. has been shown to degrade chitinous structures and synthesize metabolites including phenazine derivatives, prodigiosin, nicotinic acid, homogentisic acid, and exopolysaccharides. These compounds exhibit wide-ranging bioactivities, such as antioxidant, anticancer, anti-inflammatory, antidiabetic, and anti-Alzheimer's effects in medicine, as well as insecticidal and anti-nematode properties in agriculture. This chapter provides an overview of the biochemical composition of major chitinous substrates, microbial strains, fermentation strategies employed for valorization, and the spectrum of bioactive compounds produced. Challenges related to large-scale implementation, strain improvement, and regulatory approval are highlighted, alongside future directions involving genetic engineering, synthetic biology, and techno-economic evaluation. Overall, microbial fermentation of chitinous byproducts represents a sustainable and cost-effective strategy to reduce environmental pollution for producing high-value products via fermentation technology.

Keywords Chitinous marine byproducts · Microbial fermentation · Waste valorization · Cost-effective strategy · Bioactive compounds · Bioactivities

T. H. Nguyen · S.-L. Wang
Department of Chemistry, Tamkang University, New Taipei City, Taiwan
e-mail: sabulo@mail.tku.edu.tw

A. D. Nguyen · V. B. Nguyen (✉)
Institute of Biotechnology and Environment, Tay Nguyen University, Buôn Ma Thuột, Dak Lak, Vietnam
e-mail: nvbon@ttn.edu.vn

S.-L. Wang, M. Anpo (eds.), *Chitin and Chitosan*, Springer Series on Polymer and Composite Materials, https://doi.org/10.1007/978-981-95-7457-5_2

1 Introduction

The increasing demand for sustainable biotechnological solutions has intensified global interest in valorizing agro-industrial and marine byproducts [1]. Chitinous marine byproducts, such as shrimp heads, shrimp shells, squid pens, and crab shells, have emerged as promising byproduct sources due to their abundance, and rich biochemical composition. Traditionally discarded or used for low-value purposes like animal feed or fertilizers, these materials are now considered valuable raw resources for microbial fermentation processes to generate high-value bioactive compounds. Chitinous marine byproducts are produced in vast quantities as side streams of seafood processing industries. For instance, shrimp processing alone contributes millions of tons of solid waste annually, primarily consisting of heads and shells. These byproducts are composed predominantly of chitin as well as significant amounts of protein, minerals, and other components [2]. Thus, all these chitinous marine byproducts offer a suitable platform for biotechnological conversion via microbial fermentation.

Microbial fermentation presents an eco-friendly and effective strategy for the sustainable bioconversion of chitinous marine byproducts, offering a dual advantage, addressing environmental pollution caused by seafood processing waste and generating high-value bioactive compounds [3]. Used microbial strains, such as *Pseudomonas aeruginosa*, *Serratia marcescens*, *Paenibacillus mucilaginosus*, and *Paenibacillus* sp., have been shown to degrade chitinous structures and synthesize a range of secondary metabolites with promising applications. Recent studies, including our work, have demonstrated that microbial fermentation of these chitinous marine byproducts can yield a broad spectrum of bioactive compounds such as phenazine, prodigiosin, exopolysaccharides, nicotinic acid, and homogentisic acid. These metabolites exhibit diverse biological activities in both medicine and agriculture [4–17].

This chapter explores the microbial fermentation of chitinous marine byproducts as a sustainable platform for bioactive compound production (Fig. 1). Overview of

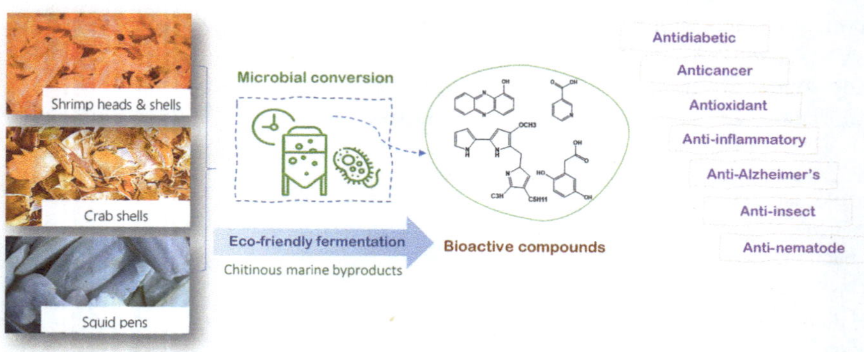

Fig. 1 Potential of chitinous marine byproducts for bioactive compounds production via microbial fermentation

major chitinous substrates and their biochemical profiles followed by a detailed discussion of microbial species, fermentation strategies, and bioconversion pathways. The latter sections focus on the types of bioactive compounds produced and potential applications. Insights into current challenges and future perspectives for industrial implementation are also presented. Special attention is given to findings from our research group to highlight the potential of microbial fermentation as a low-cost, high-impact solution for utilizing chitinous marine byproducts.

2 Chitinous Marine Byproducts: Sources and Characteristics

2.1 *Shrimp Heads and Shells*

Shrimp processing often produces a large amount of byproducts, mainly shrimp heads and shells. Shrimp heads are one of the main byproducts, accounting for about 35–45% of the total shrimp weight. With global shrimp production reaching millions of tons annually, shrimp heads byproducts are extremely large. Chemically, its composition (on a dry weight basis) includes 25–40% protein, 15–25% chitin, 20–25% minerals (mainly $CaCO_3$), and 5–10% lipids. Shrimp shells, including the body and tail, account for about 20–30% of the total shrimp weight. Similar to shrimp heads, the amount of shrimp shells discarded annually globally is also significant. Compared to shrimp heads, shrimp shells have a higher chitin content (25–40%) and a lower protein content (20–30%) on a dry weight basis. Meanwhile, the mineral content, mainly $CaCO_3$ and $Ca_3(PO_4)_2$, is higher, accounting for 30–50%. In addition, shrimp heads and shells contain valuable bioactive compounds such as astaxanthin, which have antioxidant, anticancer, anti-aging, and improve blood circulation and the immune system. Polyunsaturated fatty acids can support brain development and the immune system, reduce blood pressure and allergies, improve vision, prevent heart disease, and maintain cognitive health. In addition, many proteins in the byproduct are related to weight control and growth enhancement.

2.2 *Crab Shells*

Crab shells are a chitinous marine by-product of the seafood processing industry, usually accounting for about 60–65% of the crab weight. Regarding chemical composition, crab shells mainly consist of chitin (about 15–40%), protein (20–40%), and $CaCO_3$ accounts for about 20–50%. In addition, crab shells also contain a small amount of lipids, carotenoid pigments, and trace elements such as magnesium, phosphorus, sodium, and potassium. The minerals in crab shells are mainly $CaCO_3$, which exists in the form of calcite or aragonite crystals. In shrimp and crab shells, chitin is usually in the form of α-chitin, with a tightly arranged anti-parallel polymer chain structure that creates strong hydrogen bonds and high crystallinity. So, α-chitin has good mechanical strength and high stability.

2.3 Squid Pens

Squid pens, also known as gladius, are the long, hard endoskeleton structures of squid, accounting for about 1% of the body weight, and are also a major by-product of the squid processing industry. Chemical analysis confirmed that it consists of 30–40% chitin, 60–70% protein, and a small amount of mineral and lipid (less than 1–2%). The most prominent feature of squid pens is that they contain β-chitin, which possesses many superior structural and functional characteristics compared to α-chitin (commonly found in shrimp and crab shells). The structure of β-chitin has a parallel arrangement between polymer chains, creating a looser and more open crystalline structure, with fewer intra- and inter-molecular hydrogen bonds. This structure makes β-chitin more flexible and susceptible to enzymatic degradation. Therefore, β-chitin is an ideal raw material for microbial fermentation, without requiring strong chemical treatment like α-chitin. In addition, β-chitin also contains fewer inorganic minerals and fats, which helps reduce pretreatment costs and increase biodegradation efficiency.

In summary, chitinous marine byproducts vary significantly in their chemical composition and structural features, influencing their suitability for microbial bioconversion (Table 1 summarized chemical composition of these chitinous marine byproducts). While shrimp and crab shells are abundant sources of α-chitin with high mechanical strength, squid pens provide β-chitin with superior enzymatic accessibility and minimal pretreatment requirements. Understanding these substrate-specific characteristics is essential for selecting appropriate microbial strains and optimizing fermentation strategies to produce high-value bioactive compounds.

Table 1 The chemical composition of chitinous marine byproducts

Byproduct	Chitin (%)	Protein (%)	Minerals (%)	Other components
Shrimp head	15–25 (α-chitin)	25–40	20–25 (Mainly $CaCO_3$)	5–10% lipids, astaxanthin, polyunsaturated fatty acids, functional proteins
Shrimp shell	25–40 (α-chitin)	20–30	30–50 (Mainly $CaCO_3$ & $ca_3(PO_4)_2$)	Astaxanthin, polyunsaturated fatty acids, functional proteins
Crab shell	15–40 (α-chitin)	20–40	20–50 (Mainly $CaCO_3$)	Carotenoids, minor lipids, trace minerals (Mg, P, Na, K)
Squid pen	30–40 (β-chitin)	60–70	<1–2	Low lipid content

3 Microorganisms Involved and Their Metabolic Capacity for Chitinous Marine Byproducts Bioconversion

Marine by-products containing chitin, such as shrimp heads, shrimp shells, squid pens, and crab shells, can be metabolized by some bacteria through bio-fermentation. Our research group has also utilized these chitin-rich materials as a nutrient source for the fermentation of some bacterial strains, such as *Pseudomonas aeruginosa*, *Serratia marcescens*, *Paenibacillus mucilaginosus*, and *Paenibacillus* sp., and successfully synthesized various bioactive compounds. Microbial strains belonging to *Pseudomonas*, *Serratia*, and *Paenibacillus* species are widely distributed in soil, water, and marine environments and are recognized for their metabolic versatility and ability to utilize a wide range of organic substrates. Its potent adaptability is attributed to the production of diverse extracellular enzymes and secondary metabolites that enable it to thrive in diverse environments. In our studies, some strains have been successfully used to ferment chitin-rich marine byproducts, such as *Pseudomonas aeruginosa* (shrimp heads and squid pens) [4, 5], *Serratia marcescens* (squid pens, crab shells, shrimp shells, shrimp heads) [6–11], and *Paenibacillus* sp. (shrimp heads, squid pens) [13–17]. The efficient degradation and utilization of these complex biological materials requires a coordinated enzyme system capable of hydrolyzing chitin and proteins into simpler forms for cellular uptake. Chitinases hydrolyze chitin into monosaccharide units of N-acetylglucosamine (GlcNAc). Bacteria can utilize this GlcNAc in intracellular metabolism. After entering the cell, GlcNAc is phosphorylated to GlcNAc-6-phosphate and then deacetylated by the enzyme NagA to glucosamine-6-phosphate (GlcN-6-P). Next, the enzyme NagB deaminates GlcN-6-P to fructose-6-phosphate. This substance immediately enters glycolysis, providing energy and biosynthetic precursors for bacteria. After degradation, these steps show that chitin also becomes an essential organic carbon source for glycolysis and the cell. Proteases participate in the hydrolysis of proteins into peptides and free amino acids. These amino acids are then deaminated, releasing ammonia and the carbon skeleton. This carbon skeleton is imported into the citric acid cycle to generate energy or as a precursor for synthesizing other biological substances. The resulting ammonia can be reused to synthesize new amino acids or released into the environment. In summary, the mechanism of chitinous marine byproducts metabolism is based on an extracellular enzyme system that breaks down chitin–protein complexes into simple components to participate in the basic metabolic pathways of the cell. This allows these bacterial strains to grow well in chitin-rich byproduct substrate sources and produce valuable compounds.

4 Bioactive Compounds Produced from Chitinous Marine Byproducts Fermentation Conversion and Their Applications

Microbial fermentation technology is a technology that brings many outstanding advantages to the production of natural compounds, contributing to the trend of sustainable and environmentally friendly development. Moreover, the current trend is toward effectively utilizing renewable, low-cost raw materials such as various byproducts to make fermentation media, thereby minimizing production costs and limiting environmental pollution. With particular biological transformation capabilities, microorganisms can efficiently synthesize target compounds. In addition, the fermentation process can easily control conditions such as pH, temperature, and nutrient sources, helping to increase reproducibility and industrial-scale expansion. This technology is also highly flexible and suitable for the production of a variety of valuable biological compounds. In particular, microbial fermentation is an ideal platform for integrating other modern technologies such as genetic engineering and synthetic biology to improve the performance and effectively exploit the potential of native microbial strains. As a result, fermentation technology is increasingly becoming a strategic tool in modern bioindustrial research and application. According to this trend, utilizing chitinous marine byproducts for microbial fermentation has also produced a variety of valuable bioactive compounds such as hemi-pyocyanin, prodigiosin, nicotinic acid, exopolysaccharides, and homogentisic acid, which were mainly mentioned in this section (Table 2). The chemical structures of these bioactive compounds are shown in Fig. 2.

4.1 Metabolites from Microbial Fermentation Using Chitinous Marine Byproducts

4.1.1 Hemi-Pyocyanin

Hemi-pyocyanin (1-hydroxyphenazine) is a mono-hydroxylated derivative of phenazine with the molecular formula $C_{12}H_8N_2O$. Its structure consists of a tricyclic aromatic phenazine core, in which two nitrogen atoms occupy the 1 and 4 positions of the central diazine ring. A distinctive hydroxyl group (-OH) is substituted at the 1-position, forming the compound 1-hydroxyphenazine. It serves as a key intermediate in the biosynthetic pathway of pyocyanin, a biologically active phenazine pigment produced by *Pseudomonas aeruginosa*. Recent studies have demonstrated that chitin-rich marine byproducts, particularly shrimp heads and squid pens, serve as effective and sustainable substrates for the biosynthesis of this compound via fermentation by *P. aeruginosa* TUN03. Optimal production conditions were established using a medium containing 1.5% shrimp heads at an initial pH of 7.0–7.5, incubated at 27.5 °C for 48 h. Under these parameters, a yield of 3100 U/mL was attained in flask-scale fermentation, which was further enhanced to 4200 U/mL within only 12 hours when scaled up to a 14 L bioreactor [4]. Additionally, a fermentation medium comprising 1% squid pens, 0.05% $MgSO_4$, and 0.1% $Ca_3(PO_4)_2$ (pH 7.0) resulted in a high hemi-pyocyanin concentration of 22.73 μg/mL after 12 h

Table 2 Summary of bioactive compounds produced from chitinous marine byproducts via microbial fermentation

Compound	Microbial strain	Byproducts used	Bioactivities	Ref.
Hemi-pyocyanin	*Pseudomonas aeruginosa* TUN03	Shrimp heads, Squid pens	Antidiabetic (α-amylase inhibitor), Anti-nematode	[4, 5]
Prodigiosin	*Serratia marcescens* TKU011, TNU01	Squid pens, Shrimp heads, Shrimp shells, Crab shells, α-chitin	Anticancer, Antioxidant, Anti--inflammatory, Anti-Alzheimer's, Anti-insect	[6–11]
Nicotinic acid	*Paenibacillus* sp. TKU042	Shrimp heads	Antidiabetic (α-glucosidase inhibitor)	[12]
Exopolysaccharides	*Paenibacillus* sp. TKU023, *P. mucilaginosus* TKU032	Squid pens	Antioxidant	[13, 14]
Homogentisic acid	*Paenibacillus* sp. TKU042, TKU036	Squid pens	Antidiabetic (α-glucosidase inhibitor), Antioxidant, Anti-inflammatory	[15–17]

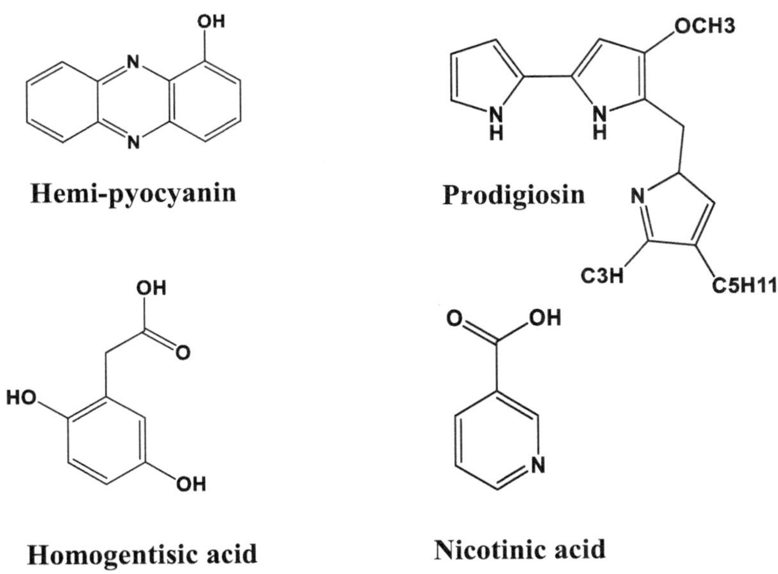

Fig. 2 The chemical structures of some bioactive compounds produced from chitinous marine byproducts via microbial fermentation

at large scale [5]. These findings underscore the remarkable bioconversion efficiency of *P. aeruginosa* TUN03 in transforming chitinous marine waste into value--added phenazine compounds, and highlight its significant potential for industrial-scale bioactive pigment production.

4.1.2 Prodigiosin

Prodigiosin (PG) is a natural pigment with a characteristic structure of three conjugated pyrrole rings (tripyrrole), in which one ring contains a methoxy group and an alkyl side chain. The molecular formula of PG is $C_{20}H_{25}N_3O$, with a molecular weight of 323.43 g/mol. This compound exists in solid form and is a characteristic bright red to purple pigment. It is odorless and slightly volatile at room temperature. The melting point of PG is in the range of 151–152 °C. PG is insoluble in water but soluble in polar organic solvents such as methanol, ethanol, chloroform, acetone, and DMSO. It can absorb light strongly at wavelengths of about 535–540 nm. However, PG is unstable in alkaline environments and is easily degraded by light, oxygen, or high temperatures.

Serratia marcescens is a primary microbial producer of PG. High-yield production of PG has been successfully achieved by utilizing β-chitin-rich marine byproducts, particularly squid pens. A fermentation medium containing 1.5% squid pens as the sole carbon and nitrogen source supported efficient PG biosynthesis by *S. marcescens* TKU011, reaching a yield of 0.978 mg/mL [6]. Further optimization using a culture medium comprising 1.75% squid pen powder (SPP), 0.05% $MgSO_4$, and 0.03% K_2HPO_4 under bioreactor conditions with *S. marcescens* TNU01 significantly enhanced PG production to 3450 mg/L [8]. Additionally, commercially available α-chitin was identified as a viable alternative substrate for PG biosynthesis [7]. Although β-chitin is easier to hydrolyze, the study showed that α-chitin is more effective in prodigiosin biosynthesis. This may be due to the stronger induction of chitinase enzyme, α-chitin has a structure that is difficult to degrade, so bacteria increase the secretion of chitinase and auxiliary enzymes, thereby releasing more chito-oligosaccharides and GlcNAc to provide signals and nutrients for the prodigiosin production process. In addition, α-chitin may have played a role in bio-induction, stimulating the prodigiosin biosynthetic gene pathway, while β-chitin is easier to degrade, so the induction effect is weaker. Moreover, α-chitin is less soluble, so it maintains a long-term solid substrate source, creating conditions for *S. marcescens* to produce prodigiosin more sustainably. Consequently, α-chitin-rich marine byproducts such as crab shells, shrimp shells, and shrimp heads were employed as fermentation substrates, resulting in high PG yields ranging from 5100 to 6310 mg/L in large-scale bioreactor systems [9–11]. These findings demonstrate that both α- and β-chitin-rich marine byproducts serve as efficient and sustainable substrates for PG production via microbial fermentation.

4.1.3 Nicotinic Acid

Nicotinic acid, also known as niacin or vitamin B3, is a heterocyclic compound of the pyridine carboxylic acid group, with the molecular formula $C_6H_5NO_2$ and a molecular weight of 123.11 g/mol. The molecular structure consists of a pyridine ring with a carboxyl group attached to the nitrogen atom. In the crystalline state, nicotinic acid is a white or off-white solid, odorless, slightly bitter, and has a melting point of about 234–237 °C. This compound is well soluble in water and ethanol, but slightly soluble in ether. Nicotinic acid is a weak acid with a pKa of about 4.9 and can form salts when reacting with weak bases such as NaOH or NH_4OH. Regarding physicochemical properties, niacin has high thermal stability, is stable under cooking conditions, and is not easily decomposed by light or oxygen at room temperature.

Shrimp heads were effectively utilized as a substrate for the production of α-glucosidase inhibitors through fermentation with *Paenibacillus* sp. TKU042 [14]. The resulting fermented shrimp heads exhibited high pH stability and remarkable inhibitory activity against yeast α-glucosidase, reaching 92%, which significantly exceeded that of the commercial drug acarbose (60%). Optimization of fermentation conditions led to a more than twofold increase in α-glucosidase inhibitors productivity, from 225 U/mL to 560 U/mL. Subsequent bioactivity-guided fractionation enabled the identification of two major active antidiabetic compounds, including nicotinic acid and adenine. These findings underscore the potential of shrimp heads bioconversion for the development of cost-effective and bioactive antidiabetic agents.

4.1.4 Exopolysaccharides and Homogentisic Acid

Exopolysaccharides are high molecular weight carbohydrate polymers. The structure of exopolysaccharides usually consists of sugar units such as glucose, galactose, mannose, rhamnose, or uronic acid, linked together by α- or β-glycosidic bonds, forming straight or branched chains. Due to the nature of polymers, the molecular formula is not fixed, but the molecular weight can range from a few kDa to millions of Da. Exopolysaccharides usually exists as a viscous glue or gel in an aqueous environment, and can be dried into an off-white or pale-yellow powder. Exopolysaccharides is soluble in water, has high viscosity, and can form biofilms. It is insoluble in organic solvents such as ethanol or acetone, and is often precipitated by adding ethanol. Some exopolysaccharides are negatively charged due to carboxyl or sulfate groups, affecting their colloidal properties and interactions with metal ions. Exopolysaccharides is stable under neutral pH conditions, susceptible to hydrolysis at high temperatures or in strong acid/alkaline environments, and degrades when exposed to enzymes such as cellulase, amylase, or protease.

Homogentisic acid has the molecular formula $C_8H_8O_4$ and a molecular weight of 168.15 g/mol. It is a phenolic acid with a structure consisting of a benzene ring with a hydroxyl group and an acetic acid (Fig. 2). Homogentisic acid exists as an off-white crystal, with a melting point of about 150–153 °C. The compound is poorly soluble in water, but more soluble in methanol and ethanol. The pKa of homogentisic acid is about 4.4, indicating a weak acidity. Chemically, homogentisic acid is susceptible to oxidation in air, especially in mildly alkaline environments, resulting in a color change from pale yellow to dark brown or black. Although stable in weakly acidic environments and at room temperature, homogentisic acid requires storage conditions protected from light and oxidation to maintain its original structure and color.

Squid pen has been effectively utilized as the sole carbon and nitrogen source for microbial fermentation by various *Paenibacillus* strains to produce bioactive compounds [13–17]. Under optimized conditions, *Paenibacillus* sp. TKU023 and *P. mucilaginosus* TKU032 successfully converted squid pen into high yields of exopolysaccharides, with maximum productivity reaching 4.55 g/L and 14.8 g/L, respectively. These exopolysaccharides were mainly composed of glucose and maltose and exhibited notable antioxidant properties. Similarly, *Paenibacillus* sp. TKU036 and TKU042 fermented squid pens to produce potent antioxidant and α-glucosidase inhibitory activities. Bioactivity-guided purification enabled the identification of homogentisic acid and tryptophan as the major antioxidant compounds, while homogentisic acid was also determined to be the primary α-glucosidase inhibitor from the fermentation broth of *Paenibacillus* sp. TKU042. These aGIs showed high stability across a wide pH (2–13) and temperature (40–100 °C) range. Collectively, these findings highlight the versatility of *Paenibacillus* spp. in bioconverting squid pens byproduct into high-value bioactive compounds such as exopolysaccharides, homogentisic acid, and tryptophan, supporting the sustainable reuse of marine byproducts in functional ingredient production.

4.2 Potential Applications of Bioactive Compounds Produced from Chitinous Marine Byproducts Fermentation Conversion

4.2.1 Potential Bioactivities Applied in Medicine

- **Antidiabetic**

Type 2 diabetes mellitus is characterized by elevated postprandial blood glucose levels, primarily regulated by two key digestive enzymes, including α-amylase and α-glucosidase [18, 19]. Inhibiting these enzymes can effectively delay carbohydrate hydrolysis and glucose absorption, offering a therapeutic glycemic control strategy. In this context, microbial fermentation of chitinous marine byproducts has emerged as a promising approach to produce natural enzyme inhibitors with antidiabetic potential. Several studies have demonstrated that shrimp heads and squid pens are efficient carbon and nitrogen sources for microbes to produce natural α-amylase inhibitors (aAIs) and α-glucosidase inhibitors (aGIs). Notably, *P. aeruginosa* TUN03 fermented shrimp heads and produced hemi-pyocyanin, which exhibited strong α-amylase inhibitory activity (maximum inhibition 92.2%, IC_{50} = 3.1 μg/mL), comparable to the commercial drug acarbose (IC_{50} = 4.2 μg/mL) [4]. Similarly, *Paenibacillus* sp. TKU042 fermentation of shrimp heads yielded potent aGIs, with inhibitory activity of 92% against yeast α-glucosidase, substantially higher than that of acarbose (60%). Through bioactivity-guided fractionation, two active non-sugar-based compounds, nicotinic acid and adenine, were identified. These showed selective inhibition against yeast and rat α-glucosidases with IC_{50} values of 22 μg/mL and 70 μg/mL, respectively [12]. In addition, squid pens were also shown to be a viable substrate for aGI production via fermentation by *Paenibacillus* strain [16, 17]. The fermentation supernatants displayed high inhibition against α-glucosidases from yeast (79–98%) and rats (76–83%), outperforming acarbose in some cases. Optimization of the fermentation process with *Paenibacillus* sp. TKU042 increased aGI yield by 3.1-fold. The main active compound isolated was homogentisic acid, a novel non-sugar-based aGI, with notable enzyme inhibition (IC_{50} = 215 μg/mL) and anti-inflammatory activity (IC_{50} = 9.8 μg/mL). Homogentisic acid also demonstrated superior pH and thermal stability (active at pH 2–13, 40–100 °C), enhancing its potential application in functional foods or pharmaceutical formulations. These findings underscore the potential of chitinous marine byproduct fermentation as a sustainable platform for producing safe and effective natural enzyme inhibitors. The bioactive compounds, particularly hemi-pyocyanin, homogentisic acid, nicotinic acid, and adenine exhibit strong antidiabetic properties, offering promising candidates for future development of functional ingredients or therapeutic agents for type 2 diabetes management.

- **Anticancer**

Natural compounds have been recognized as a safe and effective source of anticancer agents. Of those, microbial secondary metabolites produced utilizing sustainable resources, such as byproducts/wastes, are gaining increasing attention. In this context, chitinous marine byproducts have also been explored for generating high-value bioactive compounds with anticancer properties. A notable example is prodigiosin, a red tripyrrole pigment known for its strong cytotoxic effects against over 60 human cancer cell lines. In a recent study, *Serratia marcescens* TKU011 was employed to convert α-chitin for PG production. PG's anticancer activity was evaluated against four cancer cell lines, including A549 (lung), Hep G2 (liver), MCF-7 (breast), and WiDr (colon) [7]. PG exhibited potent cytotoxicity, with IC_{50} values of 0.06, 0.04, 0.04, and 0.2 μg/mL, respectively. Compared to mitomycin C (a commercial anticancer drug), PG demonstrated significantly stronger effects, being 1.67–3.25 folds more active against MCF-7, A549, and Hep G2 cells, respectively. PG was also successfully produced by various chitinous marine byproducts such as shrimp heads, shrimp shells, crab shells, and squid pens. These findings highlight the remarkable potential of marine chitins as a sustainable feedstock for the biosynthesis of prodigiosin through microbial fermentation, suggesting PG as a promising anticancer compound for pharmaceutical development.

- **Antioxidant**

Antioxidant activity is one of the most important biological properties of natural compounds, as it plays a vital role in preventing oxidative stress-related damage caused by reactive oxygen species [20]. These damages are closely associated with aging, inflammation, and various chronic diseases, including diabetes, cancer, etc. In recent years, considerable efforts have been made to explore microbial fermentation as a sustainable and eco-friendly approach for producing natural antioxidants from marine chitinous byproducts. Among the promising outcomes, prodigiosin biosynthesized by *S. marcescens* TNU01 using chitin- and protein-rich marine byproducts as carbon/nitrogen sources, demonstrated good antioxidant activity [8]. Specifically, PG exhibited a DPPH radical scavenging effect up to 98% at 234 μg/mL, comparable to α-tocopherol (98% at 117 μg/mL), with respective IC_{50} values of 79.1 and 38.5 μg/mL. This indicates PG's potential as a natural antioxidant compound. Similarly, various *Paenibacillus* species fermented on squid pen media produced diverse antioxidant compounds [13–17]. Culture supernatant of *Paenibacillus* sp. TKU023 achieved a DPPH scavenging rate of approximately 92%. Another strain, *P. mucilaginosus* TKU032, produced exopolysaccharides that showed strong, dose-dependent antioxidant activity, with a maximum scavenging rate of 80% and an IC_{50} of 157.1 μg/mL. Notably, the IC_{50} of exopolysaccharides from *Paenibacillus* sp. TKU032 was lower than that of standard ascorbic acid (191.7 μg/mL), reinforcing its efficiency as a potential natural antioxidant. Homogentisic acid is also identified as the primary antioxidant in *Paenibacillus* sp. TKU036 culture broth, exhibited superior antioxidant capacity with an IC_{50} of just 6.9 μg/mL, much lower than α-tocopherol (17.6 μg/mL). Exopolysaccharides and homogentisic acid from fermented squid pens from another *Paenibacillus* sp., both showing remarkable activities of 96% and 99% DPPH scavenging ability, with low IC_{50} values of 30 μg/mL and 5.4 μg/mL, respectively. These findings confirm that fermentation of chitinous marine byproducts using specific microbial strains is a promising route for producing potent natural antioxidants.

- **Anti-inflammatory**

Inflammation is a complex physiological response to harmful stimuli, associated with the pathogenesis of various chronic disorders. Among the key mediators in the inflammatory response is nitric oxide (NO), excessive production of NO contributes to tissue damage and promotes the progression of chronic inflammatory diseases [21]. Thus, compounds exhibiting anti-NO activity are considered promising candidates for anti-inflammatory drug development. In recent years, the microbial fermentation of chitinous marine byproducts has emerged as a sustainable approach for producing bioactive compounds with potential therapeutic applications. Several studies have highlighted the anti-inflammatory potential of such fermentation-derived products. One notable compound is prodigiosin (PG), which was biosynthesized using shrimp waste through microbial fermentation [10]. The anti-NO activity of PG was evaluated using LPS-stimulated RAW264.7 macrophage cells, a widely accepted in vitro model for screening anti-inflammatory agents. PG demonstrated a maximum NO inhibition of 91% at 80 μg/mL. Furthermore, PG exhibited a low IC_{50} value of 19.1 μg/mL, indicating strong anti-NO efficacy, nearly matching homogentisic acid (IC_{50} = 15.9 μg/mL). Importantly, these results support the feasibility of using shrimp shell waste as a low-cost, effective substrate for the large-scale microbial production of anti-inflammatory agents like PG.

In parallel, homogentisic acid, derived via microbial fermentation of squid pens, also exhibited significant anti-inflammatory activity. The IC_{50} of homogentisic acid was 10.14 μg/mL, although slightly less potent than that of the flavonoid quercetin (IC_{50} = 1.14 μg/mL) [15]. In another study, homogentisic acid also showed anti-inflammatory activity with an IC_{50} of 9.8 μg/mL [17].

- **Anti-Alzheimer's**

One of the promising biomedical applications of bioactive compounds derived from chitinous marine byproducts is their potential to manage neurodegenerative disorders, such as Alzheimer's disease. In particular, prodigiosin, a red pigment with various bioactivities, was produced by *S. marcescens* TNU01 using chitin- and protein-containing marine byproducts, including crab shells, squid pens, and shrimp shells as the carbon and nitrogen sources [8]. Among these, squid pens powder was the most effective substrate for prodigiosin biosynthesis in flask and bioreactor systems. Notably, the purified prodigiosin exhibited notable acetylcholinesterase (AChE) inhibitory activity, a key therapeutic target in Alzheimer's disease treatment. Prodigiosin demonstrated a moderate maximum inhibition of 62% with an IC_{50} value of 1.12 mg/mL. In contrast, the crude extract exhibited a lower inhibition rate of 37%, indicating the enhanced potency of the purified compound. These findings suggest that fermentation-derived prodigiosin from marine byproducts could serve as a potential natural inhibitor of AChE, contributing to developing anti-Alzheimer functional agents.

4.2.2 Potential Bioactivities Applied in Agriculture

- **Anti-nematode**

Nematode-related diseases are among the major threats to global agriculture, particularly in black pepper cultivation, where *Meloidogyne incognita* causes severe root damage and yield loss. Due to the limitations of chemical nematicides, microbial-derived compounds have emerged as eco-friendly alternatives [22]. Chitinous marine by-products were used for microbial fermentation to produce phenazine compounds, including hemi-pyocyanin [5]. Hemi-pyocyanin demonstrated strong nematicidal activity, effectively inhibiting second-stage juveniles (J2) and delaying egg hatching of *M. incognita*. At 10 mg/mL, hemi-pyocyanin reached maximum J2 inhibition of 99%. Its IC_{50} values were 0.377 μg/mL, indicating potent activity. Moreover, hemi-pyocyanin significantly reduced egg hatching with inhibition rates up to 97%. These results highlight the potential of hemi-pyocyanin as natural, scalable agents for sustainable nematode control based on chitinous marine by-products fermentation.

- **Anti-insect**

Insect pests are a major cause of agricultural and economic loss, damaging crops, reducing yields, and acting as vectors for plant and animal pathogens. Conventional chemical insecticides, though effective, often lead to environmental contamination, insect resistance, and risks to human and animal health [23]. Therefore, there is growing interest in natural, microbial-derived insecticidal compounds as sustainable alternatives. Prodigiosin, a red pigment produced by *S. marcescens* TKU011 using squid pen powder as the sole carbon and nitrogen source, has demonstrated promising insecticidal activity [6]. In bioassays using *Drosophila larvae* as a model, various fractions of the fermentation culture, especially the whole broth and purified prodigiosin, exhibited strong insecticidal effects. After 5 days of feeding, the survival rates of larvae exposed to the culture broth, broken cells, cell pellets, and cell-free supernatant were 0%, 6.25%, 12.5%, and 37.5%, respectively. Purified prodigiosin showed dose-dependent toxicity, with complete larval mortality at 1.2 g/L and an LC_{50} value of 0.23 g/L. These results indicate that prodigiosin is a key contributor to the insecticidal activity of the fermentation products and may serve as a potential biopesticide agent for integrated pest management strategies.

5 Challenges and Future Perspectives

5.1 Challenges in Scaling Up Bioprocessing Techniques

Scaling up the fermentation of chitinous materials from the laboratory to an industrial level presents significant technical challenges. The solid chitin-containing medium has high viscosity and non-Newtonian properties, which hinders mixing, oxygen exchange, and heat transfer. The heterogeneity in the flow also complicates pH and temperature control, causing an internal pH gradient in the fermentation tank. In addition, if not treated evenly, input chitin-containing materials will cause clumping, clog the stirrer, and reduce the biological contact area. The investment cost of stainless-steel industrial equipment is huge, and high operating costs for energy, cooling, aeration, and wastewater treatment. In particular, electricity consumption for stirring and aeration often accounts for the most significant proportion. Therefore, it is necessary to apply energy-saving and automation solutions, such as inverter control or innovative monitoring systems. Biological yield at a large scale is typically reduced by 10–30% compared to a small scale due to reduced oxygen diffusion efficiency and environmental inhomogeneities. Therefore, choosing the proper fermentation regime is essential, such as batch, which is challenging but straightforward to control nutrients, fed-batch, which helps maintain stable substrate concentrations, or continuous, which increases productivity but requires tight control. All require design optimization (aeration-stirring, substrate feeding, and enzyme support) to maximize the final biological yield.

5.2 Challenges in Overcoming Legal and Biosafety Constraints

Moving toward applying active ingredients from microbial fermentation requires overcoming a series of legal barriers related to food safety, pharmaceuticals, and environmental protection. Potential compounds produced from the fermentation of chitinous marine byproducts have demonstrated many outstanding biological potentials, but still need to be comprehensively evaluated for toxicity, ecological impact, and the ability to bioaccumulate in the food chain. In particular, using different bacterial strains in industrial production must also strictly comply with biosafety regulations to avoid the potential risk of causing human disease. Therefore, practical applications must comply with specific legal regulations depending on the particular application field. For example, in food, new compounds must demonstrate absolute safety (toxicity, allergenicity, and bioaccumulation) and comply with GRAS standards (USA), Novel Foods (EU), Codex Alimentarius system, and HACCP. Standards such as QCVN (Vietnam) or EC 396/2005 (EU) stipulate residue limits and assess eco-accumulation risk. The drug registration process in pharmaceuticals requires complete preclinical and clinical data, compliance with international GMP (WHO, ICH, FDA Q7A), and pharmacopoeia standards for purity and potency. Active compounds must be controlled as pharmaceutical substances, not simply as additives. Regarding environment and biology, microorganisms must be managed according to BSL-2 standards, EU Directive 2000/54/EC, and the Cartagena Protocol on genetically modified organisms. Post-fermentation waste must be treated according to the Basel Protocol. In particular, if the compound has the potential to be biopersistent or ecotoxic, a comprehensive environmental risk assessment (ERA) is required. There is also a need for specific risk control and testing strategies throughout the production cycle—from raw materials, microorganisms, processes, to the final product. Compliance with international standards appropriate to the characteristics of biological products derived from marine waste is also a challenge before aiming to expand the application and production scale of this byproduct source.

5.3 Future Perspectives

Using chitinous marine byproducts for microbial fermentation has shown great potential in producing valuable bioactive compounds. However, most current studies remain at a laboratory scale and rely on a few common microbial strains, which are insufficient for industrial-scale applications. A key direction is to improve microbial strains through modern biotechnology to optimize chitin conversion and increase the yield of target compounds. Developing stable super-producing strains will be the foundation for large-scale production. In terms of process, environmental-friendly pretreatment methods are needed to replace acid/alkali treatments. These methods should enrich chitin while preserving nutrients that are useful for microbial growth. Semi-continuous fermentation models using low-cost raw materials, with optimized pH, temperature, nutrients, and culture conditions, will enhance biological efficiency and support industrial applications. Currently, the primary raw materials used are shrimp shells, crab shells, and squid pens. Expanding to less utilized byproducts, such as other crustacean shells or seafood organs, and combining multiple chitin sources is essential. This may create a richer nutrient base, reduce input costs, and improve resource efficiency. Regarding the final products, many bioactive compounds obtained still lack a clear understanding of their structure–activity relationships. More research is needed to study their mechanisms of action at the molecular level, perform detailed structural analysis, and conduct biological tests to clarify the links between chemical structure, mechanism, and biological function. This is a critical step in developing specific applications. Simultaneously, life cycle analysis and cost–benefit assessments should be conducted to compare this technology with traditional methods. A complete evaluation of environmental impact, product stability, toxicity, and legal requirements will help guide the technology toward sustainability and international standards. Finally, collaboration between research institutes and enterprises is essential to promote technology transfer, product standardization, and commercialization. Pilot models and semi-industrial projects will help bridge the gap between research and real-world applications.

6 Conclusion

The microbial fermentation of chitinous marine byproducts represents a promising and sustainable approach to valorize seafood processing waste into high-value bioactive compounds. These byproducts, such as shrimp heads and shells, crab shells, and squid pens, are generated in large quantities globally and possess a rich biochemical composition, particularly in chitin, proteins, and minerals. Via microbial fermentation conversion, these chitinous materials can be effectively degraded and used as carbon/nitrogen sources to produce a variety of bioactive compounds, including hemi-pyocyanin, prodigiosin, nicotinic acid, exopolysaccharides, and homogentisic acid. These metabolites have demonstrated a broad range of potent biological activities, with applications in medicine such as antidiabetic, antioxidant, anticancer, anti-inflammatory, and anti-Alzheimer's functions and in agriculture, including anti-nematode and insecticidal effects. However, there are still several technical and regulatory challenges before these materials can be effectively deployed on an industrial scale. Future directions should focus on genetic engineering and synthetic biology to construct superior microbial producers, expand substrate diversity, explore novel marine byproducts, and elucidate structure–activity relationships of bioactive compounds to tailor them for specific applications. Integrating life cycle assessments, techno-economic evaluations, and collaborative industry-academia partnerships will be vital for bridging laboratory research with real-world production and commercialization. Overall, this biotechnological strategy offers an innovative solution for reducing waste and environmental pollution and provides a sustainable avenue for generating valuable functional ingredients, supporting circular economy principles, and contributing to the advancement of green bioindustries for utilizing chitinous marine byproducts.

References

1. Nguyen TH, Wang SL, Nguyen VB (2025) Comprehensive overview on preparation and diverse applications of prodigiosin nanocomposites. Res Chem Intermed 51:509–550
2. Ngo VA, Wang SL, Nguyen VB, Phan TQ, Tran THT, Doan MD, Nguyen DS, Nguyen AD (2024) Utilization of shrimp heads for scaling up of production of *Bacillus velezensis* EB.KN15, its bioactive compounds and novel anti-fungal effect against durian pathogen fungi. Res Chem Intermed 50:5061–5080
3. Wang SL, Liang TW (2017) Microbial reclamation of squid pens and shrimp shells. Res Chem Intermed 43:3445–3462
4. Nguyen TH, Wang SL, Nguyen AD, Doan MD, Tran TN, Doan CT, Nguyen VB (2022) Novel α-amylase inhibitor hemi-pyocyanin produced by microbial conversion of chitinous discards. Mar Drugs 20:283
5. Nguyen TH, Wang SL, Nguyen TH, Doan MD, Tran THT, Ngo VA, Ho ND, Tran TN, Doan CT, Do VC, Nguyen AD, Nguyen VB (2022) Utilization of fishery-processing by-product squid pens for scale-up production of phenazines via microbial conversion and its novel potential antinematode effect. Fishes 7:113
6. Wang SL, Wang CY, Yen YH, Liang TW, Chen SY, Chen CH (2012) Enhanced production of insecticidal prodigiosin from *Serratia marcescens* TKU011 in media containing squid pen. Process Biochem 47:1684–1690
7. Nguyen VB, Chen SP, Nguyen TH, Nguyen MT, Tran TTT, Doan CT, Tran TG, Nguyen AD, Kuo YH, Wang SL (2020) Novel efficient bioprocessing of marine chitins into active anticancer prodigiosin. Mar Drugs 18:15
8. Nguyen VB, Nguyen DN, Wang SL (2020) Microbial reclamation of chitin and protein-containing marine by-products for the production of prodigiosin and the evaluation of its bioactivities. Polymers 12:1328
9. Nguyen VB, Nguyen DM, Nguyen AD, Ngo VA, Ton TQ, Doan CT, Pham TP, Tran TPH, Wang SL (2020) Utilization of crab waste for cost-effective bioproduction of prodigiosin. Mar Drugs 18:523
10. Nguyen TH, Wang SL, Nguyen DM, Nguyen AD, Nguyen TH, Doan MD, Ngo VA, Doan CT, Kuo YH, Nguyen VB (2021) Bioprocessing of marine chitinous wastes for the production of bioactive prodigiosin. Molecules 26:3138
11. Nguyen VB, Wang SL, Nguyen AD, Phan TQ, Techato K, Pradit S (2021) Bioproduction of prodigiosin from fishery processing waste-shrimp heads and evaluation of its potential bioactivities. Fishes 6:30
12. Nguyen VB, Wang SL (2019) Production of potent antidiabetic compounds from shrimp head powder via *Paenibacillus* conversion. Process Biochem 76:18–24
13. Wang CL, Huang TH, Liang TW, Wang SL (2011) Production and characterization of exopolysaccharides and antioxidant from *Paenibacillus* sp. TKU023. New Biotechnol 28:559–565

14. Liang TW, Tseng SC, Wang SL (2016) Production and characterization of antioxidant properties of exopolysaccharides from *Paenibacillus mucilaginosus* TKU032. Mar Drugs 12:40–51
15. Wang SL, Li HT, Zhang LJ, Lin ZH, Kuo YH (2016) Conversion of squid pen to homogentisic acid via *Paenibacillus* sp. TKU036 and the antioxidant and anti-inflammatory activities of homogentisic acid. Mar Drugs 14:183
16. Nguyen VB, Nguyen AD, Wang SL (2017) Utilization of fishery processing for α-glucosidase inhibitors production by *Paenibacillus* sp. Mar Drugs 15:274
17. Nguyen VB, Nguyen TH, Doan CT, Tran TN, Nguyen AD, Kuo YH, Wang SL (2018) Production and bioactivity-guided isolation of antioxidants with α-glucosidase inhibitory and anti-NO properties from marine chitinous materials. Molecules 23:1124
18. Shahab N, Kong D, Ali M, Alam A, Rehman NU, Ullah S, Zainab Z, Khan M, Latif A, Shah M, Khan A, Al-Harrasi A, Ahmad M (2023) Novel polyhydroquinoline Schiff's base derivatives: synthesis, characterization, in vitro α-glucosidase inhibitory, and molecular docking studies. Res Chem Intermed 49:3005–3028
19. Le Vu TT, Wang SL, Ho TTT, Luc QT, Phan TQ, Phan TKT, Dam TBH, Phan TKP, Nguyen AD, Nguyen VB (2024) Elucidation of potent mammalian enzymes inhibitors targeting anti-diabetes drug from *Castanea mollissima* Blume, 1851: an edible herbal collected in Vietnam via experimental and computation approaching. Res Chem Intermed 50:6065–6086
20. Hussein MO, Abdulhameed AS (2025) Development of a versatile biomaterial of chitosan polymer/ginger (*Zingiber officinale* Roscoe) extract/ZnO bionanocomposite: physicochemical properties, antioxidant activity, and breast cancer therapy. Res Chem Intermed 51:551–570
21. Nguyen VB, Ton TQ, Nguyen DN, Nguyen TT, Ngu TN, Nguyen TH, Doan CT, Tran TN, Nguyen MT, Ho ND, Nguyen AD, Kuo YH, Wang SL (2020) Reclamation of beneficial bioactivities of herbal antioxidant condensed tannin extracted from *Euonymus laxiflorus*. Res Chem Intermed 46:4751–4766
22. Nguyen TH, Wang SL, Phan TQ, Nguyen TH, Nguyen AD, Nguyen VB (2024) Enhancing nematicidal effect of prodigiosin via micro-encapsulation using chitosan as a novel carrier substance. Res Chem Intermed 50:2873–2896
23. Shi JJ, Li WW, Tan CX, Hu DS, Xu TM, Liu XH (2024) Synthesis, insecticidal activity, and in silico study of novel carboxamide compounds containing benzoxazole moiety. Res Chem Intermed 50:1809–1825

Thi Hanh Nguyen

San-Lang Wang

Anh Dzung Nguyen

Van Bon Nguyen

Molecular Mechanisms of Chitosan-Based Hydrogel Formation: From Carbamate Chemistry to Polyelectrolyte Complexation and Nanocomposite Design

Hiroshi Tamura and Tetsuya Furuike

Abstract Chitosan (CS), a naturally derived cationic polysaccharide, has received extensive attention as a versatile building block for functional hydrogels. However, its limited solubility at neutral pH and relatively weak mechanical strength restrict direct applications. In this chapter, we outline a molecular-level framework for enhancing the structural and functional performance of CS-based hydrogels through carbamate chemistry, polyelectrolyte complexation (PEC), and nanocomposite design. Reversible carbamate formation at the C-2 amino group enables homogeneous solubilization of CS under weakly basic conditions, providing a mild and efficient pathway for uniform hydrogel preparation. Upon protonation, CS regains its cationic character and forms stable PEC networks with sodium alginate (SA), yielding hydrogels with tunable structural properties. To further expand functionality, silver nanoparticles (AgNPs) were synthesized in situ using microwave irradiation and stabilized by SA, producing narrow size distributions and high colloidal stability. Incorporation of AgNPs into SA/CS PEC hydrogels resulted in robust, self-supporting gels that resisted collapse, displayed moderated swelling, and exhibited sustained antimicrobial activity against both Gram-positive and Gram-negative bacteria. Spectroscopic analysis revealed that AgNPs serve not only as antimicrobial agents but also as reinforcing centers, contributing electrostatic and coordinative interactions within the matrix. These findings establish a design strategy where solubilization chemistry, electrostatic complexation, and nanoparticle incorporation act synergistically to confer mechanical robustness and biological activity. The concepts presented here provide a foundation for developing next-generation CS-based hydrogels for biomedical and environmental applications.

Keywords Chitosan hydrogels · Carbamate chemistry · Polyelectrolyte complexation · Silver nanoparticles (AgNPs) · Antibacterial biomaterials

---

H. Tamura (✉) · T. Furuike
Faculty of Chemistry, Materials and Bioengineering, Kansai University, Suita, Osaka, Japan
e-mail: tamura@kansai-u.ac.jp; furuike@kansai-u.ac.jp

S.-L. Wang, M. Anpo (eds.), *Chitin and Chitosan*, Springer Series on Polymer and Composite Materials, https://doi.org/10.1007/978-981-95-7457-5_3

1 Introduction

Chitosan (CS), a cationic polysaccharide derived from the partial deacetylation of chitin, has attracted significant attention as a versatile biomaterial due to its unique combination of properties, including excellent biocompatibility, inherent biodegradability, and intrinsic antibacterial activity. Chitin, the parent polymer, is one of the most abundant natural polysaccharides, found primarily in the exoskeletons of crustaceans, insects, and the cell walls of fungi. By selectively removing acetyl groups from chitin, CS acquires free amino groups along its backbone, which not only confer a cationic character but also enable a wide range of chemical modifications. These functional amino groups play a central role in enabling interactions with various biomolecules, polymers, and metal ions, thereby expanding the utility of CS in diverse applications. One of the most attractive features of CS is its ability to form hydrogels, which are three-dimensional networks capable of absorbing and retaining significant amounts of water while maintaining structural integrity. Such hydrogels provide a highly tunable platform for biomedical applications, including drug delivery, wound healing, tissue engineering scaffolds, and antimicrobial coatings, as well as for environmental applications, such as water purification and heavy metal adsorption [1–3]. The gelation behavior of CS can be modulated through factors such as molecular weight, polymer concentration, degree of deacetylation, pH, and the presence of crosslinkers or polyanionic partners, allowing the design of materials with tailored mechanical and functional properties. Despite its wide-ranging potential, the practical utilization of CS is often limited by its solubility characteristics. Being a cationic polymer, CS is soluble primarily under acidic conditions, typically below pH 6, where protonation of the amino groups occurs. This limited solubility restricts its processing and functionalization under neutral or basic conditions, and conventional approaches often require harsh acidic solutions that can induce depolymerization and negatively affect molecular weight and material performance. Recent studies have revealed that CS can undergo reversible carbamate formation in weakly basic aqueous media, providing a pathway to temporarily mask the amino groups and enhance solubility without extensive depolymerization [4]. This chemical modification allows for more homogeneous processing and broader versatility in hydrogel formation, nanoparticle incorporation, and composite material design. Furthermore, the reversible nature of carbamate formation ensures that the original cationic character of CS can be restored under controlled conditions, preserving its bioactive functions and enabling sophisticated material engineering. These advances highlight the potential of CS not only as a functional biomaterial but also as a chemically tunable platform for next-generation biomedical and environmental technologies.

The incorporation of sodium alginate (SA), an anionic polysaccharide obtained from brown algae, into the CS system further broadens the functional versatility of chitosan-based materials. This is because electrostatic interactions take place between the protonated amino groups of CS and the carboxyl groups of SA, resulting in the formation of polyelectrolyte complex (PEC) hydrogels [5]. In practice,

PEC hydrogels can be prepared simply by mixing a weakly acidic solution of CS with an aqueous solution of SA. However, because gelation in this case initiates at the interface between the two solutions, the gels produced tend to be structurally heterogeneous. By contrast, when carbamoylated CS is employed, a clear and uniform solution can be obtained with SA. If glucono-δ-lactone (GDL) is then added, the pH gradually decreases, allowing gelation to proceed in a controlled manner and ultimately leading to the formation of homogeneous hydrogels [4]. Such hydrogels exhibit mechanical properties and swelling behaviors that can be systematically tuned by adjusting the pH of the solution as well as the mixing ratio of the components. Spectroscopic and rheological studies have clarified that controlled carbamate dissociation and stepwise protonation govern gelation, while molecular orbital interactions—specifically HOMO–LUMO overlap between ammonium and carboxylate groups—stabilize the PEC network at the electronic level.

Beyond polysaccharide–polysaccharide assemblies, incorporation of metallic nanoparticles offers additional functional advantages. Silver nanoparticles (AgNPs) are of particular interest because of their broad-spectrum antimicrobial activity, sustained Ag^+ release, and high surface reactivity at low concentrations [6]. Traditional AgNPs synthesis often relies on toxic reductants, but polysaccharide-assisted routes provide a greener and more biocompatible alternative [7]. SA acts not only as a stabilizer of AgNPs via carboxylate coordination but also as a structural component in the PEC matrix, while microwave-assisted synthesis promotes uniform nucleation and prevents aggregation [8]. Embedding AgNPs within SA-CS hydrogels yields hybrid materials with enhanced mechanical strength, reduced swelling, and controlled Ag^+ release. At the molecular scale, incorporation of AgNPs induces electron density redistribution at the polymer–nanoparticle interface, strengthening ionic and coordinative interactions. These orbital-level effects rationalize the observed improvements in structural integrity and antibacterial efficacy. The released Ag^+ further exert bactericidal activity by binding to thiol groups in proteins and enzymes, disrupting metabolic pathways and ultimately leading to cell death.

This study integrates spectroscopic, rheological, and molecular orbital analyses to provide a comprehensive framework for understanding CS hydrogel formation [9], SA-CS PEC stabilization [4], and AgNPs incorporation [10]. By correlating macroscopic material properties with molecular-level interactions, we establish a mechanistic basis for the design of multifunctional CS-based hydrogels with significant potential in biomedical and environmental applications.

2 Hydrogel State Chitosan: Neutral-Condition Processing and Molecular Insights into Solubilization

This section summarizes and expands upon our previous work [9], providing additional molecular-level interpretation. Dissolution of CS in its powdered form typically necessitates the use of excess acids. Under such conditions, the medium becomes highly acidic, which promotes hydrolytic cleavage of glycosidic bonds and markedly decreases the molecular weight of CS. This degradation reduces its functional integrity and complicates long-term storage. To circumvent these limitations, our previous work demonstrated that CS can be transformed into a hydrogel under nearly neutral conditions, thereby preserving polymer integrity and enabling stable aqueous storage [11]. The water content of these hydrogels can be finely adjusted (85–98%) by controlling processing parameters such as the starting polymer concentration, centrifugation conditions, or the molecular weight of the precursor CS.

When weak acids are gradually introduced, these hydrogels become transparent, which can be followed by monitoring pH and optical transmittance [9]. For acetic acid, the pH remained constant (~5.8) until ~0.4 molar equivalents per sugar unit were added. Beyond this threshold, the pH decreased steadily while transmittance sharply increased, exceeding 95% at the same acid concentration. Similar trends were observed with other monocarboxylic acids, while dicarboxylic acids such as oxalic or malonic acid achieved dissolution at only ~0.2 molar equivalents, highlighting the correlation between the number of carboxyl groups and dissolution efficiency. Considering the degree of deacetylation (~80%) of the CS employed [FL-80 (DAC 84.7, 5 mPa s) and FM-80 (DAC 77.9, 50 mPa s) were supplied from Koyo Chemical Co., Ltd. (Japan)], approximately 0.5 equivalents of anions per amino group are sufficient to achieve complete dissolution.

Viscosity measurements revealed distinct stability differences between CS solutions prepared from powder and from hydrogels. Powdered CS failed to dissolve at pH 5, while hydrogel-derived CS yielded stable solutions. The hydrogel's partially disrupted crystalline domains permit enhanced hydration and acid accessibility, thereby facilitating dissolution. For hydrogel-derived CS solutions, viscosity decreased gradually over 2 days at room temperature before stabilizing, with smaller reductions at higher pH (30% at pH 3, 20% at pH 5). By contrast, solutions from powdered CS showed severe molecular weight loss (~60% viscosity reduction after 7 days), emphasizing the protective role of the hydrogel state.

The dissolution of chitosan in organic acids can be rationalized in terms of counterion binding, a process that becomes accessible to direct observation through ^{1}H NMR spectroscopy. To minimize contributions from excess acid, the acetate form of chitosan was obtained as a dry salt by freeze-drying and then examined in D_2O. The resulting spectra (Fig. 1) allowed resonances from both glucosamine (GlcN) and N-acetylglucosamine (GlcNAc) units to be distinguished. Thus, three separate components for methyl group were apparent: unassociated acetic acid (a), the N-acetyl methyl protons of GlcNAc (b), and acetate signals attributable to ion pairing with protonated amino groups of GlcN residues (c).

When incremental amounts of acetic acid were introduced, approaching one equivalent per sugar residue, characteristic changes were seen in the carbohydrate resonances. Both the anomeric (H-1) and C2 (H-2) protons shifted progressively downfield, with the larger displacement occurring at H-2. Simultaneously, the methyl resonance ascribed to ion-paired acetate grew in intensity in proportion to the amount of acid present. Quantitative evaluation of this trend confirmed that the

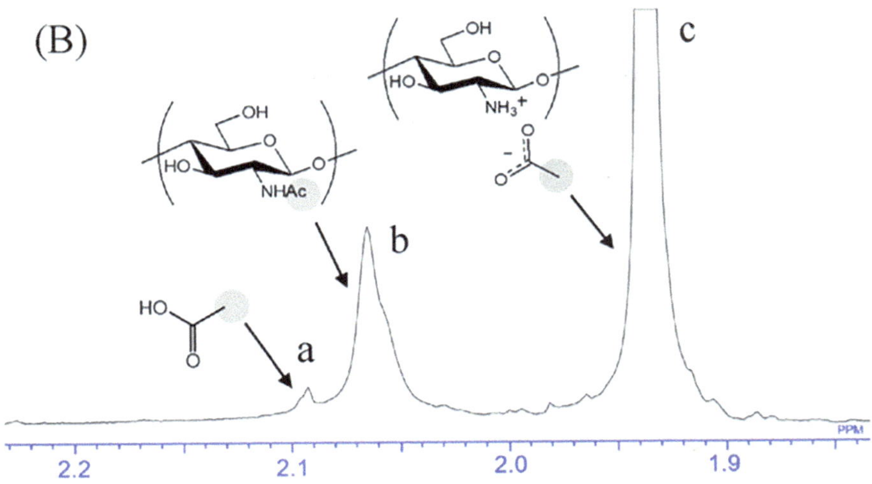

Fig. 1 ^{1}H-NMR spectra of CS acetate (addition of 1.0 molar equivalent of acetic acid) in D_2O. The spectrum shows the peaks of CH_3 groups due to (a) free acetic acid, (b) the acetamide of GlcNAc, and (c) the acetate salt of $GlcNH_2$. Reproduced with permission from Elsevier [9]

extent of acetate association closely tracked the degree of protonation of amino groups, providing molecular-level evidence for anion–cation interactions as the basis of solubilization.

The dynamic nature of these complexes was revealed by counterion substitution experiments (Fig. 2). Introduction of trifluoroacetic acid into the chitosan acetate solution led to a disappearance of the resonance assigned to bound acetate, while new signals from liberated acetate became evident. This exchange demonstrates that the ionic pairing of acetate is reversible and that the counterion environment can be replaced by stronger acids such as trifluoroacetate. Changes in the chemical shifts of sugar protons further confirmed this substitution process.

From a molecular orbital perspective, these results can be rationalized by considering both electronic redistribution and hydration effects. In the crystalline state, the lone-pair electrons on CS amino groups are engaged in intra- and intermolecular hydrogen bonding, stabilizing the dense packing. Upon protonation, the HOMO of the neutral amino group (localized lone pair) is lowered in energy and interacts with the LUMO of incoming anions (carboxylate or acetate). This orbital overlap facilitates charge delocalization, stabilizing the ionic complex. At the same time, the hydration shell surrounding protonated amino groups and associated counter anions redistributes the local electron density, weakening interchain hydrogen bonds. This dual effect—electrostatic stabilization via HOMO–LUMO interactions and disruption of crystalline domains through hydration—synergistically enhances dissolution. Notably, the requirement of only ~0.5 equivalents of acid per amino group can be explained by cooperative effects: once local protonation disrupts hydrogen bonding clusters, water penetration increases, further lowering the energy barrier for dissolution.

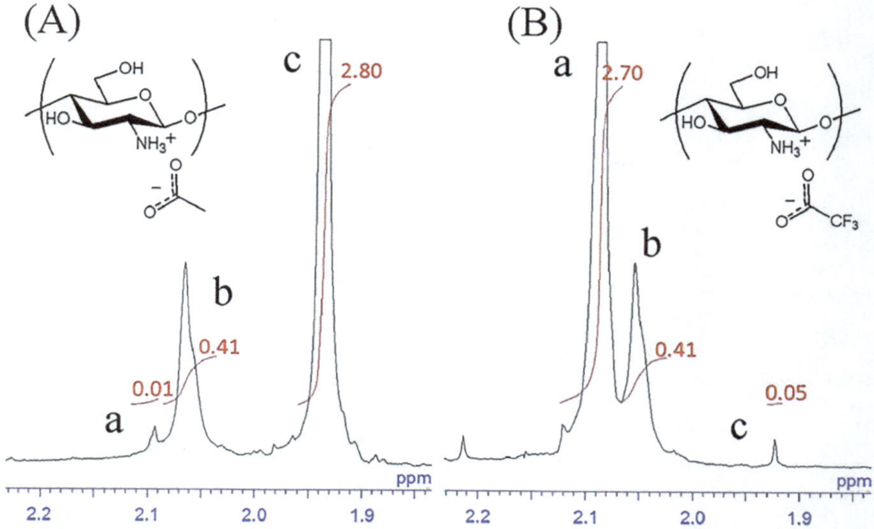

Fig. 2 ^{1}H-NMR spectra of CS acetate in (A) D_2O and (B) D_2O containing 1% TFA-d_1. Each spectrum shows peaks from the CH_3 group due to (a) free acetic acid, (b) the acetamide of GlcNAc, and (c) the acetate salt of $GlcNH_2$. Reproduced with permission from Elsevier [9]

In summary, CS hydrogels offer a structurally pre-conditioned state in which polymer chains are hydrated and partially decrystallized, thereby reducing the acid input required for solubilization. The dissolution process is thus driven not only by ionic complex formation but also by molecular orbital interactions and electron density redistribution, providing a comprehensive electronic level rationale for the enhanced solubility of hydrogel-derived CS compared with powdered CS.

3 Homogeneous Gels of SA–CS Polyelectrolyte Complexes: Reversible Carbamate Formation Enables Molecularly Uniform SA–CS Hydrogels

This section summarizes and expands upon our previous work [4], providing additional molecular-level interpretation. Polyelectrolyte complexes (PECs) are assemblies of polymers that are formed through electrostatic interactions between positively charged polycations and negatively charged polyanions. These interactions occur when oppositely charged polymers come into contact in solution, leading to the spontaneous formation of a physically or chemically crosslinked network. The resulting PECs often display unique physicochemical properties that differ substantially from those of the individual polymers, including changes in solubility, mechanical strength, and thermal stability. Importantly, the structure and behavior of PECs are highly sensitive to environmental conditions such as pH, ionic strength, and temperature. Variations in these parameters can induce conformational changes, phase separation, or swelling, which frequently culminate in the formation of hydrogels with tunable mechanical and transport properties. Because of these distinctive features, PECs have attracted considerable attention as versatile materials in the fields of biomaterials, drug delivery, and tissue engineering [5]. Their ability to encapsulate bioactive molecules, respond to external stimuli, and form self-supporting hydrogels makes them particularly attractive for controlled release systems, wound dressings, and scaffolds for cell growth. Furthermore, by carefully selecting the type of polycation and polyanion, as well as controlling the mixing ratios and environmental conditions, researchers can finely tune the degree of cross-linking, network density, and gel stiffness. In particular, the combination of sodium alginate (SA), a negatively charged polysaccharide, with CS, a positively charged polysaccharide, has been extensively studied due to the biocompatibility, biodegradability, and bioactivity of both components. Such SA–CS PEC hydrogels exhibit homogeneous network structures that are advantageous for uniform drug distribution, consistent mechanical properties, and reliable biological performance. Overall, PECs, and specifically SA–CS complexes, represent a class of materials in which molecular-level electrostatic interactions can be harnessed to design gels with precise structural and functional properties. This level of control enables the development of advanced biomedical and environmental applications, bridging fundamental polymer science with practical material engineering.

CS is nearly the sole naturally occurring cationic biopolymer, whereas numerous anionic natural polysaccharides exist. In this study, we selected SA, a well-characterized anionic polysaccharide derived from brown algae, as the complementary component. SA consists of copolymeric chains of D-mannuronic acid and L-guluronic acid linked through 1,4-glycosidic bonds. The ionic nature of both polymers makes them suitable candidates for PEC gel formation.

Previous methods for preparing SA-CS gels often resulted in rapid gelation and heterogeneous structures when CS and SA solutions were simply mixed. Alternative strategies, such as using D-glucono-δ-lactone (GDL) to dissolve CS powder in SA solution, slowed the gelation process but failed to achieve molecular-level homogeneity [12]. Low-molecular-weight chitosan oligomers (CSOs) can dissolve in basic solutions, producing more homogeneous gels; however, due to insufficient chain entanglement with SA, these gels exhibit limited mechanical strength [13].

CS exhibits a p*K*a of approximately 6.5 and precipitates when the pH exceeds 7. Nevertheless, it has recently been demonstrated that CS can dissolve in weakly basic solutions of sodium hydrogen carbonate (pH 7–8) [4]. SA, as a typical anionic polysaccharide, also dissolves readily in this pH range. By exploiting the solubility of CS under these conditions, homogeneous and transparent SA-CS solutions can be prepared. Gradual acidification via GDL addition triggers the decomposition of carbamate-modified CS, regenerating protonated amino groups and enabling uniform PEC gelation through electrostatic interactions with SA.

To clarify how CS behaves in weakly basic media, both ^{1}H and ^{13}C NMR spectroscopy were applied, with signal assignments aided by two-dimensional COSY experiments. Figure 3 summarizes the ^{1}H NMR spectra obtained for CS acetate and for solutions treated with different amounts of sodium hydrogen carbonate. In the acetate salt (Fig. 3A), resonances attributable to the polysaccharide backbone were readily identified: the C-2 proton of GlcN appeared near 3.15 ppm, while multiplets from H-3 through H-6 were clustered between 3.5 and 4.0 ppm [14, 15]. A partial anomeric signal was observed at approximately 4.5 ppm, but much of the H-1 resonance was obscured by the residual D_2O peak around 4.9 ppm. Two distinct methyl signals were also resolved: the acetamide group from N-acetylglucosamine (GlcNAc) at 2.07 ppm (a) and the acetate associated with protonated glucosamine ($GlcNH_3^+$) at 1.97 ppm (b) [9]. Integration of these signals indicated a degree of deacetylation of ~86%. Notably, a resonance from unbound acetic acid at 2.09 ppm was absent, consistent with stoichiometric salt formation (1:1 acid-to-sugar ratio). Marked changes emerged upon addition of sodium hydrogen carbonate. At four molar equivalents (Fig. 3B), the methyl resonance of $GlcNH_3^+$ acetate at 1.97 ppm disappeared and was replaced by a peak at 1.92 ppm, attributable to sodium acetate produced in situ. The H-2 resonance split into two components at ~2.7 and 3.4 ppm, whose combined integral was smaller than that of the H-3–6 protons (integral ratio ~ 1:4). The anomeric region also showed new doublets between 4.4 and

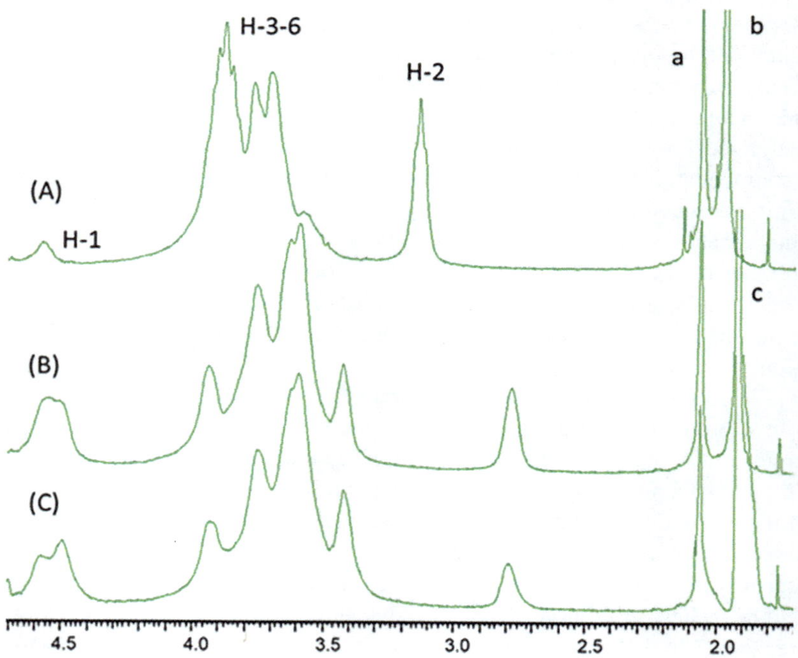

Fig. 3 ^{1}H-NMR spectra of CS acetate and its basic solution (addition of 1.0 molar equivalent of acetic acid) in D_2O; (A) CS solution, basic CS solution prepared by addition of (B) 4.0 and (C) 8.0 molar equivalent of sodium hydrogen carbonate, respectively. Each spectrum shows the peaks of CH_3 group due to (a) the acetamide of GlcNAc, (b) the acetate salt of $GlcNH_2$, and (c) the sodium acetate. Reproduced with permission from Elsevier [4]

4.6 ppm. Increasing the bicarbonate concentration (Fig. 3C) enhanced these splitting patterns, suggesting modification of the C-2 amino group of glucosamine.

To obtain clearer information, ^{13}C NMR spectra were recorded using a CSO consisting of dimers and short oligomers (2–6 units), which gave sharper signals than the high-molecular-weight CS polymer. In the acetate form (Fig. 4A), signals from acetic acid carbons were visible at 179.6 ppm (C=O) and 25.1 ppm (CH_3), confirming salt formation. Upon addition of 1.2 equivalents of sodium hydrogen carbonate (Fig. 4B), carbon dioxide was generated but the solution remained transparent. The acetate signals shifted slightly to 184.1 ppm and 26.1 ppm, while no sodium hydrogen carbonate peaks were observed. When sodium hydrogen carbonate was added in excess of 1.4 equivalents (Fig. 4 C and D), two new carbon resonances appeared near 162.9 and 166.6 ppm. The former was assigned to carbonate/bicarbonate species undergoing rapid proton exchange, whereas the latter was attributed to carbamate carbonyl carbons ($-NH-CO_2^-$) formed by reaction of the C-2 amino group with carbonate species [16]. A further resonance around 105 ppm

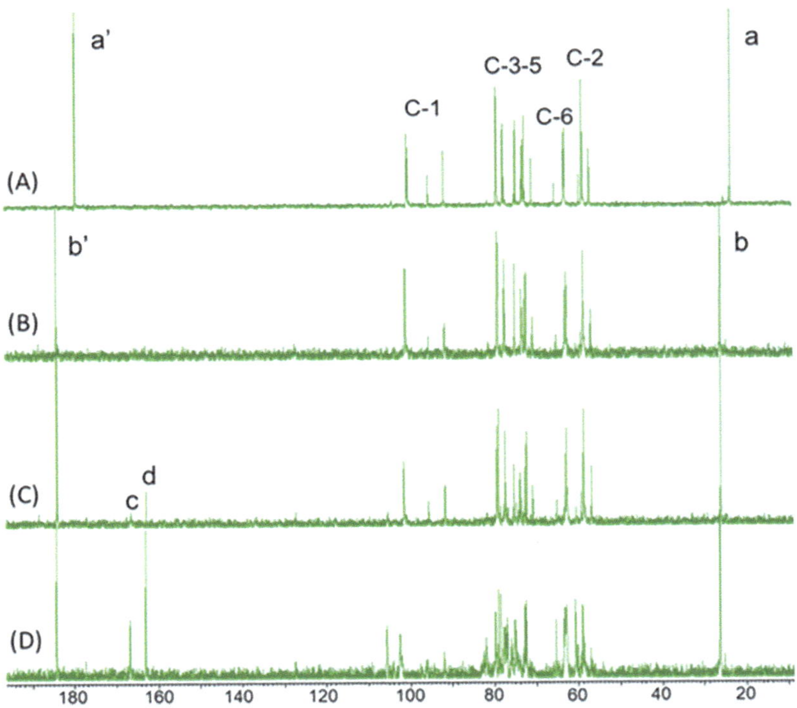

Fig. 4 ^{13}C NMR spectra of CSO acetate and its basic solution (addition of 1.0 M equivalent of acetic acid) in D_2O; (A) CSO solution, basic CSO solution prepared by addition of (B) 1.2, (C) 1.4 and (D) 2.0 M equivalent of sodium hydrogen carbonate, respectively. Each spectrum shows the peaks of CH_3 group due to (a) the acetic acid and (b) the sodium acetate; carbonyl carbon due to (a′) the acetic acid, (b′) the sodium acetate, (c) the carbamate moiety of $GlcNHCO_2^-$ and (d) the carbonate and bicarbonate ions. Reproduced with permission from Elsevier [4]

was tentatively linked to internal C-1 carbons adjacent to carbamate-modified amino groups. The intensities of these new peaks increased progressively with additional sodium hydrogen carbonate. Because carbamate formation is known to be reversible with a small equilibrium constant, the N–C linkage can be readily cleaved by subsequent acidification.

Based on this evidence, stepwise mechanism is thus proposed:

1. Neutralization of $-NH_3^+$ by sodium hydrogen carbonate, producing free $-NH_2$ and CO_2.
2. Nucleophilic addition of the liberated amine to CO_2 or HCO_3^-, yielding a transient carbamic acid intermediate (–NH–COOH).
3. Deprotonation to form the carbamate anion ($-NH-CO_2^-$).

This scheme accounts for the appearance of the 166.6 ppm carbonyl resonance and the 105 ppm C-1 shift observed in the oligomer spectra.

^{1}H NMR of the same CSO solutions (Fig. 5) further corroborated this mechanism. In the acetate salt (Fig. 5A), the reducing-end H-2α and H-2β resonances appeared at ~3.4 and 3.1 ppm, while the internal GlcN H-2 protons resonated between 3.1 and 3.2 ppm. After addition of 1.2 equivalents of bicarbonate (Fig. 5B), the acetate was consumed and glucosamine residues were present as free amines. With further sodium hydrogen carbonate (Fig. 5 C and D), the H-2 signals shifted upfield toward ~3.0 ppm, consistent with electronic changes at C-2 due to carbamate formation. Additional resonances at ~4.5 ppm (H-1) and 3.4 ppm (H-2) were observed in the carbamate form, and their assignments were verified by COSY correlations. Importantly, the H-1 signals of the unmodified residues (H-1α: 5.4 ppm; H-1β: 4.9 ppm) were largely unaffected, highlighting that the most significant spectral perturbations occurred at C-2.

From a molecular orbital standpoint, these findings can be rationalized as the interaction between the nitrogen lone pair (HOMO) and the π* orbital of carbon dioxide (LUMO). This overlap drives N–C σ-bond formation, accompanied by bending of the CO_2 structure and redistribution of electron density within the C=O bonds. The result is a stabilized carbamate anion, whose formation produces the upfield shift of the H-2 resonance observed in ^{1}H NMR.

Taken together, the combined ^{1}H and ^{13}C NMR evidence establishes that under weakly basic conditions, the amino group at C-2 of glucosamine undergoes reversible carbamate formation with bicarbonate/carbonate species. This reversible chemical modification accounts for the enhanced solubility of CSOs at near-neutral to slightly basic pH.

Based on these insights, the formation of SA-CS PEC gels can be described in Fig. 6, which was schematic illustration based on our previous study [4]. Under weakly basic conditions, the C-2 amino groups of CS are carbamate-modified and remain soluble. Mixing with SA produces a clear, homogeneous solution. Gradual pH reduction by GDL addition triggers carbamate dissociation, sequential protonation of amino groups, and electrostatic interactions with SA carboxylates, resulting in uniform gel formation. Electron density redistribution and HOMO–LUMO interactions at the molecular level contribute to the uniformity of the gel network.

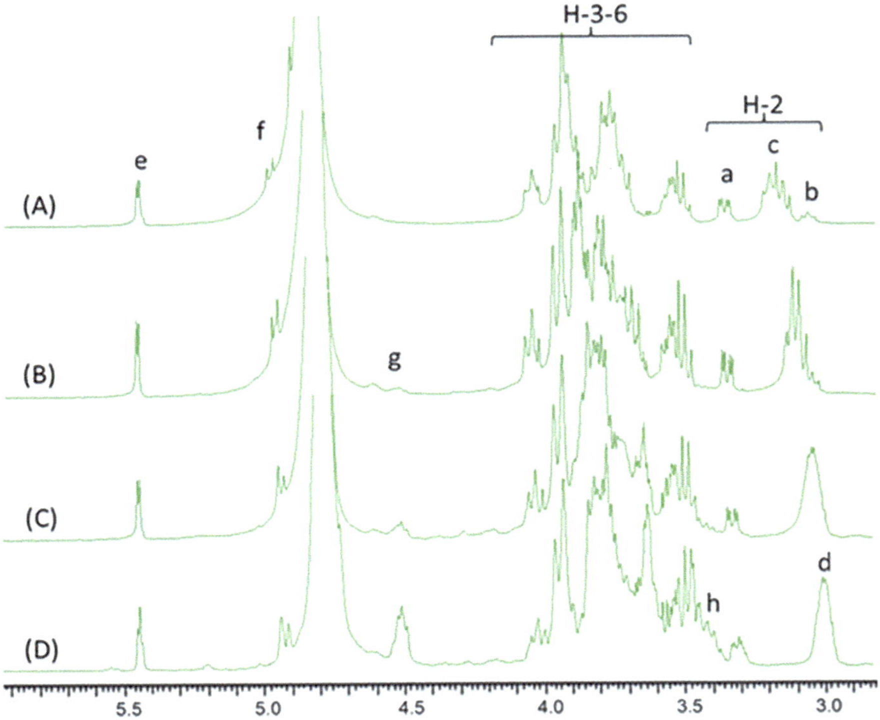

Fig. 5 ^{1}H-NMR spectra of CSO acetate and its basic solution (addition of 1.0 molar equivalent of acetic acid) in D_2O; (A) CSO solution, basic CSO solution prepared by addition of (B) 1.2, (C) 1.4 and (D) 2.0 molar equivalent of sodium hydrogen carbonate, respectively. The assignments were as follows: (a) H-2a; (b) H-2b; (c) H-2 of inner sugar units; (d) H-2 of $GlcNH_2$; (e) H-1a; (f) H-1b; (g) H-1 of $GlcNHCO_2^-$; (h) H-2 of $GlcNHCO_2^-$. Reproduced with permission from Elsevier [4]

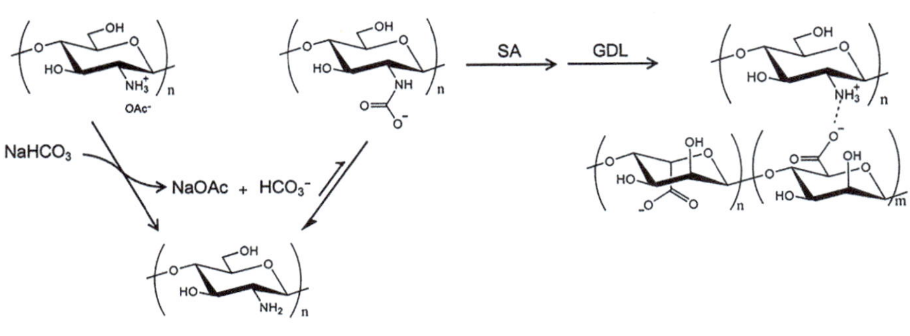

Fig. 6 Schematic illustration of SA–CS PEC gelation mechanism based on our previous study [4]

FT-IR analysis further corroborated ionic interactions: asymmetric and symmetric stretching vibrations of COO^- shifted from 1608 cm^{-1} and 1418 cm^{-1} to 1615 cm^{-1} and 1414 cm^{-1}, respectively, while the NH_3^+ shoulder at 1572 cm^{-1} shifted to lower wavenumber [17]. These shifts indicate electrostatic stabilization and partial redistribution of electron density between the COO^- and the NH_3^+ group. Rheological measurements showed that SA:CS molar ratio of 1:1 yielded maximum storage modulus ($G' \approx 7000$ Pa), approximately tenfold higher than SA-CSO gels. This enhancement is attributed to both polymer chain entanglement and strengthened electrostatic interactions, consistent with the molecular orbital–driven stabilization.

In summary, the stepwise process—carbamate formation under basic conditions, followed by protonation during acidification—induces electron density redistribution and HOMO–LUMO interactions, which are key molecular-level factors enabling the formation of homogeneous and mechanically robust SA-CS PEC gels. This molecular-level understanding not only clarifies the fundamental solubility behavior of CS but also provides a rational basis for designing advanced hydrogels with predictable and tunable properties for biomedical and environmental applications.

4 AgNP-Loaded SA/CS PEC Gels from Carbamoyl Chitosan: A Strategy for Self-Standing Nanocomposite Hydrogels

This section summarizes and expands upon our previous work [10], providing additional molecular-level interpretation. Silver nanoparticles (AgNPs) have attracted considerable interest in biomedical, food, and environmental applications owing to their nanoscale physicochemical properties [18]. Compared with bulk silver or ionic silver, AgNPs provide sustained release of Ag^+, a high surface-to-volume ratio, and strong antimicrobial efficacy at low concentrations. Conventional synthesis methods, however, often rely on toxic reducing agents, prompting the development of greener, polysaccharide-assisted strategies [19, 20]. Despite these advances, cytotoxicity remains a challenge, as AgNPs can penetrate cells and disrupt DNA integrity, induce oxidative stress, and trigger inflammatory responses. Stabilization with natural polymers such as SA and CS offers a biocompatible route to control aggregation, modulate Ag^+ release, and enhance overall safety for biological applications. Based on previous studies mentioned in Sect. 3, the dimensions of AgNPs stabilized by SA (SA-AgNPs) are strongly influenced by the electrostatic interactions between the carboxylate groups (COO^-) of SA and Ag^+. In typical AgNPs synthesis, both a reducing agent and a stabilizer are required. While the reducing agent initiates the formation of AgNPs by electron transfer to Ag^+, SA functions as a stabilizer, preventing particle aggregation through electrostatic repulsion of carboxyl groups and steric hindrance provided by its polysaccharide chains [8, 21]. This stabilization not only maintains colloidal dispersion but also allows for precise control over particle size and morphology, which are critical determinants of antimicrobial efficacy and cytocompatibility. The formation of AgNPs occurs in a sequential manner. Initially, a fraction of Ag^+ is reduced by reducing agents present in the solution, generating primary AgNPs. These initial nanoparticles then act as nucleation centers, accelerating the reduction of the remaining Ag^+ and promoting growth on the particle surfaces. The use of microwave irradiation has been shown to further facilitate homogeneous nucleation, enhance the reaction rate, and limit undesired aggregation [8, 22]. Moreover, microwave-assisted synthesis provides rapid, energy-efficient heating that contributes to uniform particle distribution and reproducibility, which is particularly advantageous when integrating AgNPs into SA/CS PEC gels for biomedical or environmental applications.

Transmission electron microscopy (TEM) analysis of the silver nanoparticle–alginate (SA–AgNP) system revealed that the particles formed under alkaline conditions (pH 9, 80 °C, 8 min reaction time) displayed a nearly spherical morphology with a narrow size distribution, averaging approximately 9 ± 2 nm (Fig. 7). The presence of SA was found to be critical, as it served not only as a stabilizing agent but also as a mild reductant, thereby preventing aggregation and yielding a homogeneous nanoparticle dispersion. In contrast, particles generated without SA protection tended to undergo self-association, resulting in larger mean diameters (around

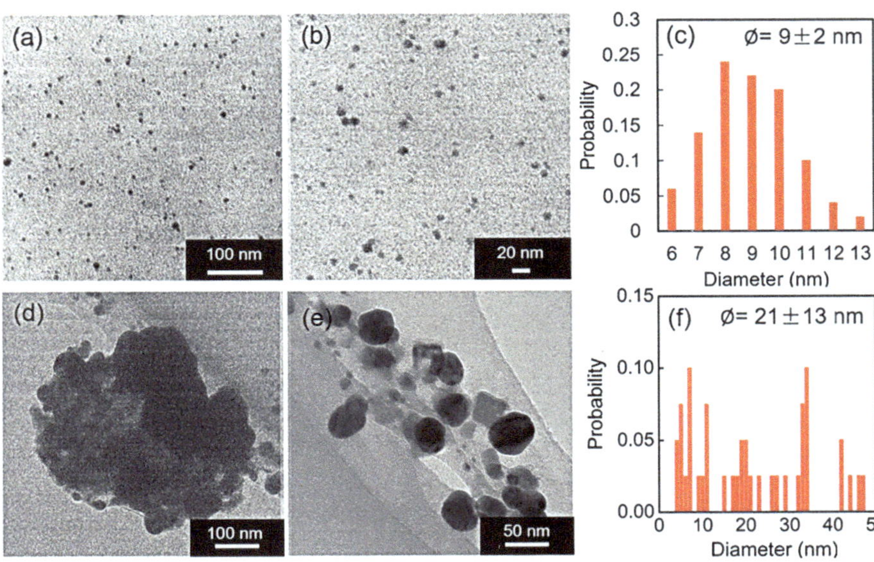

Fig. 7 TEM images of SA-AgNPs fabricated using 0.5% SA and 50mM $AgNO_3$ at pH 9 and 80°C, after 8 min of reaction: (**a**) with a scale bar of 100 nm and (**b**) with a scale bar of 20 nm. (**c**) The size distribution of the SA-AgNPs. TEM images of the AgNPs fabricated without the SA coating using 50 mM $AgNO_3$ at pH 9 and 80°C, after 8 min of reaction: (**d**) with a scale bar of 100 nm and (**e**) with a scale bar of 50 nm. (**f**) The size distribution of the AgNPs without the SA coating. Reproduced from MDPI [10] under CC BY 4.0

20 nm) with broad polydispersity. Literature reports consistently point out that AgNPs below 10 nm exhibit enhanced bactericidal effects, particularly toward *E. coli*, and that spherical geometry provides superior antimicrobial efficacy compared with anisotropic forms. Taken together, these observations suggest that the alginate-mediated synthesis route favors the formation of small, stable, and spherical AgNPs, characteristics that are closely associated with high antimicrobial performance.

However, the resulting SA-AgNPs remain dispersed in a sol state, which restricts their direct applicability in practical biomedical or environmental contexts. Incorporation of AgNPs into polymer matrices represents an effective strategy to expand their functionality, improve handling, and broaden potential applications [23]. Hydrogels, in particular, offer an advantageous platform, as they can enable gradual Ag^+ release, absorb large amounts of water, provide structural stability, and serve as biocompatible scaffolds for cells or bioactive molecules. To prepare a homogeneous SA-CS PEC hydrogel, the carboxylate groups exposed on the surface of SA-AgNPs are mixed with carbamoyl chitosan dissolved under the weakly basic conditions described in Sect. 3. Subsequent addition of glucono-δ-lactone (GDL) gradually reduces the pH, promoting the transformation of carbamoyl moiety into chitosan, which then becomes protonated under acidic conditions. Below the p*K*a of CS, electrostatic interactions between protonated CS and SA lead to the formation of a stable, crosslinked SA-CS PEC gel (Fig. 8). This carefully controlled approach produces hydrogels exhibiting favorable characteristics, including sustained release of Ag^+, excellent water absorption, tunable mechanical properties, and enhanced colloidal stability of embedded nanoparticles, making them highly suitable for diverse biomedical and environmental applications.

The hydrogel incorporating SA–AgNPs was visually distinguishable from the control system, exhibiting a self-standing, stable shape and a characteristic yellow coloration, whereas gels lacking AgNPs did not show these features (Fig. 9). Interestingly, mixtures of SA–AgNPs with CS alone were unable to form a gel matrix unless both sodium hydrogen carbonate and GDL were present, highlighting the essential role of these components in initiating gelation. The incorporation of SA–AgNPs into the CS matrix not only altered the macroscopic gelation behavior but also provided clear spectroscopic evidence of molecular-level interactions. The characteristic carboxylate stretching bands of SA at 1298 and 1412 cm^{-1} converged into a single, upshifted band at 1414 cm^{-1} upon nanoparticle formation, suggesting coordination of Ag^+ species with the COO^- groups. This shift toward higher wavenumbers can be interpreted as a strengthening of the C–O bond due to partial electron donation from the carboxylate oxygen atoms to silver atoms, thereby reducing the degree of charge delocalization within the carboxylate moiety. Similarly, in the composite gels, the COO^--related band initially observed at 1371–1406 cm^{-1} shifted further to 1410 cm^{-1}, indicating that AgNPs acted as crosslinking centers that reinforced the polyelectrolyte complexation between SA and CS. These results are consistent with a stabilization mechanism in which the AgNPs surface interacts via both electrostatic attraction and coordination bonding with alginate carboxylates,

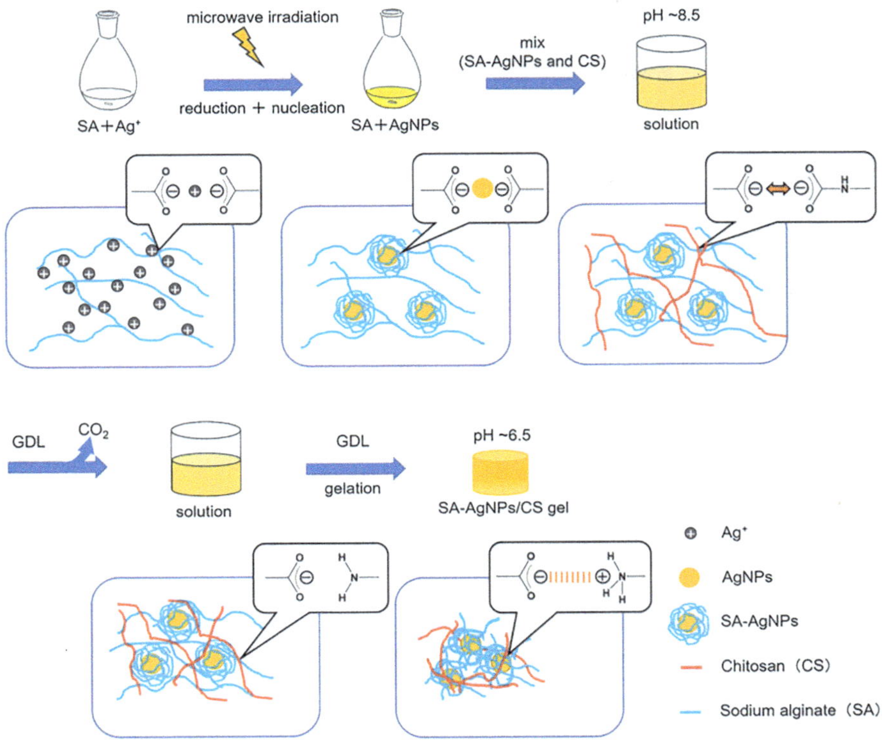

Fig. 8 Schematic illustration of the SA-AgNPs/CS gel fabrication. Reproduced from MDPI [10] under CC BY 4.0

effectively anchoring the nanoparticles within the gel network and enhancing the structural integrity of the hybrid material.

For biomedical applications, hydrogels must combine strong antimicrobial performance and minimal cytotoxicity with adequate mechanical robustness. While our earlier studies employed a fixed 1:1 molar ratio of CS to SA, subsequent investigations revealed that the balance between these two components markedly influences gel strength. Incorporation of SA–AgNPs into the CS/SA matrix significantly enhanced the Young's modulus compared with gels lacking nanoparticles, indicating that the AgNPs contribute to a denser and more stable network structure (Fig. 10). Systematic variation of the CS-to-SA molar ratio (0.5, 0.75, and 1.0) demonstrated that the intermediate ratio (0.75) yielded the stiffest gels, consistent with the observed trend in crosslinking density. These findings suggest that not only the presence of AgNPs but also the fine-tuning of the polyelectrolyte ratio governs the mechanical integrity of the hybrid gels, a key parameter for ensuring reliable performance in biomedical settings.

Swelling experiments revealed that SA/CS gels without AgNPs rapidly expanded and collapsed within 72 h, while SA-AgNPs/CS gels maintained integrity with only

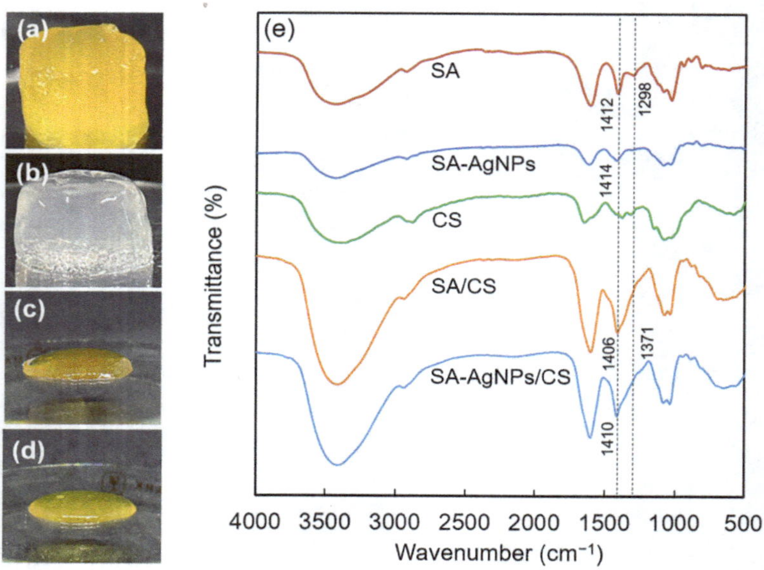

Fig. 9 Digital photographs of (**a**) the SA-AgNPs/CS, (**b**) the SA/CS gel, (**c**) the mixture of SA-AgNPs and CS without $NaHCO_3$ and GDL, and (**d**) the mixture of SA-AgNPs and CS with the addition of $NaHCO_3$ without GDL. (**e**) The FTIR spectra of SA, the SA-AgNPs, CS, the SA/CS gel, and the SA-AgNPs/CS gel. Reproduced from MDPI [10] under CC BY 4.0

moderate swelling (Fig. 11). This enhanced stability is attributed to restricted chain mobility caused by ionic interactions between carboxylate groups exposed on the surface of SA-AgNPs and protonated CS, which reduce water uptake. Additionally, the embedded AgNPs act as nanoscale crosslinking points, further reinforcing the gel network and contributing to improved mechanical resilience and prolonged structural durability under aqueous conditions.

Silver nanoparticles are well known for their broad-spectrum antimicrobial activity. To assess the antibacterial performance of the hybrid materials, inhibition zone tests were performed against *E. coli* and *B. subtilis* (Fig. 12). Gels containing SA–AgNPs exhibited clear inhibitory zones, whereas the AgNP-free SA/CS control gel showed little effect, likely due to the relatively high molecular weight of the chitosan used (~5.3×10^4 Da), which restricts its bactericidal contribution mainly to limiting nutrient transport across the cell envelope. By contrast, the combined action of AgNPs and CS in the composite gels resulted in pronounced antimicrobial effects. In particular, samples prepared with increasing concentrations of silver precursor displayed a concentration-dependent enhancement of activity, with inhibition zones extending up to ~27 mm for *E. coli* and ~ 22 mm for *B. subtilis*. These results indicate that both the intrinsic antibacterial properties of AgNPs and their synergy with the CS matrix contribute to the observed efficacy. It is also noteworthy that the present microwave-assisted synthesis route enabled the preparation of highly active SA–AgNP/CS gels within minutes, whereas conventional

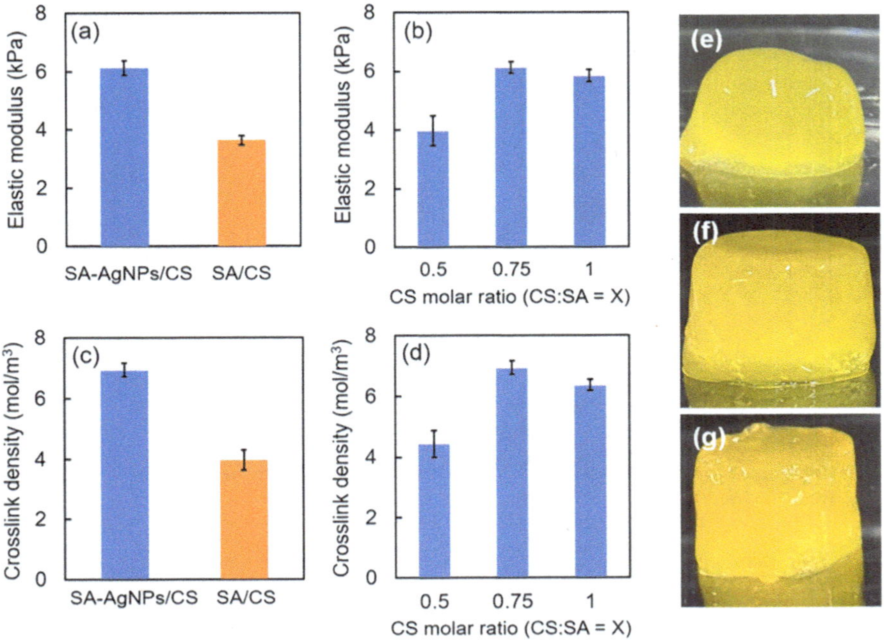

Fig. 10 Young's modulus of (**a**) the SA-AgNPs/CS and SA/CS gels (CS:SA = 0.75:1) and (**b**) the SA-AgNPs/CS gel with different molar ratios of CS:SA. Crosslinking densities of (**c**) the SA-AgNPs/CS and SA/CS/gels (CS:SA=0.75:1) and (**d**) the SA-AgNPs/CS gel with different molar ratios of CS:SA. Digital photographs of the SA-AgNPs/CS gel with different molar ratios of CS:SA: (**e**) CS:SA = 0.5:1, (**f**) CS:SA = 0.75:1, and (**g**) CS:SA = 1:1. Reproduced from MDPI [10] under CC BY 4.0

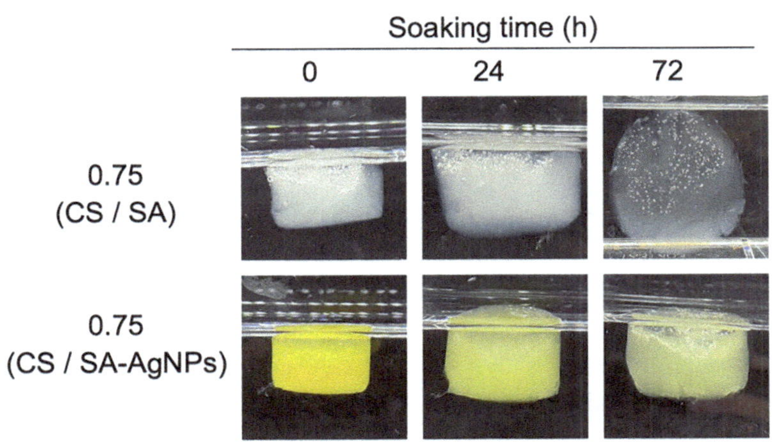

Fig. 11 Swelling experiments for SA/CS and SA-AgNPs/CS gels

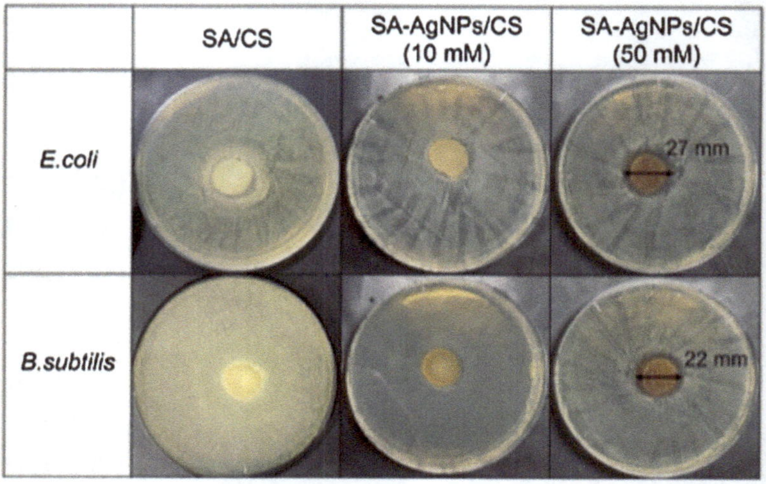

Fig. 12 Inhabitation zones of the SA/CS and SA-AgNPs/CS (fabricated with 10 and 50 mM $AgNO_3$) gels. Reproduced from MDPI [10] under CC BY 4.0

hydrothermal or solution-based routes typically require several hours. This highlights the advantage of microwave irradiation not only in shortening reaction times but also in producing gels with robust antimicrobial performance comparable to or exceeding those reported in other SA–AgNP-containing hydrogel systems.

In summary, this study demonstrates that the incorporation of AgNPs into SA/CS PEC gels formed via CS carbamate chemistry produces exceptionally robust, self-standing hydrogels. The combined effects of carbamate-mediated stabilization and polyelectrolyte complexation with SA enable these gels to maintain structural integrity under swelling stress while exhibiting pronounced antimicrobial activity. Such self-standing nanocomposite gels not only highlight the utility of carbamate chemistry in hydrogel design but also underscore their potential as advanced biomaterials. Beyond antibacterial functionality, they offer durability, controlled swelling behavior, and tunable physicochemical properties, which are relevant for wound dressings, tissue scaffolds, and stimuli-responsive systems. Furthermore, the microwave-assisted fabrication strategy provides scalability and reproducibility, supporting the future translation of these gels toward biomedical and industrial applications.

5 Conclusion and Future Perspective

This chapter has sought to provide a comprehensive account of the molecular mechanisms underlying the formation and functionalization of chitosan (CS)-based hydrogels, with emphasis on reversible carbamate chemistry, polyelectrolyte complexation (PEC) with sodium alginate (SA), and silver nanoparticle (AgNP) incorporation. Taken together, these investigations establish a molecular-to-macroscopic framework for understanding how chemical modifications, electronic interactions, and nanoscale organization converge to produce hydrogels with distinct structural and functional attributes.

A key outcome of this study is the demonstration that CS, traditionally limited by its solubility under neutral or weakly basic conditions, can be effectively solubilized through reversible carbamate formation at the C-2 amino position. This temporary masking strategy not only expands the processing window of CS but also enables homogeneous mixing with SA, thereby overcoming the long-standing issue of structural heterogeneity in directly prepared SA–CS hydrogels. Subsequent controlled acidification regenerates protonated CS and drives uniform PEC formation, which is further stabilized at the electronic level through HOMO–LUMO interactions between ammonium and carboxylate groups. These findings highlight the importance of transient chemical modifications in broadening the versatility of naturally derived polymers and in fine-tuning their gelation pathways.

Equally significant is the successful incorporation of AgNPs into SA–CS matrices using an alginate-mediated, microwave-assisted route. The resulting SA-AgNPs/CS gels not only displayed uniform nanoparticle dispersion but also exhibited enhanced mechanical integrity, swelling resistance, and sustained silver release. Importantly, the gels were self-supporting—a property rarely achieved in polysaccharide-based hydrogels without the addition of synthetic crosslinkers. This outcome underscores the synergistic effect of polymer–nanoparticle interactions, where carboxylate coordination and electrostatic stabilization act in concert with nanoparticle-mediated crosslinking to reinforce the gel network. The pronounced antibacterial activity against representative Gram-negative and Gram-positive species further validates the biomedical promise of such nanocomposite gels.

From a broader perspective, the advances described here suggest several promising future directions. First, the concept of reversible carbamate chemistry may be extended to other amino-containing biopolymers, thereby generalizing the strategy for neutral-condition processing and homogeneous PEC formation. Second, the electronic level insights into PEC stabilization invite further exploration using computational chemistry and spectroscopy, which could uncover additional design principles for tailoring gel networks with predictable mechanical and biological functions. Third, the incorporation of AgNPs represents only one example of functional nanoparticle integration. Other nanomaterials—such as gold, zinc oxide, or bioactive glass nanoparticles—could be similarly embedded, potentially endowing the gels with multifunctionality, including catalytic, osteogenic, or photothermal properties.

Biomedical translation remains a critical challenge and opportunity. For wound healing, tissue engineering scaffolds, and controlled drug delivery, the balance between mechanical robustness, biocompatibility, and degradation kinetics must be carefully optimized. In particular, while AgNPs confer strong antibacterial activity, their potential cytotoxicity at higher concentrations necessitates systematic studies of dose, release kinetics, and long-term biocompatibility in vivo. Advanced strategies, such as surface modification of AgNPs or co-encapsulation with antioxidant molecules, may mitigate adverse effects while preserving antimicrobial potency.

Finally, the sustainability and scalability of CS-based hydrogel production should not be overlooked. Chitin, the parent polymer of CS, is abundantly available as a byproduct of seafood processing, yet efficient valorization pathways remain underdeveloped. Integrating green chemistry principles, such as solvent-free carbamate modification or microwave-assisted gelation, could render CS-derived hydrogels more economically and environmentally viable. Moreover, extending the application scope beyond biomedical contexts—toward environmental remediation, water purification, or agriculture—would further leverage the inherent biodegradability and ionic responsiveness of these materials.

In conclusion, the integration of carbamate chemistry, PEC formation, and nanocomposite design provides a coherent and versatile framework for engineering next-generation CS-based hydrogels. By uniting molecular-level understanding with macroscopic material performance, this work not only advances fundamental polymer science but also lays a foundation for practical applications in healthcare, biotechnology, and environmental engineering. The future development of CS hydrogels will likely hinge on continued efforts to bridge chemical innovation, nanoscale control, and translational applicability—ensuring that this abundant natural polymer fulfills its potential as a cornerstone of sustainable and multifunctional biomaterials.

Acknowledgement This work was financially supported by Private University Research Branding Project, MEXT, 2016–2020 and in part by the Kansai University Outlay Support for Establishing Research Centers, 2016–2017 "Development and application of biocompatible polymer materials having sol-gel transition".

References

1. Jayakumar R, Menon D, Manzoor K, Nair SV, Tamura H (2010) Biomedical applications of chitin and chitosan based nanomaterials—a short review. Carbohydr Polym 82:227–232
2. Nagahama H, Maeda Y, Kashiki T, Jayakumar R, Furuike T, Tamura H (2009) Preparation and characterization of novel chitosan/gelatin membranes using chitosan hydrogel. Carbohyd Polym 76:255–260
3. Madhumathi K, Tamura H, Nagahama H, Shalumon KT, Selvamuruga N, Nair SV, Jayakumar R (2009) Wet chemical synthesis of chitosan hydrogel-hydroxyapatite composite membranes for tissue engineering applications. Int J Biol Macromol 45:12–15
4. Komoto D, Furuike T, Tamura H (2019) Preparation of polyelectrolyte complex gel of sodium alginate with chitosan using basic solution of chitosan. Int J Biol Macromol 126:54–59
5. Hamman JH (2010) Chitosan based polyelectrolyte complexes as potential carrier materials in drug delivery systems. Mar Drugs 8:1305–1322
6. Ocsoy I, Temiz M, Celik C, Altinsoy B, Yilmaz V, Duman F (2017) A green approach for formation of silver nanoparticles on magnetic graphene oxide and highly effective antimicrobial activity and reusability. J Mol Liq 227:147–152

7. Sathiyaseelan A, Saravanakumar K, Mariadoss A, Wang M (2020) Biocompatible fungal chitosan encapsulated phytogenic silver nanoparticles enhanced antidiabetic, antioxidant and antibacterial activity. Int J Biol Macromol 153:63–71
8. Demchenko V, Riabov S, Sinelnikov S, Radchenko O, Kobylinskyi S, Rybalchenko N (2020) Novel approach to synthesis of silver nanoparticles in interpolyelectrolyte complexes based on pectin, chitosan, starch and their derivatives. Carbohydr Polym 242:116431
9. Furuike T, Komoto D, Hashimoto H, Tamura H (2017) Preparation of chitosan hydrogel and its solubility in organic acids. Int J Biol Macromol 104:1620–1625
10. Oe T, Dechojarassri D, Kakinoki S, Kawasaki H, Furuike T, Tamura H (2023) Microwave-assisted incorporation of AgNP into chitosan-alginate hydrogels for antimicrobial applications. J Funct Biomater 14:199
11. Tamura H, Wada K, Rujiravanit R, Tokura S (2005) Preservation of chitosan aqueous gel under neutral conditions. J Metal Mater Miner 15:19–21
12. Khong TT, Aarstad OA, Kjåk–Bræk G, Draget KI, Vårum KM (2013) Gelling concept combining chitosan and alginate–proof of principle. Biomacromol 14:2765–2771
13. Yury S, Nadya I, Vladimir S (2009) Bionanocomposites formed by in situ charged chitosan with clay. Green Chem 11:1758–1761
14. Kasaai MR (2010) Determination of the degree of N-acetylation for chitin and chitosan by various NMR spectroscopy techniques: a review. Carbohyd Polym 79:801–810
15. Hirai A, Odani H, Nakajima A (1991) Determination of degree of deacetylation of chitosan by ^{1}H-NMR spectroscopy. Polym Bull 26:87–94
16. Mani F, Peruzzini M, Stoppioni P (2006) CO_2 absorption by aqueous NH_3 solutions: speciation of ammonium carbamate. Bicarbonate and carbonate by a ^{13}C NMR study. Green Chem 8:995–1000
17. Lawrie G, Keen I, Drew B, Chandler-Temple A, Rintoul L, Fredericks P, Grøndahl L (2007) Interactions between alginate and chitosan biopolymers characterized using FT-IR and XPS. Biomacromol 8:2533–2541
18. Qin Z, Zheng Y, Wang Y, Du T, Li C, Wang X, Jiang H (2021) Versatile roles of silver in Ag-based nanoalloys for antibacterial applications. Coord Chem Rev 449:214218
19. Rao SS, Saptami K, Venkatesan J, Rekha PD (2020) Microwave–assisted rapid synthesis of silver nanoparticles using fucoidan: characterization with assessment of biocompatibility and antimicrobial activity. Int J Biol Macromol 163:745–755
20. Ran L, Zou Y, Cheng J, Lu F (2019) Silver nanoparticles in situ synthesized by polysaccharides from Sanghuangporus sanghuang and composites with chitosan to prepare scaffolds for the regeneration of infected full-thickness skin defects. Int J Biol Macromol 125:392–403
21. Azizi-Lalabadi M, Garavand F, Jafari S (2021) Incorporation of silver nanoparticles into active antimicrobial nanocomposites: release behavior, analyzing techniques, applications and safety issues. Adv Colloid Interf Sci 293:102440
22. Zhao X, Xia Y, Li Q, Ma X, Quan F, Geng C, Han Z (2014) Microwave-assisted synthesis of silver nanoparticles using sodium alginate and their antibacterial activity. Colloids Surf A Physicochem Eng Asp 444:180–188
23. Dallas P, Sharma VK, Zboril R (2011) Silver polymeric nanocomposites as advanced antimicrobial agents: classification, synthetic paths, applications, and perspectives. Adv Colloid Interf Sci 166:119–135

Hiroshi Tamura

Tetsuya Furuike

Chemical Synthesis of Chito-Hetero-Oligosaccharides

Yuko Yoneda and Toshinari Kawada

Abstract Chito-oligosaccharides (COS) are β-1,4-linked oligosaccharides composed of *N*-acetyl-D-glucosamine (GlcNAc) and/or D-glucosamine (GlcN) residues. Based on their monosaccharide composition, COS are typically classified into chito-homo-oligosaccharides (homo-COS), which consist solely of either GlcNAc or GlcN. In contrast, chito-hetero-oligosaccharides (CHOS) contain both GlcNAc and GlcN residues. The diverse biological activities of COS are closely linked to their chemical structures. However, elucidation of the structure–activity relationships of COS is still in its early stages, primarily due to the lack of established methods for preparing COS with well-defined chemical structures. In this context, chemical structure refers to two key parameters: the degree of polymerization and the pattern of GlcNAc/GlcN arrangement. In high-molecular-weight chitin and chitosan, the GlcNAc/GlcN ratio—referred to as the fraction of acetylation—is frequently discussed. However, at the oligosaccharide level, the precise arrangement of GlcNAc and GlcN residues is generally of greater importance. Various methods have been explored for the preparation of COS, including hydrolysis and enzymatic degradation of high-molecular-weight chitin, chemoenzymatic synthesis, and organic synthesis. This chapter focuses on the organic synthesis of COS and is intended for readers who may not be specialists in organic synthesis. It begins with an overview of general oligosaccharide synthesis, followed by methodologies for homo-COS, challenges in CHOS synthesis, and future perspectives.

Y. Yoneda
College of Agriculture, Academic Institute, Shizuoka University, Shizuoka, Japan
e-mail: yoneda.yuko@shizuoka.ac.jp

T. Kawada (✉)
Graduate School of Life and Environmental Sciences, Kyoto Prefectural University, Kyoto, Japan
e-mail: kawada@kpu.ac.jp

S.-L. Wang, M. Anpo (eds.), *Chitin and Chitosan*, Springer Series on Polymer and Composite Materials, https://doi.org/10.1007/978-981-95-7457-5_4

1 Introduction

1.1 Definition of Chito-Hetero-Oligosaccharide

Chitin, a polysaccharide composed of numerous *N*-acetyl-D-glucosamine (GlcNAc) residues linked via β-1,4-glycosidic bonds, serves as the precursor of chitosan. Chitosan is the de-*N*-acetylated derivative of chitin and consists primarily of D-glucosamine (GlcN) residues. Traditionally, the term "*chitin*" refers to polysaccharides that retain a substantial portion of *N*-acetyl (Ac) groups, whereas "*chitosan*" refers to polysaccharides with a significant degree of deacetylation, although some residual Ac groups may remain [1]. Collectively, chitin and chitosan are referred to as chitinous materials.

Chito-oligosaccharides (COS) are produced by partial depolymerization of chitinous materials (Fig. 1). That is, COS are oligosaccharides derived from β-1,4-glucan composed of GlcNAc, GlcN, or a combination of both. COS composed solely of GlcNAc or GlcN are termed chito-homo-oligosaccharides (homo-COS), whereas those containing both GlcNAc and GlcN residues are referred to as chito-hetero-oligosaccharides (CHOS). COS have attracted considerable attention due to their notable biological activities, including antimicrobial, antitumor, elicitor, and immunomodulatory properties [2]. These activities are attributed to the chemical structure of COS, particularly the degree of polymerization (DP) and the arrangement of GlcNAc and GlcN residues, also referred to as the pattern of acetylation (PA). When only the proportion of GlcNAc residues or *N*-Ac groups is considered, terms such as degree of acetylation (DA), degree of deacetylation (DDA), or fraction of acetylation (FA) are typically used.

In the early stages of COS research, the concept of structure–activity relationships was not widely recognized, and the use of PA notation was either unnecessary or impractical. This was primarily due to the difficulty of obtaining COS with precisely defined DP and PA. In recent years, however, the need to clarify the details of biological activity has made the use of well-defined COS—characterized by both DP and PA—essential. This chapter discusses the precise preparation of structurally defined COS via organic synthesis, along with an overview of the historical background of this research field.

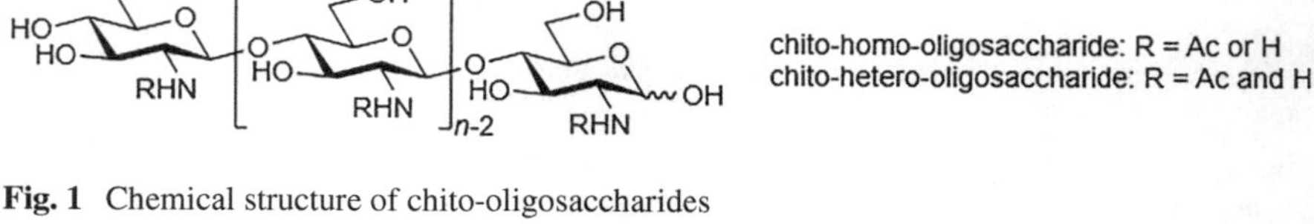

Fig. 1 Chemical structure of chito-oligosaccharides

1.2 Preparation Methods of Chito-Oligosaccharides

To date, several approaches have been developed for the preparation of COS, with the most recognized methods including the depolymerization (or degradation) method, enzymatic synthetic method, organic synthetic method, and automated synthetic method.

Depolymerization involves cleaving glycosidic bonds in high-molecular-weight chitinous materials using chemical reagents, resulting in low-molecular-weight COS [3]. Reported examples include the synthesis of $(GlcNAc)_4$ and $(GlcN)_4$ via acid hydrolysis with hydrochloric acid (HCl), $(GlcNAc)_4$ via acetolysis with acetic acid, and $(GlcNAc)_{12}$ via fluorohydrolysis with hydrogen fluoride (HF). Additionally, studies have investigated the DP and DA of COS with DP up to dodecasaccharides, prepared using concentrated HCl. Although relatively straightforward, this method typically yields a mixture of low-molecular-weight fractions and monosaccharides, requiring advanced purification techniques to isolate specific oligosaccharide species.

Enzyme-based depolymerization methods offer greater specificity, as the resulting COS depend on the substrate preferences of the enzymes employed [4]. Endo-chitinase (EC 3.2.1.14) cleaves internal glycosidic bonds at random sites, while exo-chitinase (EC 3.2.1.29) targets terminal residues. Notable examples include the production of COS up to $(GlcNAc)_6$ and various tetrasaccharides bearing a reducing end GlcNAc residue. A significant milestone in this area was the successful isolation of all mono-acetylated tetrasaccharides—GlcN-GlcN-GlcN-GlcNAc, GlcN-GlcN-GlcNAc-GlcN, GlcN-GlcNAc-GlcN-GlcN, and GlcNAc-GlcN-GlcN-GlcN—achieved through precise enzymatic specificity and advanced chromatographic techniques [5].

Chemoenzymatic synthesis is a hybrid approach that takes advantage of the reversibility of enzymatic reactions. Glycosidic hydrolases, which typically catalyze hydrolysis, can be used under suitable conditions to promote glycosylation in the reverse direction [6]. During catalysis, an enzyme–substrate complex is formed and proceeds through a transition state. By carefully designing substrate structures and optimizing reaction conditions, the glycosylation pathway can be selectively favored over hydrolysis. Drawing on insights from chemical glycosylation and enzymology, a strategy using 1,2-oxazoline derivatives as donor substrates was developed. This method has enabled the synthesis of COS with defined DP, including $(GlcNAc)_6$ and $(GlcNAc)_7$, and offers promise for the flexible synthesis of CHOS with specific PA.

Organic synthesis methods, discussed in detail in Sect. 2 and beyond, employ a bottom-up strategy. This approach entails the stepwise assembly of COS by linking monosaccharide units—typically GlcN derivatives—via β-1,4-glycosylation reactions. Although this method enables precise control over DP and PA, it involves numerous reaction steps and sophisticated experimental techniques. Consequently, it is often regarded as labor-intensive and technically demanding. Nevertheless, extensive research has yielded substantial knowledge, particularly concerning the reactivity and protection strategies of GlcNAc and GlcN derivatives.

More recently, advances in organic synthesis have facilitated the development of automated glycan assembly (AGA). In solid-phase synthesis, the reducing end monosaccharide derivative is immobilized on a solid support [7]. The synthesis cycle involves repeated removal of temporary protecting groups (PGs) to expose a free hydroxyl group (-OH) at the glycosylation site, followed by addition of the next monosaccharide derivative. This method has enabled the synthesis of COS with defined DP and specific PA, extending up to hexasaccharides. In contrast, automated electrochemical synthesis extends the reducing end via anodic oxidation [8]. This technique employs a thioglycoside derivative at the non-reducing end, which reacts with an acceptor bearing a free hydroxyl group at the glycosylation site. COS such as $(GlcNAc)_4$ and $(GlcNAc)_5$ have been synthesized successfully using this approach.

In summary, COS were initially prepared by depolymerizing polysaccharides using chemical or enzymatic methods. Subsequently, bottom-up organic synthesis from monosaccharide units was explored. The extensive knowledge accumulated through these studies has laid the foundation for chemoenzymatic synthesis and automated glycan assembly, leading to significant achievements in the field. Although purely organic synthetic methods have not yet reached their full potential, their significance is expected to grow with the advancement of new strategies and technologies.

This chapter focuses on purely organic synthesis approaches, offering an overview of past efforts and perspectives on future development in COS and CHOS synthesis.

2 Outline of Organic Synthesis of Oligosaccharide

This section provides a comprehensive overview of oligosaccharide synthesis via organic synthetic methods, with particular emphasis on the synthesis of COS.

2.1 Formation of Glycosidic Bond

The organic synthesis of oligosaccharides involves the stepwise assembly of carbohydrate units to construct structurally defined glycans. The key reaction in this process is glycosylation, which connects carbohydrate units through glycosidic bonds. This reaction may occur between two monosaccharides, between a monosaccharide and an oligosaccharide, or between two oligosaccharides, enabling both stepwise and convergent construction of complex glycans.

Formally, glycosylation is a dehydration condensation between the hydroxyl group at the binding site of the glycosyl acceptor, which does not provide an

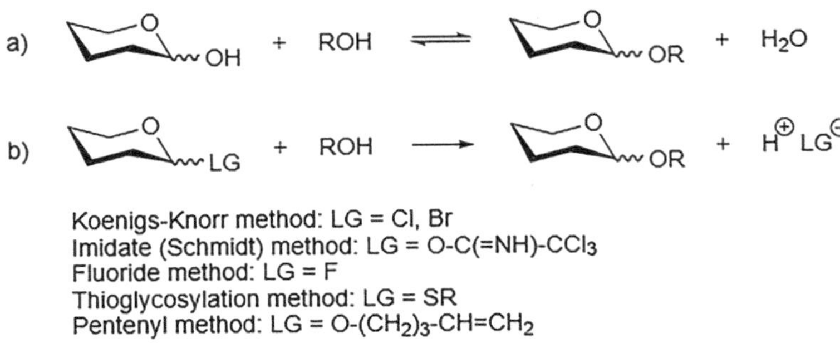

Fig. 2 Mechanism of glycosidic bond formation

anomeric carbon, and the hydroxyl group at *C*-1 of the glycosyl donor, which contributes the anomeric carbon of the glycosidic bond (Fig. 2a). Dehydration of two hydroxyl groups using sulfuric acid (H_2SO_4) or similar reagents is reversible, and shifting the equilibrium toward product formation is considered difficult [9]. Therefore, a strategy involving the introduction of a leaving group (LG) at the anomeric position of the glycosyl donor is used to facilitate irreversible glycosylation reactions (Fig. 2b) [10].

The Koenigs–Knorr method, which employs bromine (Br) as the leaving group, is the earliest glycosylation technique and remains in use. The imidate method (or Schmidt method), which utilizes the trichloroacetamide group (O-C(=NH)-CCl_3) as the leaving group, is currently among the most widely adopted. Other examples relevant to COS synthesis include the fluoride method (LG = F), the thioglycosylation method (LG = SR), and the pentenyl method (LG = O-$(CH_2)_3$-CH=CH_2).

In oligosaccharide synthesis, three types of control are required during the glycosylation reaction: regioselective control, degree of polymerization (DP) control, and stereoselective control. These aspects are discussed in detail in the following sections.

2.2 *Regioselective Control*

In the organic synthesis of oligosaccharides, regioselective control is essential to ensure that glycosidic bonds form at the intended position—specifically, the 1,4-linkage in the case of COS. Since the glycosyl donor provides the anomeric carbon (*C*-1), positional control is primarily achieved by preparing the glycosyl acceptor with a free hydroxyl group at the target bonding site. For COS synthesis, this typically requires that the acceptor possesses a free hydroxyl group exclusively at *C*-4.

Without appropriate positional control, glycosyl acceptor **2**, which possesses free hydroxyl groups at *C*-2, *C*-3, and *C*-4, may undergo glycosylation at unintended positions. As a result, undesired 1,2- (**3**) or 1,3-linkages (**4**) may form

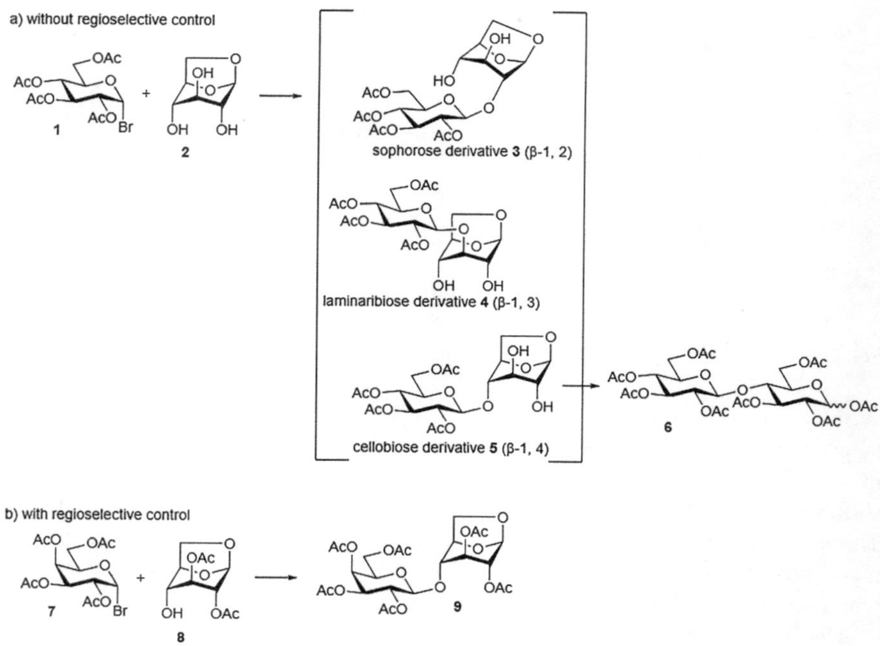

Fig. 3 Regioselective control in glycosylation reactions

alongside the desired 1,4-glycosidic bond (**5**) (Fig. 3a). In contrast, when glycosyl acceptor **8** is used—with the hydroxyl groups at *C*-2 and *C*-3 selectively protected—glycosylation proceeds regioselectively to afford the desired 1,4-linkage (**9**) (Fig. 3b) [11].

Thus, positional control is typically achieved by masking all hydroxyl groups not involved in the glycosylation reaction using persistent protecting groups. In COS synthesis, this includes the hydroxyl groups at *C*-3 and *C*-6, as well as the amino group (-NH_2) at *C*-2, which must also be protected with an appropriate group. The selection and function of protecting groups for the amino group at *C*-2 are discussed in detail in Sect. 3.1.

2.3 Degree of Polymerization Control

To achieve precise control over the DP in oligosaccharide synthesis, glycosylation reactions must be designed so that each step forms a single glycosidic bond. If the same compound serves as both the glycosyl donor and acceptor, uncontrolled polymerization may occur, resulting in a mixture of products with varying DP and polydispersity (Scheme 1) [12].

To avoid this, glycosyl donors and acceptors are typically prepared as distinct compounds [13]. However, in the absence of specific design strategies, multiple

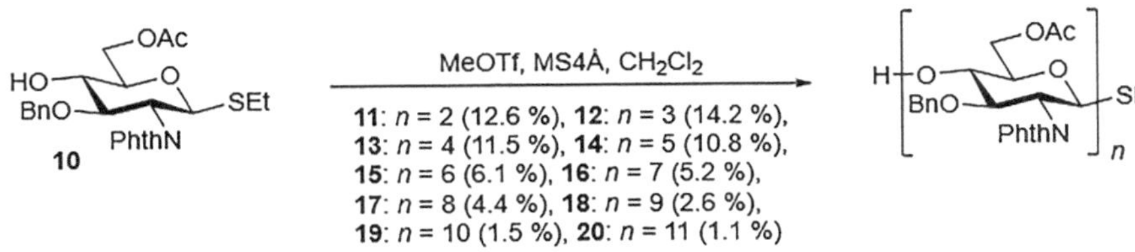

Scheme 1 Synthesis of chito-oligosaccharide derivatives up to undecasaccharide by polycondensation

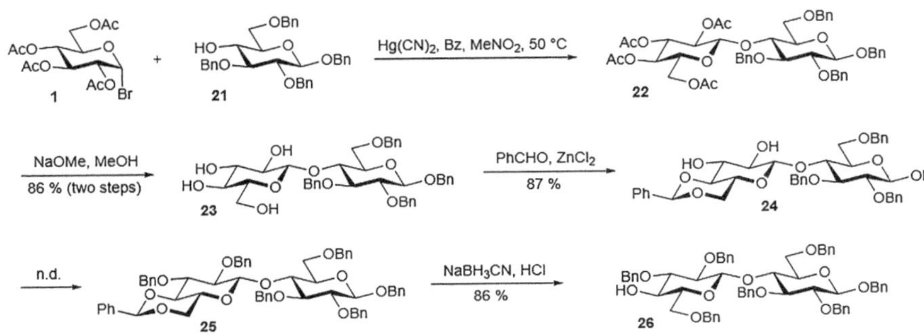

Scheme 2 Conversion of compound **22** to regioselectively controlled acceptor **26**

synthetic steps are often required to convert a disaccharide into a suitable acceptor for further elongation. For example, one approach requires four additional steps to convert a disaccharide intermediate (**22**) into a glycosyl acceptor (**26**) for the synthesis of tri- and tetrasaccharide derivatives (Scheme 2).

To streamline this process, a convergent synthesis strategy has been developed as a more efficient alternative (Fig. 4) [14]. In this method, functional groups are introduced in advance at *C*-1 and *C*-4—corresponding to the glycosylation sites—while the remaining hydroxyl groups (*C*-2, *C*-3, and *C*-6) are protected with appropriate substituents (R groups). The glycosyl donor (**29**) is generated in two steps: removal of the X group at *O*-1, followed by introduction of a leaving group. The corresponding acceptor (**30**) is prepared in a single step by removing the Y group at *O*-4. The resulting disaccharide (**31**) retains the same protecting group pattern as the starting monosaccharide (**27**), enabling iterative application of the same synthetic sequence to construct oligosaccharides with higher DP in a modular and efficient manner.

2.4 Stereoselective Control

Stereoselective control is a critical aspect of glycosylation in oligosaccharide synthesis, as the configuration of the glycosidic bond—whether α or β—directly influences the biological and physicochemical properties of the resulting oligosaccharide. The glycosylation reaction involving a leaving group, as described in Sect. 2.1,

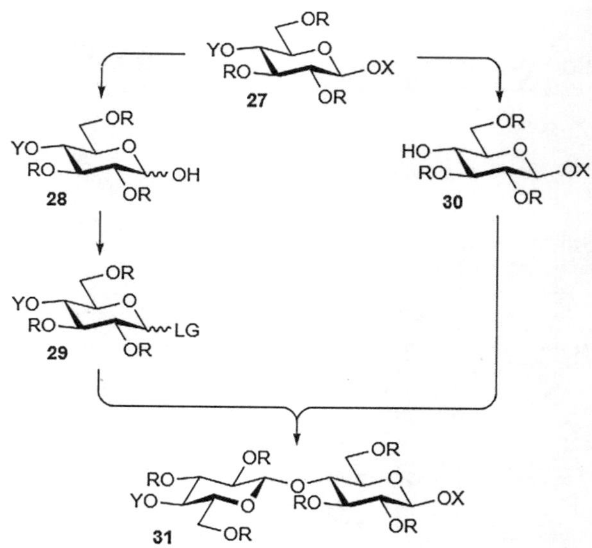

Fig. 4 Convergent synthetic strategy for oligosaccharide

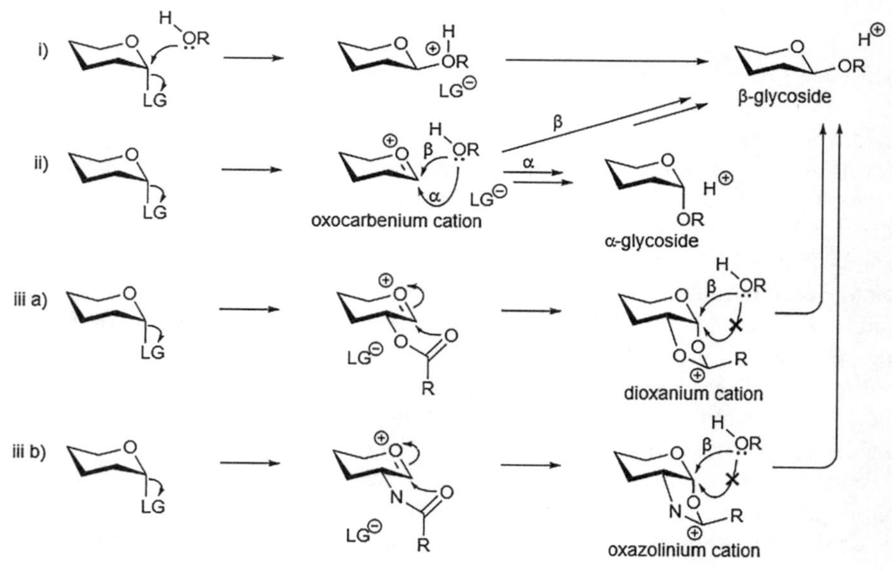

Fig. 5 Reaction mechanism of glycosylation reaction

proceeds via distinct mechanistic pathways that determine the stereochemical outcome (Fig. 5).

If the leaving group is displaced by a direct S_N2-type attack from the glycosyl acceptor, the reaction proceeds with inversion of configuration, resulting in a stereospecific glycoside (Fig. 5i). Alternatively, if the leaving group departs before nucleophilic attack, an oxocarbenium cation intermediate is formed (Fig. 5ii). In this case, the glycosyl acceptor may attack either face of the planar intermediate, leading to a mixture of anomers.

To achieve stereoselective control, neighboring group participation is commonly employed. For example, when an ester group (Fig. 5iiia) or an amide group (Fig. 5iiib) is present at *C*-2 of the glycosyl donor, it can stabilize the oxocarbenium cation by forming a 1,2-cyclic intermediate—such as a dioxanium or oxazolinium cation. This restricts the approach of the glycosyl acceptor to one face, thereby favoring the formation of a specific anomer.

Extensive research has explored stereocontrol strategies, including the selection of appropriate leaving groups, the use of persistent protecting groups, and the optimization of reaction conditions such as solvent and temperature. In COS synthesis, the choice of protecting group for the amino group at *C*-2 is particularly critical, as it determines the potential for neighboring group participation and thus affects both the efficiency and stereoselectivity of glycosylation. The next section provides a detailed overview of protecting groups and latent protecting groups for the amino group at *C*-2, which are essential for achieving both positional and stereoselective control in COS synthesis.

3 Synthesis of Chito-Oligosaccharide

3.1 Protecting Group and Latent Protecting Group for Amino Group

In the chemical synthesis of COS, selecting an appropriate protecting group for the amino group at *C*-2 is of paramount importance. This group plays a decisive role in both positional control, as discussed in Sect. 2.2, and stereoselective control, as elaborated in Sect. 2.4. Although a wide variety of protecting and latent protecting groups have been reported for the amino group at *C*-2, this section focuses on those with proven utility in COS synthesis or in the synthesis of GlcN derivatives—particularly those that support efficient glycosylation and reliable deprotection. A latent protecting group is not a conventional protecting group, but rather one that can be readily converted into a free amino group. For example, azide groups (-N_3) and nitro groups (-NO_2) are not conventional protecting groups for amino groups, but both can be readily converted into free amino groups by reduction.

GlcNAc inherently possesses an *N*-Ac group, and although it may seem convenient to use this group directly as a protecting group for the free amino group, it presents significant challenges in glycosylation reactions. Under the mildly acidic conditions typically employed, the Ac group participates in neighboring group interactions, forming a stable 1,2-oxazoline intermediate via a 1,2-cyclic oxonium cation. This intermediate inhibits glycosylation by diverting the reaction pathway. Glycosylation using 1,2-oxazoline intermediates—sometimes referred to as the oxazoline method—has shown high efficiency in certain contexts [15], but no successful examples have been reported in the purely organic synthesis of COS. In contrast, methyloxazoline derivatives have been used effectively in chemoenzymatic synthesis [6] (a), with considerable success.

Representative protecting groups other than the Ac group can be broadly classified into Ac group-based, carbonyl group-based, and cyclic group-based systems (Table 1). Ac group-based systems include the monochloroacetyl (MCA) group (-C(=O)-CH_2Cl) [15], the dichloroacetyl (DCA) group (-C(=O)-$CHCl_2$) [16], the trichloroacetyl (TCA) group (-C(=O)-CCl_3), and the trifluoroacetyl (TFA) group (-C(=O)-CF_3) [17]. These groups offer varying degrees of electron-withdrawing capacity and steric hindrance, which influence both the reactivity and stability of the protected intermediates. Carbonyl group-based systems, such as the trichloroethoxycarbonyl (Troc) group (-C(=O)-O-CH_2-CCl_3) [18] and the benzyloxycarbonyl (Cbz or Z) group (-C(=O)-O-CH_2-Ph) [19], have also demonstrated reliable performance in COS synthesis, particularly due to their compatibility with standard deprotection protocols. Cyclic group systems, notably the phthaloyl (Phth) group [20] (a) and the dimethylmaleoyl (DMM) group [20] (b), have shown exceptional effectiveness in COS synthesis. These groups not only provide robust protection during glycosylation but also facilitate stereoselective control through neighboring group participation. Latent protecting groups include the azide group, as mentioned earlier.

The choice of protecting group must be made with consideration of the entire synthetic sequence, including the compatibility of deprotection steps and the potential for neighboring group participation. In subsequent sections, the application of these groups in the synthesis of homo-COS and CHOS will be discussed in detail, highlighting their influence on yield, stereoselectivity, and overall synthetic efficiency.

Table 1 Commonly used protecting and latent protecting groups for amino group at *C*-2

Acetyl group-based	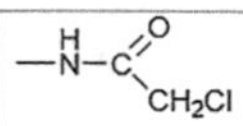 *N*-monochloroacetyl (MCA)	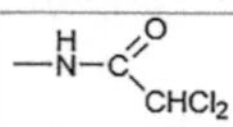 *N*-dichloroacetyl (DCA)	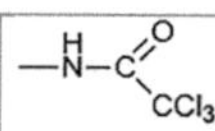 *N*-trichloroacetyl (TCA)	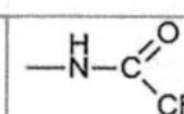 *N*-trifluoroacetyl (TFA)
Carbonyl group-based	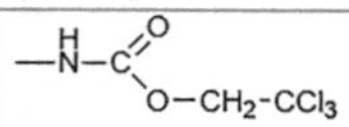 *N*-trichloroethoxycarbonyl (Troc)		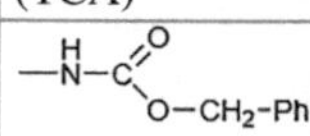 *N*-benzyloxycarbonyl (Cbz or Z)	
Cyclic group-based	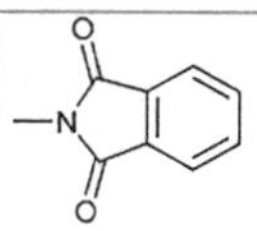 *N*-phthaloyl (Phth)		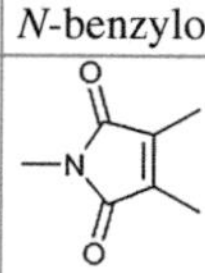 *N*-dimethylmaleoyl (DMM)	
Latent protecting group	—N_3 Azide		—NO_2 Nitoro	

3.2 Synthesis of Chito-Homo-Oligosaccharide

An overview of the synthesis of homo-COS is shown in Fig. 6. The process begins with GlcN derivative **32**, bearing appropriate protecting groups. Glycosyl donor **34** is generated by removing the temporary protecting group at *O*-1 (X group) and introducing a leaving group, while the corresponding glycosyl acceptor **35** is prepared by removing the temporary protecting group at *O*-4 (Y group).

The glycosylation reaction between **34** and **35** affords disaccharide **36**, which retains the same protecting group pattern as the starting monosaccharide **32**. This enables iterative elongation using the same synthetic sequence until the desired DP is reached. Global deprotection of the disaccharide **36** yields the target COS **37**.

Representative examples are summarized in Table 2. In these syntheses (see Fig. 6), cyclic protecting groups such as the Phth group and the DMM group were employed as protecting groups for the amino group at *C*-2. Kuyama et al. synthesized a series of homo-COS with varying DP, up to a dodecasaccharide (DP = 12), using the Phth group as the amino protecting group (R^1), the *p*-methoxyphenyl (MP) group as the temporary protecting group at *O*-1 (X), the Ac group as the temporary

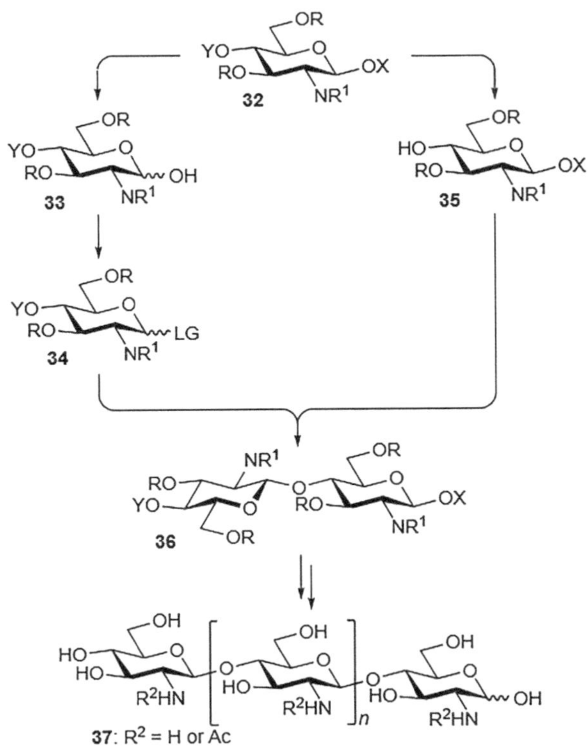

Fig. 6 Synthetic overview of chito-homo-oligosaccharides

Table 2 Examples of chito-oligosaccharides synthesis

Kuyama et al. [21]

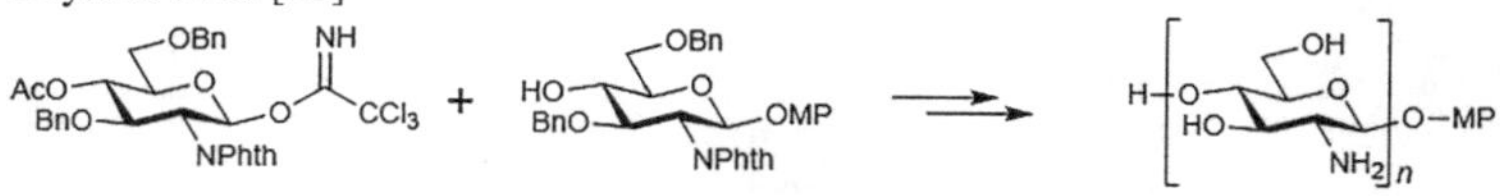

X (PG at *O*-1)	R^1 (PG at *N*)	Y (PG at *O*-4)	R (PG at *O*-3 and 6)	LG (glycosylation method)	max. DP
MP (methoxyphenyl)	Phth	Ac	Bn	O-C(=NH)-CCl_3 (Schmidt method)	12

Aly et al. [20] (c)

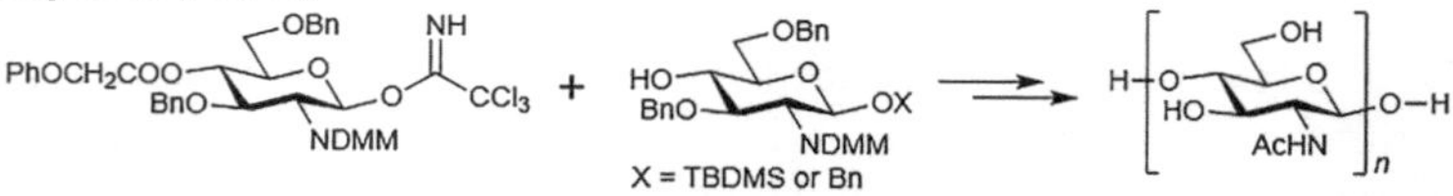

X (PG at *O*-1)	R^1 (PG at *N*)	Y (PG at *O*-4)	R (PG at *O*-3 and 6)	LG (glycosylation method)	max. DP
TBDMS or Bn	DMM	$PhOCH_2CO$	Bn	O-C(=NH)-CCl_3 (Schmidt method)	6

protecting group at *O*-4 (Y), and the benzyl (Bn) group as the persistent protecting group (R) [21]. Aly et al. constructed a series of homo-COS with varying DP, reaching up to a hexasaccharide (DP = 6), by employing the DMM group as the amino protecting group (R^1), the *tert*-butyldimethylsilyl (TBDMS) or Bn group as the temporary protecting group at *O*-1 (X), the Ac group at *O*-4 (Y), and the Bn group as the persistent protecting group (R) [20] (c).

4 Synthesis of Chito-Hetero-Oligosaccharide

4.1 Basic Strategy for Synthesis of Chito-Hetero-Oligosaccharide

The synthetic strategy for CHOS is illustrated in Fig. 7 [22, 23]. In contrast to the synthesis of homo-COS (Fig. 6), CHOS synthesis necessitates the preparation of two structurally distinct monosaccharide derivatives (**38a** and **38b**): one intended to be converted into a GlcNAc residue and the other into a GlcN residue. Protecting groups for the amino group at *C*-2—R^{1a} and R^{1b}—must be selected to allow orthogonal removal, thereby enabling selective deprotection.

Sequential glycosylation between intermediates **40** and **41** establishes the desired PA and DP. Selective removal of either the R^{1a} or R^{1b} group yields a free amino group. This is then acetylated to form an acetamide group (-NHAc), thereby producing the GlcNAc residue. Finally, the remaining amino protecting group (R^{1a} or R^{1b}), along with all other protecting groups (R, X, Y), is removed to afford the target CHOS **43**.

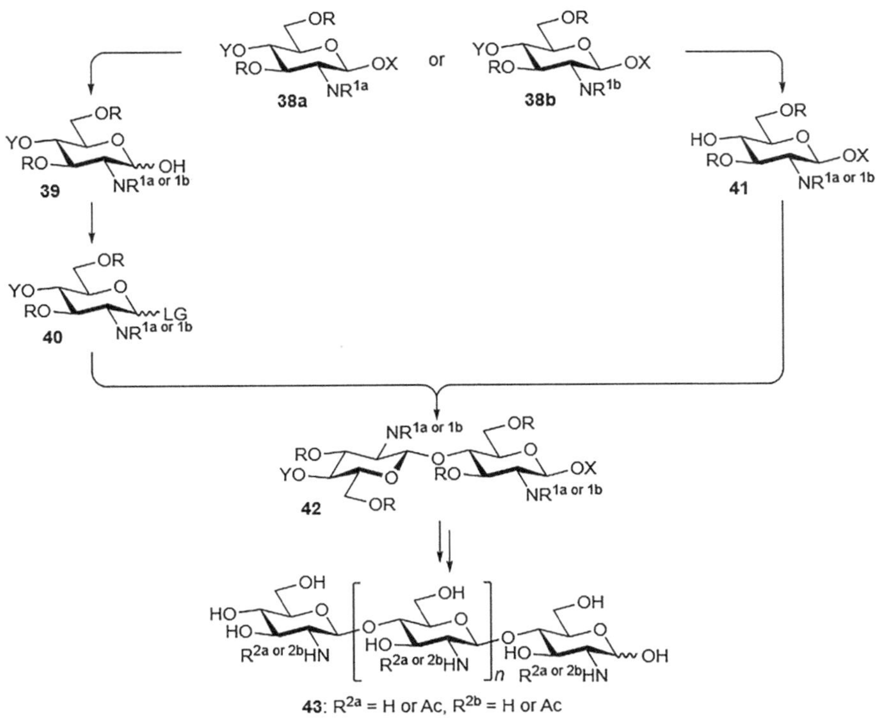

Fig. 7 Synthetic strategy for chito-hetero-oligosaccharide synthesis

This strategy enables the synthesis of CHOS with any desired PA, including non-repetitive sequences. The compatibility and orthogonality of R^{1a} and R^{1b} are crucial, as they influence both glycosylation efficiency and the practicality of selective deprotection. These aspects are discussed in Sect. 4.3.

4.2 Synthesis of Chito-Hetero-Oligosaccharide

The first synthesis of CHOS is outlined in Schemes 3 and 4 [22]. In this example, the azide (a latent protecting group) and Phth (protecting group) are used as R^{1a} and R^{1b}, respectively, for the amino group at *C*-2 (see Fig. 7).

Scheme 3 outlines the synthetic route for the preparation of the monosaccharide derivatives **51a/51b** and **52a/52b**. Initially, the free amino group is protected, and the hydroxyl groups are acetylated to yield compounds **45a/45b**. Selective removal of the Ac group at *O*-1 affords **46a/46b**. The reactive hemiacetal hydroxyl group is temporarily protected with a TBDMS group, yielding **47a/47b**. Subsequent removal of the three Ac groups at *O*-3, *O*-4, and *O*-6 produces **48a/48b**. Selective protection

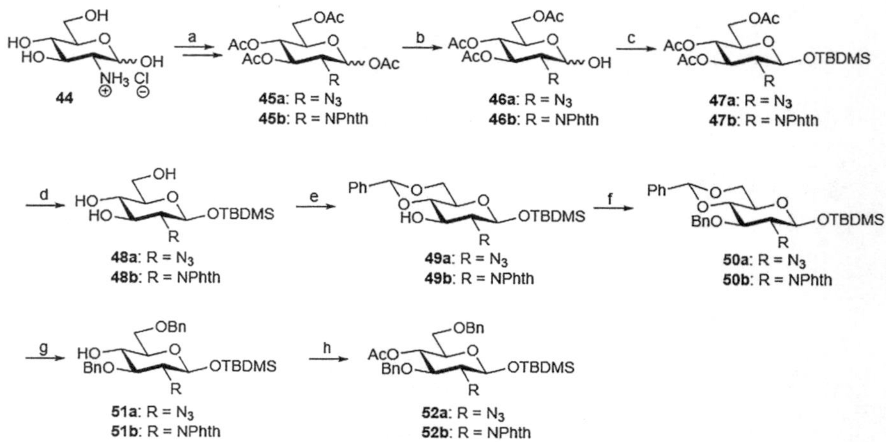

a: for **45a**: (i) imidazole-1-sulfonyl azide, K_2CO_3, $CuSO_4$, MeOH, 0 °C → rt, 5 d, (ii) Ac_2O, Pyr, DMAP, 0 °C → rt, overnight (overall 41 %); for **45b**: (i) MeONa, MeOH, rt, 12 min, (ii) phthalic anhydride, Et_3N, 0 °C → rt, 80 min, (iii) Ac_2O, Pyr, DMAP, 0 °C → rt, overnight (overall 73 %), b: $H_2NNH_2 \cdot AcOH$, DMF, rt, 30 min (90 % for **46a**; 73 % for **46b**), c: TBDMSCl, imidazol, $CHCl_3$, 0 °C → rt, 1 h (74 % for **47a**; 95 % for **47b**), d: MeONa, MeOH, 0 °C → rt, 2.5 h (quant. for **48a**; 93 % for **48b**), e: $PhCH(OMe)_2$, *p*-TsOH·H_2O, dioxan, 30 °C, 70 mmHg, 1 h (78 % for **49a**; 71 % for **49b**), f: BnBr, NaH, TBAI, THF, 0 °C → rt, overnight (78 % for **50a**; 56 % for **50b**), g: TES, TFA, $CHCl_3$, 0 °C → rt, 2.5 h (55 % for **51a**; 64 % for **51b**), h: Ac_2O, Pyr, DMAP (quant. for **52a** and **52b**).

Scheme 3 Synthetic routes for monosaccharide derivatives

of the *O*-3 position is achieved by forming a cyclic acetal between *O*-4 and *O*-6, resulting in **49a/49b**. The free hydroxyl group at *C*-3 is then protected with a Bn group to obtain **50a/50b**. Regioselective reduction of the acetal affords a free hydroxyl group at *C*-4 (**51a/51b**), which acts as the glycosyl acceptor for 1,4-glycosidic bond formation. Protection of the *C*-4 hydroxyl group with an Ac group, followed by removal of the TBDMS group, affords the glycosyl donor **52a/52b**.

Scheme 4 outlines the synthesis of a tetrasaccharide derivative **58** and a hexasaccharide derivative **60**. In this example, CHOS with a GlcNAc-GlcN repeating structure is synthesized. However, the methodology is applicable to any desired PA. Glycosylation of the Phth acceptor **54** with the azide donor **53** yields the disaccharide derivative **55**. This intermediate retains the same protecting group pattern as the monosaccharide derivatives **52a/52b**—specifically, Bn as the R group, TBDMS as the X group, and Ac as the Y group (see Fig. 7). Therefore, the same reaction sequence can be iteratively applied to extend the DP.

In this manner, the disaccharide imidate **56** was prepared by removal of the TBDMS group followed by imidoylation. The corresponding disaccharide acceptor **57** was obtained by removal of the Ac group. The glycosylation reaction between **56** and **57** affords the tetrasaccharide derivative **58**. The tetrasaccharide acceptor **59** was prepared by deacetylation of **58**. It was glycosylated with the disaccharide imidate **56**, yielding the hexasaccharide derivative **60**.

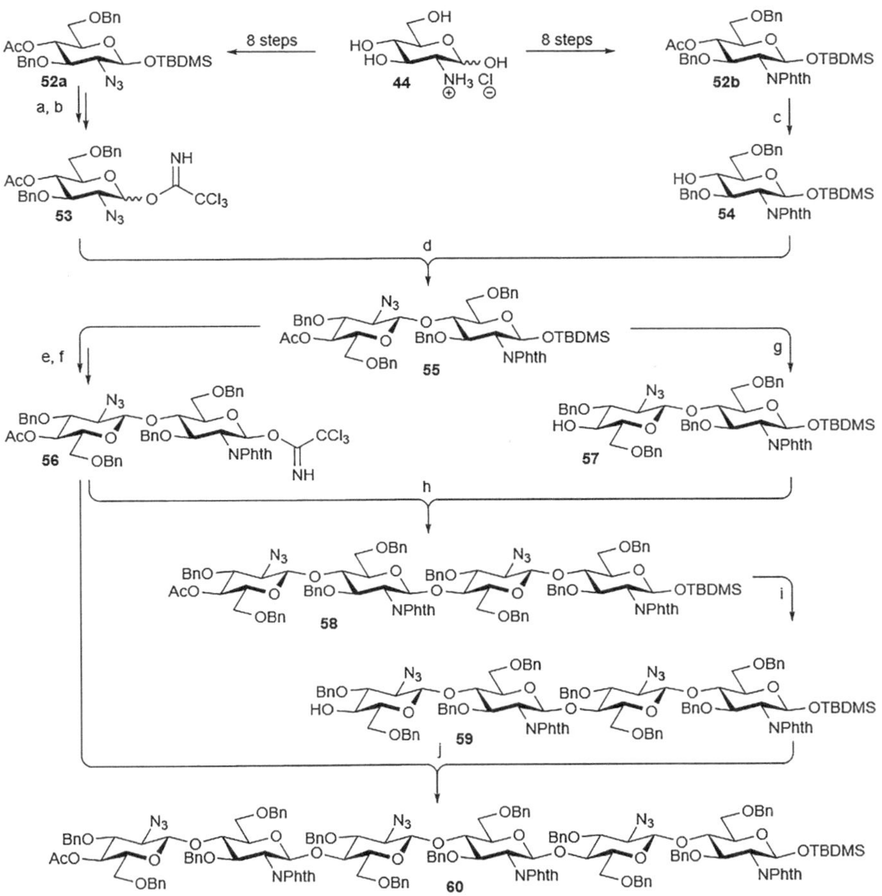

a: Bu_4NF, AcOH, THF, 0 °C, 1 h (80 %), b: CCl_3CN, DBU, CH_2Cl_2, rt, 40 min (69 %), c: MeONa, MeOH, rt, 1 h (quant.), d: TMSOTf, $CHCl_3$, −20 °C, 10 min (26 %), e: Bu_4NF, AcOH, THF, 0 °C → rt, 1.5 h (82 %), f: Cl_3CCN, DBU, $CHCl_3$, rt, 30 min (81 %), g: NaOMe, MeOH, 0°C → rt, 90 min (78 %), h: TMSOTf, $CHCl_3$, −20 °C (74 %), i: MeONa, MeOH, 0 °C → rt, 48 h (77 %), j: TMSOTf, $CHCl_3$, −80 °C (35 %).

Scheme 4 Synthesis of chito-hexasaccharide derivative **60**

A key insight from this synthesis is the influence of the amino protecting group on glycosylation efficiency and stereoselectivity. The azide donor **53**, lacking neighboring group participation, results in a low glycosylation yield. In contrast, the Phth donor **56** provides nearly stereospecific β-anomer formation and improved yields. In this strategy, the azide monosaccharide **53** is intentionally used as the donor in the initial glycosylation (reaction d in Scheme 4) to construct the repeating disaccharide unit. This choice prioritizes its performance as a glycosyl donor. As a result, the glycosylation yield for [disaccharide + disaccharide] (76%) is significantly higher than that for [monosaccharide + monosaccharide] (26%).

Table 3 Modified selection of amino protecting group

Kawada and Yoneda [22]

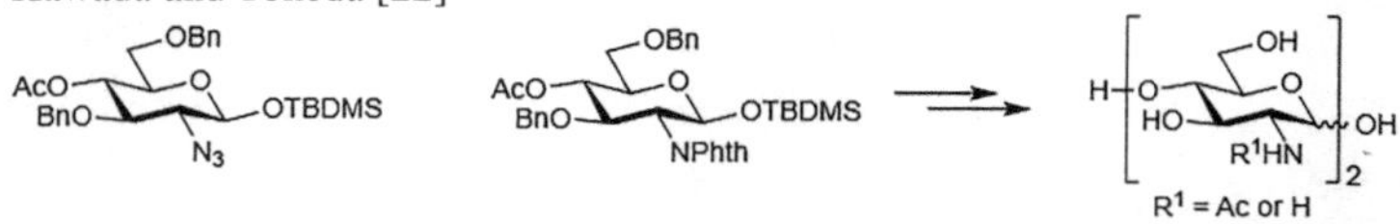

X (PG at *O*-1)	R^1 (PG at *N*)	Y (PG at *O*-4)	R (PG at *O*-3 and 6)	LG (glycosylation method)
TBDMS	N_3/Phth	Ac	Bn	O-C(=NH)-CCl_3 (Schmidt method)

Barroca-Aubry et al. [24]

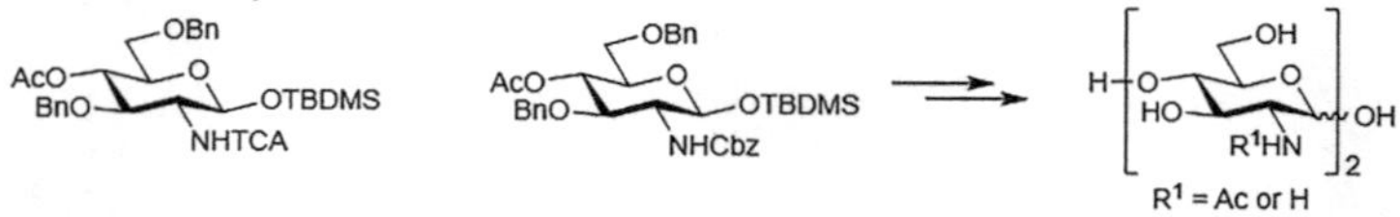

X (PG at *O*-1)	R^1 (PG at *N*)	Y (PG at *O*-4)	R (PG at *O*-3 and 6)	LG (glycosylation method)
TBDMS	TCA/Cbz	Ac	Bn	O-C(=NH)-CCl_3 (Schmidt method)

To address the low glycosylation yield associated with azide donors, an improved method employing TCA and Cbz groups as alternative amino protecting groups has been proposed (Table 3) [24].

4.3 Deprotection Process

Following the construction of the CHOS derivative, the final target compound is obtained by removing all protecting groups and latent protecting groups. At this stage, a strategic decision must be made as to whether the R^{1a} or R^{1b} groups (see Fig. 7) will be converted into an acetamide or a free amine group. In the synthesis described above, the deprotection sequence was carried out according to the pathway shown in Scheme 5 [22].

The combination of the R^{1a} and R^{1b} groups—specifically, azide and Phth—was selected not only to mitigate the low glycosylation yield associated with azide donors, but also to provide a synthetic advantage referred to as "reverse deprotection." In the pathway illustrated in Scheme 5, the Phth group is first removed to generate a free amino group, which is then acetylated to form an acetamide. The azide group is then reduced to yield another amino group. Alternatively, converting the Phth group to an amino group and the azide group to an acetamide group allows the synthesis of two distinct CHOS structures from a single synthetic intermediate. This approach significantly reduces the synthetic workload and enhances the efficiency of sequence-defined CHOS synthesis.

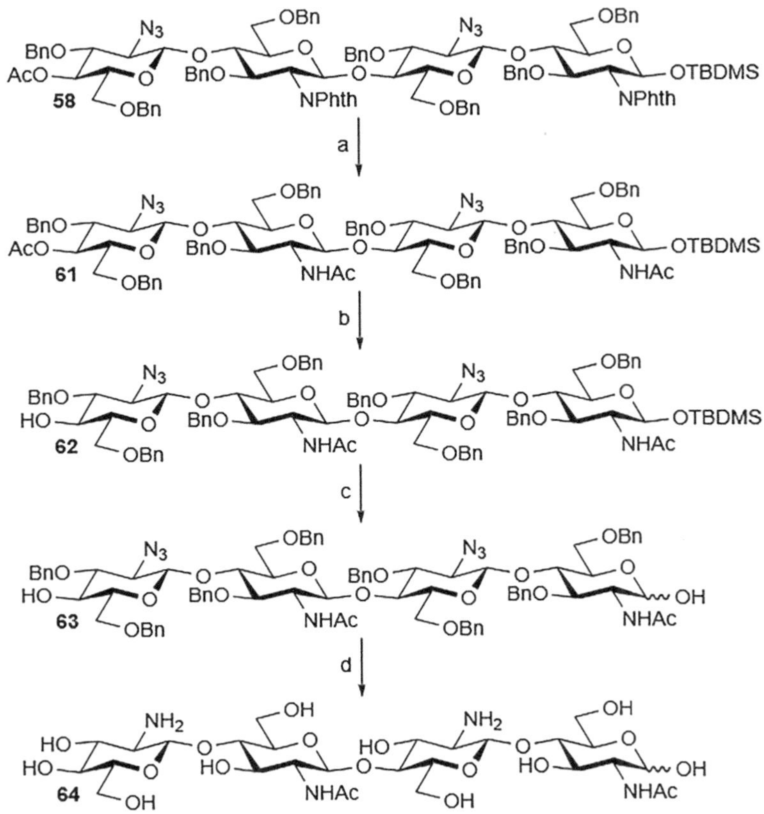

a: (i) $H_2NCH_2CH_2NH_2$, *n*-BuOH, 90 °C, 6 h, (ii) Ac_2O, Pyr, DMAP, 0 °C → rt, overnight (overall 95 %), b: NaOMe, MeOH, 0 °C → rt, overnight (97 %), c: Bu_4NF, AcOH, THF, 0 °C → rt, 32 h (93 %), d: 10 % Pd-C, MeOH, 50 °C, 30 min (72 %).

Scheme 5 Synthesis of chito-hetero-pligosaccharide **64** by deprotection of compound **58**

4.4 Automated Synthesis

Regardless of the target structure, the organic synthesis of oligosaccharides requires extensive experimentation, including iterative trials to optimize reaction conditions. The development of automated synthesis platforms for glycans is therefore a natural and important progression in this field. This section highlights two representative examples of recent progress in the automated synthesis of CHOS.

Seeberger et al. reported the synthesis of four types of CHOS using a solid-phase synthesis approach, in which the reducing-end monosaccharide is immobilized on a solid support. The synthesis cycle involves iterative removal of the 9-fluorenylmethyloxycarbonyl (Fmoc) group, employed as a protecting group at *O*-6. This is followed by glycosylation using a dibutylphosphoryloxy group as the leaving group, catalyzed by trifluoromethanesulfonic acid (TfOH) [7].

In contrast, Nokami et al. developed an electrochemical method utilizing a thioaryl group as the protecting group at *O*-1. This group was converted to an α-triflate via anodic oxidation, enabling catalyst-free glycosylation. Through iterative cycles, a CHOS derivative was synthesized in a one-pot reaction [8].

In automated synthesis, multiple glycosylation steps are performed in sequence without intermediate purification. As a result, if a glycosidic bond with incorrect stereochemistry is formed during the process, it is extremely difficult to remove or correct. Therefore, achieving complete stereocontrol in glycosylation—ideally 100% selectivity—is essential for the efficient synthesis of CHOS with defined PA and high DP, while avoiding labor-intensive purification of anomeric mixtures.

5 Conclusion

Over the past five decades, the preparation of COS has undergone significant advancements. Initial efforts focused on depolymerizing high-molecular-weight chitin to obtain COS with a DP of approximately 6 or 7. These early studies revealed correlations between biological activity and DP and were later extended to explore the influence of the PA.

Subsequent advances enabled the development of purely organic and enzymatic synthetic methods, with early research emphasizing the synthesis of homo-COS. Building on this foundation, the organic synthesis of CHOS has been both theoretically and practically realized. Enzymatic approaches have also successfully produced CHOS with defined PA. More recently, the concept of automated synthesis has been established, enabling the synthesis of CHOS through both solid-phase and electrochemical methods.

Further advancements in automated chemical synthesis—particularly electrochemical and solid-phase techniques—is expected. Integration with enzymatic chemistry is anticipated to play an increasingly important role. However, this does not mean that classical organic synthesis methods have lost relevance. On the contrary, foundational studies, including those involving traditional chemical approaches, continue to be indispensable. A detailed understanding of protecting group stability, stereocontrol strategies, glycosylation catalysts, solvent effects, temperature dependencies, and deprotection protocols continues to play a critical role in driving future breakthroughs.

In closing, we express our sincere respect and gratitude to all researchers who have contributed to the advancement of COS synthesis.

References

1. (a) Piekarska K, Sikora M, Owczarek M, Jóźwik-Pruska J, Wiśniewska-Wrona M (2023) Chitin and chitosan as polymers of the future—obtaining, modification, life cycle assessment and Main directions of application. Polymers 15:793; (b) Giraldo JD, García Y, Vera M, Garrido-Miranda KA, Andrade-Acuña D, Marrugo KP, Rivas BL, Schoebitz M (2024) Alternative processes to produce chitin, chitosan, and their oligomers. Carbohydr Polym 332:121924
2. (a) Zhou J, Wen B, Xie H, Zhang C, Bai Y, Cao H, Che Q, Guo J, Su Z (2021) Advances in the preparation and assessment of the biological activities of chitosan oligosaccharides with different structural characteristics. Food Funct 12:926–951; (b) Liaqat F, Eltem R (2018) Chitooligosaccharides and their biological activities: a comprehensive review. Carbohydr Polym 184:243–259
3. (a) Chang K-LB, Tai M-C, Cheng F-H (2001) Kinetics and products of the degradation of chitosan by hydrogen peroxide. J Agric Food Chem 49:4845–4851; (b) Horowitsza ST, Roseman U, Blumenthal HJ (1957) The preparation of glucosamine oligosaccharides. I Separation. J Am Chem Soc 79:5046; (c) Rupley JA (1964) The hydrolysis of chitin by concentrated hydrochloric acid, and the preparation of low-molecular-weight substrates for lysozyme. Biochim Biophys Acta 83:245–255; (d) Inaba T, Ohgushi T, Iga Y, Hasegawa E (1984) Synthesis of 4-methylcoumarin-7-yloxy tetra-*N*-acetyl-β-chitotetraoside, a novel synthetic substrate for the fluorometricassay of lysozyme. Chem Phram Bull 32:1597–1603; (e) Bosso C, Defaye J, Domard A, Gadelle A, Pedersen C (1986) The behavior of chitin towards anhydrous hydrogen fluoride. Preparation of β-(1-4)-linked 2-acetamido-2-deoxy-D-glucopyranosyl oligosaccharides. Carbohydr Res 156:57–68; (f) Trombotto S, Ladavière C, Delolme F, Domard A (2008) Chemical preparation and structural characterization of a homogeneous series of chitin/chitosan oligomers. Biomacromol 9:1731–1738
4. (a) Pechsrichuang P, Lorentzen SB, Aam BB, Tuveng TR, Hamre AG, Eijsink VGH, Yamabhai M (2018) Bioconversion of chitosan into chito-oligosaccharides (CHOS) using family 46 chitosanase from *Bacillus subtilis* (*Bs*Csn46A). Carbohydr Polym 186:420–428; (b) Ohtakara A, Matsunaga H, Mitsutomi M (1990) Action pattern of *Streptomyces griseus* chitinase on partially *N*-acetylated chitosan. Agric Biol Chem 54:3191–3199
5. Basa S, Nampally M, Honorato T, Das SN, Podile AR, El Gueddari NE, Moerschbacher BM (2020) The pattern of acetylation defines the priming activity of chitosan tetramers. J Am Chem Soc 142:1975–1986
6. (a) Kobayashi S (2007) New developments of polysaccharide synthesis via enzymatic polymerization. Proc Jpn Acad Ser B 83:215–247; (b) Umemoto N, Saito N, Noguchi M, Shoda S, Ohnuma T, Watanabe T, Sakuda S, Fukamizo T (2022) Plant chitinase mutants as the catalysts for chitooligosaccharide synthesis using the sugar oxazoline derivatives. J Agric Food Chem 70:12897–12906; (c) Ogata M, Umemoto N, Ohnuma T, Numata T, Suzuki A, Usui T, Fukamizo T (2013) A novel transition-state analogue for lysozyme, 4-*O*-β-tri-*N*--acetylchitotriosyl moranoline, provided evidence supporting the covalent glycosyl-enzyme intermediate. J Biol Chem 288:6072–6082
7. (a) Tuck OT, Sletten ET, Danglad-Flores J, Seeberger PH (2022) Towards a systematic understanding of the influence of temperature on glycosylation reactions. Angew Chem Int Ed 61:e202115433; (b) Tyrikos-Ergas T, Bordoni V, Fittolani G, Chaube MA, Grafmeller A, Seeberger PH, Delbianco M (2021) Systematic structural characterization of chitooligosaccharides enabled by automated glycan assembly. Chem Eur J 27:2321–2325
8. (a) Nokami T, Isoda Y, Sasaki N, Takaiso A, Hayase S, Itoh T, Hayashi R, Shimizu A, Yoshida J (2015) Automated electrochemical assembly of the protected potential TMG-chitotriomycin precursor based on rational optimization of the carbohydrate building block. Org Lett 17:1525–1528; (b) Isoda Y, Sasaki N, Kitamura K, Takahashi S, Manmode S, Takeda-Okuda N, Tamura J, Nokami T, Itoh T (2017) Total synthesis of TMG-chitotriomycin based on an automated electrochemical assembly of a disaccharide building block. Beilstein J Org Chem 13:919–924

9. Fischer E (1894) Ueber die Glucoside der Alkohle. Ber Dtsch Chem Ges 1894(A 27, T 4):2478–2486
10. (a) Königs W, Knorr E (1901) Ueber einige Derivate des Traubenzuchers und der Galactose. Ber Dtsch Chem Ges 1901/01(A 34, T 1):957–981; (b) Schmidt RR, Michel J (1980) Facile synthesis of α- and β-*O*-glycosyl imidates; preparation of glycosides and disaccharides. Angew Chem Int Ed Engl 19:659–748
11. (a) Freudenberg K (1933) Die Synthese der Cellobiose. Ber Dtsch Chem Ges (A and B Series) 66:27–29; (b) Shapiro D, Rabinsohn Y, Acher AJ, Diver-Habe A (1969) A new micro-scale synthesis of lactose. Biochem Chiophys Res Commun 37:28–30
12. Hashimoto H, Abe Y, Horito S, Yoshimura J (1989) Synthesis of chitooligosaccharides derivatives by condensation polymerization. J Carbohydr Chem 8:307–311
13. Takeo K, Yasato T, Kuge T (1981) Synthesis of α- and β-cellotriose hendecaacetates and of several 6,6′,6″-trisubstituted derivatives of methyl β-cellotrioside. Carbohydr Res 93:148–156
14. Nakatsubo F, Takano T, Kawada T, Someya H, Harada T, Shiraki H, Murakami K (1985) Synthetic studies of cellulose I: synthetic design and selection of the protective groups. Mem Coll Agric Kyoto Univ 127:37–47
15. Kiso M (1985) Protected glycosides and disaccharides of 2-amino-2-deoxy-D-glycopyrnose by ferric chloride-catalyzed coupling. Carbohydr Res 136:309–323
16. Shapiro D, Acher AJ, Rachaman ES (1967) Studies in the ganglioside series. I. Synthesis of 4-*O*-(2-acetamido-2-deoxy-β-D-glucopyranosyl)-D-galactopyranose. J Org Chem 32:3767–3771
17. Wolfrom ML, Bhat HB (1967) Trichloroacetyl and Trifluoroacetyl as N-blocking groups in nucleoside synthesis with 2-amino sugars. J Org Chem 32:1821–1823
18. Dullenkopf W, Castro-Palomino JC, Manzoni L, Schmidt RR (1996) N-Trichloroethoxycarbonyl-glucosamine derivatives as glycosyl donors. Carbohydr Res 296:135–147
19. Lessard J, Driguez H, Vermes JP (1970) The chromous chloride promoted addition of N-chlorocarbamates to enol ethers: the synthesis of N-alkoxycarbonyl derivatives of 2-amino sugars and of α-amino ketones and aldehydes. Tetrahedr Lett 56:4887–4890
20. (a) Lemieux RU, Takeda T, Chung BY (1977) Synthesis of 2-Amino-2-deoxy-β-D-glucopyranosides: properties and use of 2-deoxy-2-phthalimidoglycosyl halides. In: El Khadem H (ed) Synthetic Methods for Carbohydrates. ACS symposium series, vol 39. American Chemical Society, Washington, DC; (b) Aly MRE-S, Ashry E-SHE (2016) Recent advances toward robust N-protecting groups for glucosamine as required for glycosylation strategies. Adv Carbohydr Chem Biochem 73:117–223; (c) Aly MRE, Ibrahim E-SI, El Ashry E-SH, Schmidt RR (2001) Synthesis of chitotetraose and chitohexaose based on dimethylmaleoyl protection. Carbohydr Res 331:129–142
21. Kuyama H, Nakahara Y, Nukada T, Ito Y, Nakahara Y, Ogawa T (1993) Stereocontrolled synthesis of chitosan dodecamer. Carbohydr Res 243:C1–C7
22. (a) Kawada T, Yoneda Y (2009) Selection of protecting groups and synthesis of a β-1,4-GlcNAc-β-1,4-GlcN unit. Monatsh Chem 140:1245–1250; (b) Kawada T, Yoneda Y (2009) Synthesis of a chito-tetrasaccharide β-1,4-GlcNAc-β-1,4-GlcN repeating unit. Monatsh Chem 140:1251–1256
23. Fittolani G, Tyrikos-Ergas T, Vargová D, Chaube MA, Delbianco M (2021) Progress and challenges in the synthesis of sequence controlled polysaccharides. Beilstein J Org Chem 17:1981–2025
24. Barroca-Aubry N, Pernet-Poil-Chevrier A, Domard A, Trombotto S (2010) Towards a modular synthesis of well-defined chitooligosaccharides: synthesis of the four chitodisaccharides. Carbohydr Res 345:1685–1697

Yuko Yoneda

Toshinari Kawada

Utilization of Azide–Alkyne Cycloaddition, "Click Chemistry" for the Synthesis of Chitosan Derivatives and Conjugates

Sankar Rathinam, Aathira Pradeep, Maonian Xu, Rangasamy Jayakumar, and Már Másson

Abstract Chitosan, a natural polysaccharide derived from chitin, has garnered significant attention in the biomedical and materials sciences due to its biocompatibility, biodegradability, and reactive functional groups. However, its limited solubility and chemical versatility frequently necessitate functional modifications. Click chemistry, particularly the copper(I)-catalyzed azide–alkyne cycloaddition (CuAAC), offers a potent and selective tool for chitosan functionalization. This chapter examines published research on the application of click chemistry for the efficient and site-specific modification of chitosan, providing previously published and new examples of its use in facilitating the introduction of various functional moieties under mild conditions. Through azide or alkyne pre-functionalization of chitosan backbones, followed by click conjugation, novel derivatives with enhanced physicochemical and biological properties have been synthesized. The reaction conditions can be optimized to preserve the polymer's structural integrity and ensure high reaction yields. The resulting materials demonstrated improved solubility, targeted bioactivity, and potential for applications in drug delivery, tissue engineering, and biosensing. This work highlights the utility of click chemistry as a robust platform for tailoring chitosan-based materials with precise functionality and reproducibility, especially the conversion to chitotriazolan derivatives with antimicrobial properties.

Keywords Chitosan · Click reaction · Copper(I)-catalyzed azide–alkyne cycloaddition · Drug conjugates

---

S. Rathinam · M. Xu · M. Másson (✉)
Faculty of Pharmaceutical Sciences, School of Health Sciences, University of Iceland, Reykjavík, Iceland
e-mail: sankar@hi.is; maonian@hi.is; mmasson@hi.is

A. Pradeep · R. Jayakumar
Polymeric Biomaterials Lab, School of Nanosciences and Molecular Medicine, Amrita Vishwa Vidyapeetham, Kochi, India

S.-L. Wang, M. Anpo (eds.), *Chitin and Chitosan*, Springer Series on Polymer and Composite Materials, https://doi.org/10.1007/978-981-95-7457-5_5

1 Introduction

1.1 Click Chemistry

The rise of *click chemistry* at the beginning of the twenty-first century marked a major shift in synthetic strategies. Traditional organic synthesis often required multistep processes, harsh conditions, and laborious purification, which limited its utility in creating complex molecular systems. In 2001, K. Barry Sharpless and co-workers introduced the term "click chemistry" to describe reactions that are modular, high-yielding, selective, and operationally simple. The vision was to create a small set of "ideal" reactions that could be applied broadly, in the same way nature uses a few fundamental transformations to build complex biomolecules [1].

Among the various click reactions, the copper-catalyzed azide–alkyne cycloaddition (CuAAC) emerged as the flagship transformation. Reported independently by Sharpless and Meldal in 2002, CuAAC quickly gained recognition as the most reliable click reaction. The profound impact of click chemistry is highlighted by its recognition in the 2022 Nobel Prize in Chemistry, awarded to K. Barry Sharpless, Morten Meldal, and Carolyn R. Bertozzi for their contributions to "the development of click chemistry and bio-orthogonal chemistry. "Its ability to produce 1,4-disubstituted 1,2,3-triazoles with remarkable regioselectivity, speed, and tolerance of diverse functional groups made it immediately useful in polymer chemistry, bioconjugation, drug discovery, and materials science [2].

At the same time, chitosan, a naturally occurring polysaccharide obtained by deacetylation of chitin, was attracting increasing interest as a sustainable biomaterial [3]. Its biodegradability, biocompatibility, and abundance made it highly suitable for biomedical and environmental applications. However, chitosan's limited chemical reactivity constrained the diversity of derivatives achievable by conventional reactions [4].

The integration of CuAAC with chitosan chemistry has provided a powerful solution. By introducing azide or alkyne groups into the chitosan backbone, CuAAC enables precise, reproducible, and modular modification of the polymer. This chapter provides a detailed discussion of the fundamental aspects of CuAAC and its application in chitosan modification, with an emphasis on the reaction mechanism, synthetic strategies, and characterization methods. It specifically provides new and previously published examples of its use for converting chitosan and chitosan derivatives to chitortriazolan. Special emphasis is placed on chemistry and chemical characterization, whereas information on antimicrobial activity and applications can be found in the cited publications.

2 Chitosan: Structure and Chemical Properties

Chitosan is a naturally derived linear polysaccharide obtained by partial or complete deacetylation of chitin, the structural component found in crustacean shells, insect exoskeletons, and fungal cell walls. Structurally, it is composed of β-(1 → 4)-linked D-glucosamine and N-acetyl-D-glucosamine units, where the proportion of free amino groups is defined by the degree of deacetylation (DD) [4, 5]. When DD exceeds ~50%, the polymer is generally regarded as chitosan and acquires solubility in dilute acidic solutions due to protonation of the amino groups, which makes it unique among natural polysaccharides in exhibiting a cationic character. This property enables electrostatic interactions with negatively charged biomolecules, polyanions, and cell membranes, imparting mucoadhesive behavior and facilitating complex formation in biomedical and pharmaceutical systems [6, 7]. The polymer's molecular weight, which can vary from a few kilodaltons to several thousand, together with DD, strongly governs its solubility, viscosity, degradation rate, and mechanical properties. Chemically, chitosan presents three main reactive functional groups: amino groups at the C-2 position and hydroxyl groups at the C-3 and C-6 positions. These functionalities provide versatile platforms for derivatization, including *N*-acylation, *N*-alkylation [8, 9], *O*-acylation, *O*-alkylation, graft copolymerization, cross-linking, and click chemistry reactions such as copper-catalyzed azide–alkyne cycloaddition (CuAAC), which extend its range of applications. Hydrogen bonding between chains confers semi-crystalline behavior, with acetylated units promoting crystallinity and deacetylated regions contributing to amorphous character, thereby influencing solubility and mechanical strength. Furthermore, chitosan is biodegradable, primarily through enzymatic hydrolysis by lysozyme, producing oligomers and monomers that can also be biologically active. Collectively, the structural features of chitosan, including its cationic nature, tunable solubility, reactive functional groups, and biodegradability, make it an exceptionally versatile biopolymer with potential applications across biomedical, pharmaceutical, agricultural, and environmental fields [10].

3 Copper-Catalyzed Azide–Alkyne Cycloaddition (CuAAC) Reaction

3.1 *Historical Perspective*

The 1,3-dipolar cycloaddition between azides and alkynes was first described by Rolf Huisgen in the early 1960s, now widely known as the "Huisgen cycloaddition [11]." This thermal process required high temperatures (often >100 °C) and resulted in a mixture of 1,4- and 1,5-disubstituted 1,2,3-triazoles, which limited its practical utility. Despite its conceptual significance, the lack of regioselectivity and harsh reaction conditions confined it mainly to academic interest.

The true breakthrough occurred in 2002, when two groups independently reported that the presence of a copper(I) catalyst enabled the reaction to proceed under mild conditions with exceptional regioselectivity. K. Barry Sharpless coined the term "click chemistry" and highlighted CuAAC as the paradigmatic click reaction, while Morten Meldal demonstrated its utility in peptide coupling on solid supports [2]. These studies transformed the Huisgen reaction into a highly selective, fast, and broadly applicable tool, now recognized as one of the most widely used transformations in chemistry.

3.2 General Click Reaction

The CuAAC reaction involves three essential components. Overall, the combination of a terminal alkyne (–C ≡ CH), an organic azide ($-N_3$), and a copper(I) catalyst (see Fig. 1) represents a powerful strategy for creating complex molecular architectures in a straightforward manner.

Catalyst sources include $CuSO_4$ combined with sodium ascorbate (for in situ reduction) and CuI, CuBr, or copper nanoparticles. The solvent systems can be water, mixtures of t-butanol and water, DMSO, DMF, or ionic liquids. Ligands such as tris(triazolylmethyl)amine (TTMA) and bathophenanthroline disulfonate (BPDS) help to stabilize Cu(I) and accelerate the reaction [12]. The regioselective formation of 1,4-disubstituted 1,2,3-triazoles is particularly notable for its stability. The triazole ring is chemically stable and resistant to hydrolysis, oxidation, and enzymatic cleavage.

3.3 Reaction Mechanism

The mechanism of the CuAAC reaction is described in the following steps. The copper-catalyzed azide–alkyne cycloaddition (CuAAC) is a prototypical "click" reaction that efficiently links terminal alkynes with organic azides to yield highly

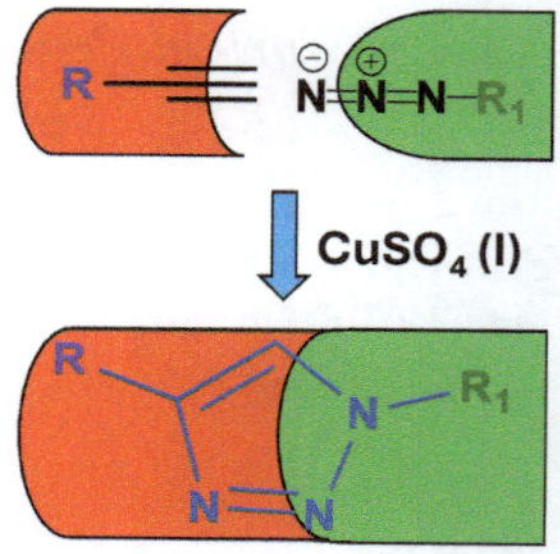

Fig. 1 Copper-Catalyzed Azide–Alkyne Cycloaddition Reaction (CuAAC)

stable 1,4-disubstituted 1,2,3-triazoles. The mechanism begins with Cu(I) coordination to the terminal alkyne (see in Fig. 2), which significantly lowers its pKa by nearly ten units, making deprotonation under mild conditions feasible.

This generates a reactive copper acetylide species that serves as the true nucleophilic partner in the cycloaddition. The organic azide then coordinates to the copper center (or to a second copper in the dinuclear pathway), aligning its terminal nitrogen with the acetylide carbon in a geometry that favors regioselective bond formation. A key metallacyclic intermediate, often described as a six-membered Cu(III) complex, is formed, stabilizing the transition state and enforcing the exclusive production of the 1,4-triazole regioisomer. Subsequent ring contraction collapses the metallacycle into a triazolyl copper complex, which undergoes protolysis to release the final triazole product while regenerating the copper catalyst for subsequent cycles. Computational studies combined with experimental evidence, indicate that both mononuclear and dinuclear copper pathways are possible, with the latter being kinetically more favorable and often dominant under aqueous conditions [13].

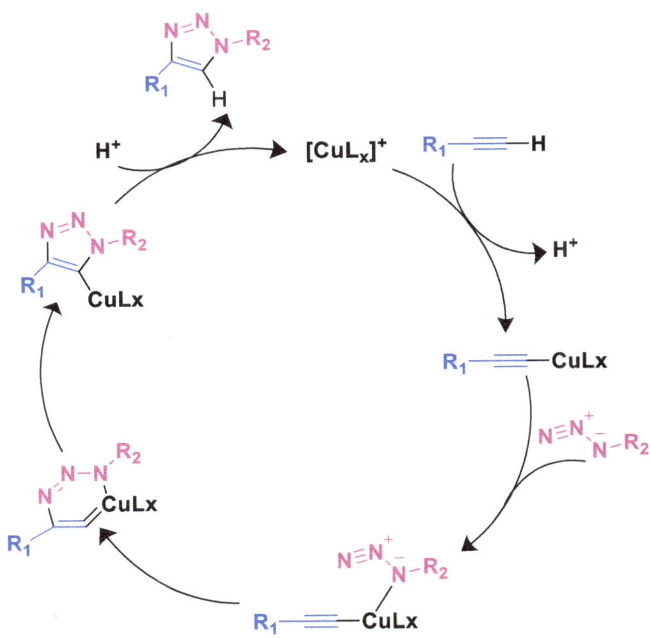

Fig. 2 Mechanism of CuAAC reaction

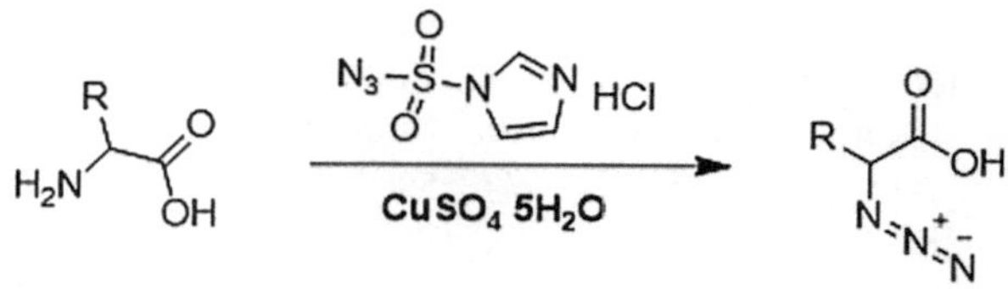

Fig. 3 Diazo-transfer reaction

3.4 Diazo-Transfer Reaction

The diazotransfer reaction is a method for converting primary amines ($-NH_2$) into azides ($-N_3$) using a diazotransfer reagent, such as sodium azide, sodium nitrite, triflyl azide, or imidazole-1-sulfonyl azide hydrochloride [14] (Fig. 3).

3.5 Chitosan Derivatives by CuAAC Reaction

This reaction can be specifically utilized in chitosan modification to transform C2-amino groups into azides, which can then undergo click chemistry (CuAAC) for bioconjugation [15]. Kulbokaite and colleagues reported that *N*-azidated chitosan was synthesized to introduce azide groups onto the amino functionalities of chitosan. The methods used included reactions with sodium azide and sodium nitrite (a diazotransfer approach), trifluoromethanesulfonyl azide, and imidazole-1-sulfonyl azide hydrochloride. Among these, the use of trifluoromethanesulfonyl azide and imidazole-1-sulfonyl azide hydrochloride yielded the highest degrees of azidation, reaching up to 40% and 65%, respectively. These higher efficiencies underscore the suitability of modern diazotransfer reagents for the effective *N*-azidation of chitosan, thereby enabling further functionalization through click chemistry [16].

Our previous work demonstrated an efficient synthesis, whereby the diazotransfer reaction and CuAAC could be used to obtain water-soluble chitotriazolan derivatives, a term used to describe structures where all the amino groups in the glucosamine monomers have been replaced with triazole moieties [17]. In the primary approach, the C-2 amino groups of chitosan were first converted into azides via diazo transfer, followed by copper(I)-catalyzed azide–alkyne cycloaddition (CuAAC) with various alkynes, yielding stable 1,2,3-triazole units at each sugar residue. This method enabled fine-tuning of the polymer's physicochemical properties by varying the alkyne substituents, with clear conversion profiles demonstrated (Figs. 4 and 5).

A second route involved protecting the hydroxyl groups with TBDMS, followed by conversion of the amine to azide using imidazole sulfonyl azide, and subsequent CuAAC to form chitosan triazole derivatives. Functionalized products bearing

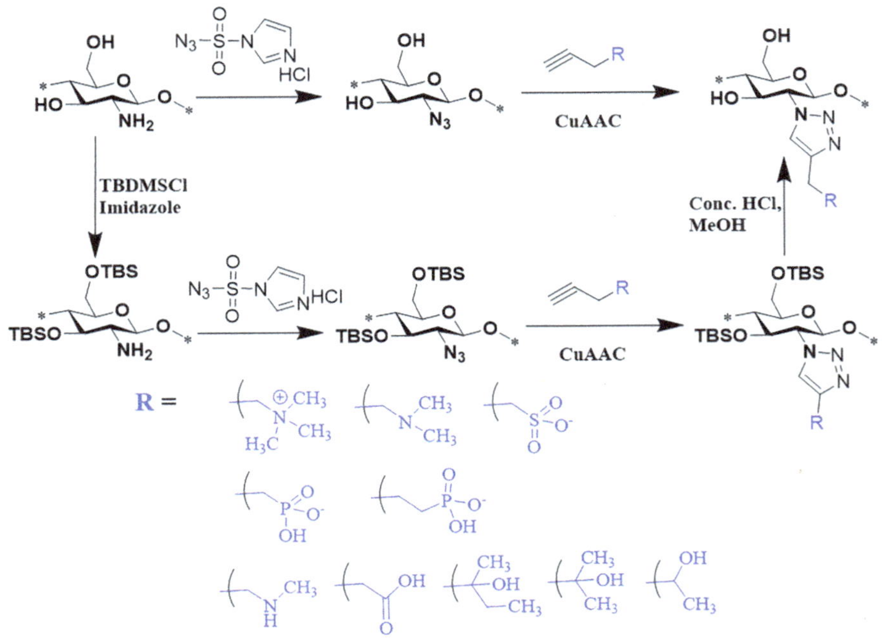

Fig. 4 Diazo-transfer reaction and Chitosan derivatives by using the Click reaction

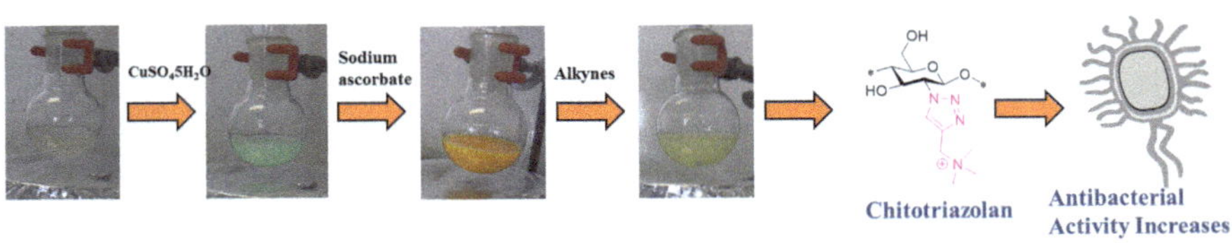

Fig. 5 CuAAC reactions profile by color changes

quaternary ammonium, sulfonyl, or phosphonyl groups exhibited markedly improved water solubility and potent antibacterial activity against *Staphylococcus aureus* and *Escherichia coli*, attributed to the cationic charge-mediated interaction with bacterial membranes. Conversely, neutral or anionic triazole derivatives showed poor solubility and negligible antimicrobial activity, highlighting the critical influence of substituent charge on biofunctionality. Structural confirmation was obtained through FTIR and advanced ^{1}H NMR analyses, validating the formation of the triazole. Collectively, this modular CuAAC-based approach positions chitotriazolans as a versatile class of water-soluble, bioactive polysaccharides with strong potential in antimicrobial and broader biomedical applications [17].

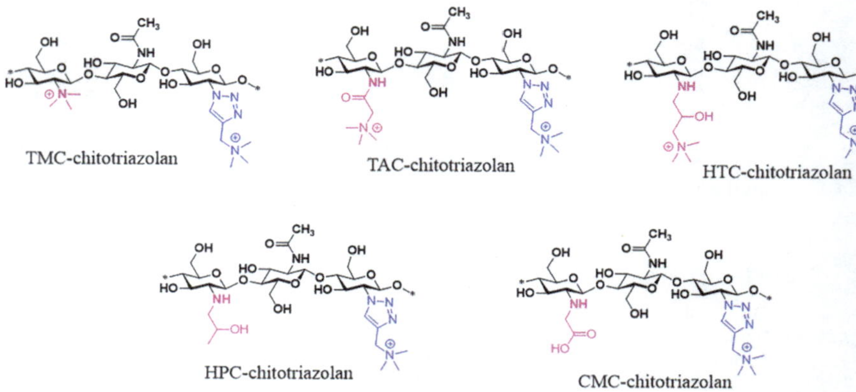

Fig. 6 Mixed chitotriazolan derivatives by click reactions

The efficient synthesis of water-soluble "mixed" chitotriazolan derivatives was achieved through a two-step strategy involving the diazo-transfer of azide groups onto different chitosan backbones, including cationic derivatives (TMC, TAC, HTC), a neutral derivative (HPC), and an anionic derivative (CMC) [8, 18], followed by CuAAC with a quaternary ammonium alkyne (Fig. 6). Structural characterization by ^{1}H NMR spectroscopy confirmed near-complete conversion of amino groups to triazoles in most derivatives, with the notable exception of TAC, which exhibited only ~50% substitution due to steric and electronic limitations. Importantly, the introduction of quaternary ammonium groups through triazole linkages significantly improved aqueous solubility in all cationic and neutral derivatives, while simultaneously imparting potent antibacterial activity (*S. aureus, E. faecalis, E. coli, P. aeruginosa*), while the anionic CMC-based derivative remained insoluble and inactive, highlighting the role of backbone charge in bioactivity [19].

In another study, a systematic library of chitotriazolan derivatives was synthesized by converting the C-2 amino groups of chitosan into azides, followed by CuAAC with various quaternary ammonium and protonable amine-containing alkynes, yielding both water-soluble quaternary derivatives and protonable analogs [20]. A total of nine distinct derivatives were obtained, and their chemical structures were thoroughly analyzed using FTIR and ^{1}H NMR analyses, which confirmed the formation of triazoles and their structural integrity. All derivatives demonstrated significantly enhanced aqueous solubility compared to native chitosan and exhibited substantial, broad-spectrum antibacterial activity, particularly against *S. aureus* and *E. coli*. These results demonstrate the effectiveness of CuAAC functionalization in producing chitosan-based biomaterials with enhanced physicochemical and antimicrobial properties [20].

4 Conjugation of Peptide and Drugs on Chitosan Biopolymer

4.1 Conjugation of CRAMP Peptide to Chitosan via CuAAC

The diazo-transfer reaction and controlled CuAAC reaction have also been utilized for the selective conjugation of the $CRAMP_{18-35}$ peptide to chitosan biopolymers, while preserving the integrity of both biomolecules. To minimize extensive modification of the chitosan backbone, the degree of azide conversion of the C-2 amino groups was deliberately kept low. Specifically, only 0.4 equivalents of imidazole sulfonyl azide hydrochloride were used relative to the available amino groups on chitosan. This mild degree of substitution prevented excessive structural disruption of the polysaccharide backbone, maintaining its inherent solubility and biological properties.

The $CRAMP_{18-35}$ peptide sequence (KLKKIGQKIKNFFQKLVP) was synthesized separately using standard peptide synthesis protocols [21]. To introduce an alkyne functionality required for click chemistry, the N-terminus of $CRAMP_{18-35}$ was selectively coupled with 4-pentynoic acid, producing the pentynoyl–$CRAMP_{18-35}$ derivative. This modification provided the terminal alkyne moiety necessary for CuAAC conjugation, while preserving the peptide's overall structural integrity and bioactivity.

The azidated chitosan and hydroxypropyl chitosan (HPC) intermediate presented challenges in terms of solubility. It was insoluble in both aqueous and most common organic solvents, reflecting the reduced flexibility of the polymer upon introduction of azide groups. Limited solubility was achieved only in dimethyl sulfoxide (DMSO) at 50 °C, though this was still insufficient for direct characterization by proton nuclear magnetic resonance (^{1}H NMR) spectroscopy. The poor solubility emphasized the importance of optimizing reaction conditions for successful bioconjugation.

The CuAAC reaction between the chitosan azide, hydroxypropyl chitosan azide intermediate and pentynoyl-$CRAMP_{18-35}$ (see in Fig. 7) was carried out under mild conditions in the presence of tris(3-hydroxypropyltriazolylmethyl)amine (THPTA) as a stabilizing ligand [9]. THPTA plays a critical role in such bioconjugation reactions it coordinates to the copper(I) center, maintaining its reduced oxidation state and thereby sustaining catalytic activity. In addition, THPTA prevents oxidative side reactions, which are particularly detrimental when working with sensitive biomolecules such as peptides. The ligand also accelerates the cycloaddition process, ensuring efficient and regioselective formation of the 1,4-disubstituted 1,2,3-triazole linkage between chitosan and the peptide.

The overall strategy thus ensured the site-specific attachment of $CRAMP_{18-35}$ to chitosan through a stable triazole bond, producing a hybrid biopolymer that combines the cationic and antimicrobial properties of the peptide with the biocompatibility, biodegradability, and film-forming properties of chitosan. This conjugation approach highlights the versatility of CuAAC chemistry in generating peptide polysaccharide conjugates for advanced biomedical applications.

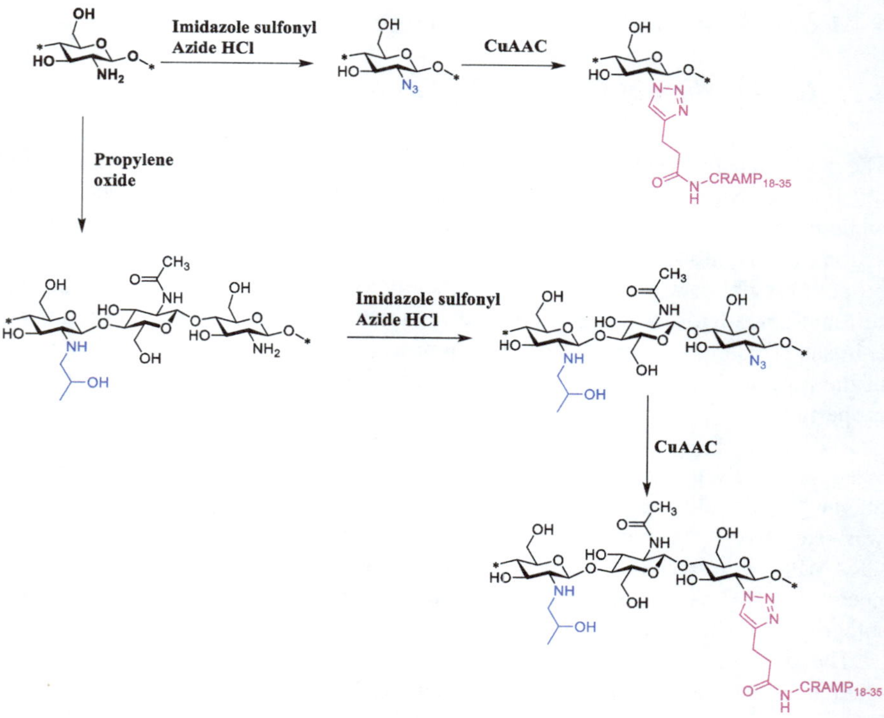

Fig. 7 $CRAMP_{18-35}$ peptide conjugation to chitosan through CuAAC reaction

4.2 Conjugation of Ciprofloxacin to Chitosan via CuAAC

To design an effective drug delivery system capable of targeting intracellular pathogens such as *Staphylococcus aureus* and *Mycobacterium tuberculosis* [22]. Novel chitosan-ciprofloxacin conjugates are being developed using CuAAC, a prototypical click chemistry reaction. In this approach, the C-2 amino groups of chitosan are selectively converted into azide functionalities, providing reactive handles along the polysaccharide backbone. In parallel, ciprofloxacin (see Fig. 8B) derivatives are synthesized by introducing terminal alkyne groups onto the drug molecule without disrupting its pharmacophoric regions, which are essential for antibacterial activity. Subsequent CuAAC coupling between the drug alkyne derivatives and the chitosan azides yields stable, covalent 1,2,3-triazole linkages, resulting in polymer drug conjugates with high efficiency. The conjugates are not soluble in aqueous solution so alternatively started synthesized *N*-(2-hydroxyl) propyl-3-trimethyl ammonium chitosan (HTC) was synthesized and then followed by azidation and click reaction to obtain HTC-CIP chitotriazolan conjugates (see Fig. 8A). Thus, the CuAAC-mediated synthesis of chitosan-ciprofloxacin conjugates represents a rational and versatile platform for developing next-generation antimicrobial delivery systems tailored to combat challenging intracellular infections.

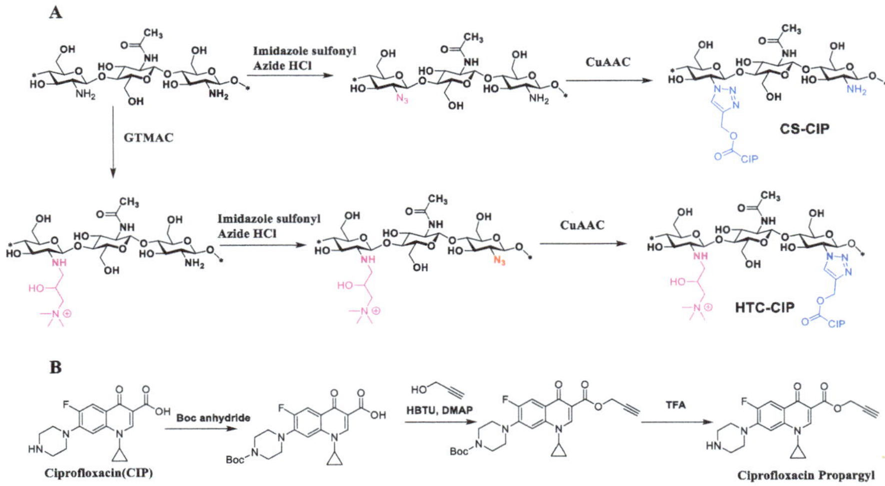

Fig. 8 Chitosan-Ciprofloxacin conjugation by using CuAAC strategy (A) and Ciprofloxacin propargyl preparation(B)

5 Characterization of Chitotriazolan Derivatives and Peptide Conjugates

5.1 ^{1}H NMR Characterization of Chitosan Triazole Derivatives

The unmodified chitosan biopolymer was characterized by proton NMR (see Fig. 9A) for comparison with the modified chitotriazolan derivatives. The structural confirmation of chitosan derivatives obtained via CuAAC can be effectively achieved by ^{1}H NMR spectroscopy, as demonstrated for the water-soluble derivatives 4-(*N,N,N*-trimethylammoniummethyl)-chitotriazolan derivative and 4-sulfomethylchitotriazolan derivative, whose spectra are illustrated in Fig. 9. In the case of chitotriazolan derivative (see Fig. 9B), the successful formation of the 1,2,3-triazole moiety is clearly evidenced by the appearance of a distinct aromatic proton resonance at $\delta = 8.59$ ppm, which is diagnostic of triazole substitution. Additionally, the presence of the quaternary trimethylammonium group is confirmed by a sharp singlet at $\delta = 3.20$ ppm, while the associated methylene ($-CH_2-$) group resonates at $\delta = 4.80$ ppm, overlapping with the residual D_2O peak. A significant chemical shift perturbation is observed for the C-2 proton of chitosan. Upon conversion of the primary amino group at C-2 into the triazole substituent, the resonance shifts dramatically from its typical position near $\delta = 2.80$ ppm in unmodified chitosan to $\delta = 4.58$ ppm in the derivative, providing strong evidence of successful functionalization.

Other protons of the chitosan backbone also exhibit notable downfield shifts as a result of electronic effects introduced by triazole substitution. The C-6 protons

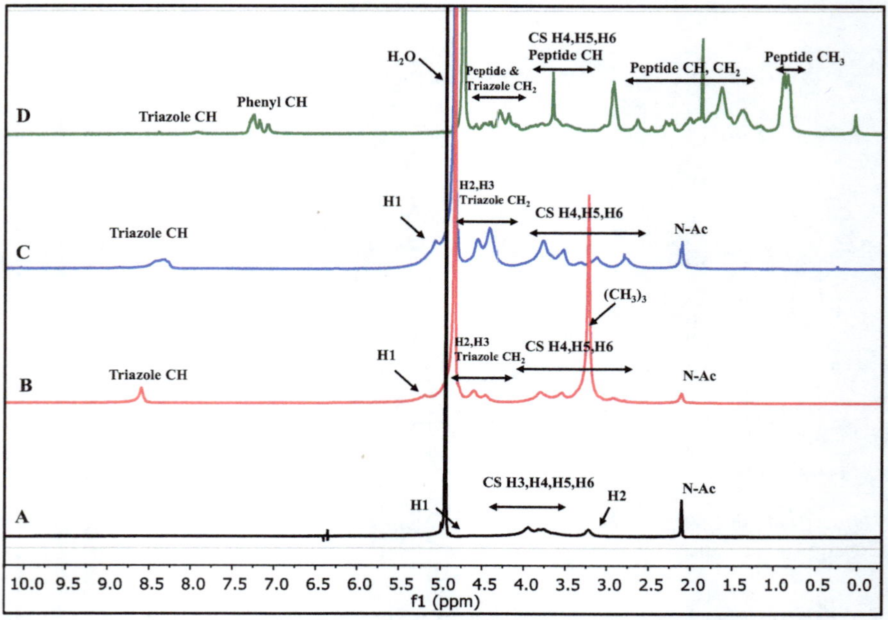

Fig. 9 Proton NMR characterization of Chitosan (A), Chitotriazolan derivative (B), Chitotriazolan sulfonate derivative (C), and Chitotraizolan $CRAMP_{18-35}$ peptide conjugates (D)

resonate at δ = 2.90 ppm and δ = 3.20 ppm (the latter coinciding with the $N(CH_3)_3$ signal). The C-5, C-4, and C-3 protons appear at δ = 3.52 ppm, 3.78 ppm, and 4.44 ppm, respectively, further confirming the systematic perturbation of the polymer's proton environment.

For chitotriazolan sulfonate derivative (see Fig. 9C) (bearing the sulfonated substituent), the aromatic triazole proton is still observed in the downfield region but appears broadened and slightly upfield, between δ = 8.13–8.43 ppm, relative to the chitotriazolan derivative. The C-6, C-5, C-4, C-3, and C-2 protons display nearly identical chemical shift values to those in derivative 3, consistent with similar backbone modifications. However, an additional feature is characteristic of the sulfonated derivative: the methylene protons adjacent to the sulfonate group resonate between δ = 4.27–4.42 ppm, overlapping with the signals of the C-2 and C-3 protons.

Together, these spectral data provide unambiguous evidence of successful triazole ring formation at the C-2 position of chitosan and highlight the distinctive electronic and steric influences exerted by different substituents (trimethylammonium vs. sulfonate) on the local proton environments. The pronounced chemical shift changes, particularly at the C-2 proton, serve as reliable markers for confirming CuAAC-mediated functionalization of chitosan.

5.2 ^{1}H NMR Evidence of CRAMP18–35 Conjugation to Chitosan

The successful covalent conjugation of the antimicrobial peptide $CRAMP_{18–35}$ to the chitosan backbone via the CuAAC reaction can be clearly verified by ^{1}H NMR spectroscopy. A defining feature of the spectrum is the presence of a triazole proton resonance at δ = 7.94 ppm, which serves as a direct indicator of triazole ring formation and, by extension, of the click-mediated linkage between chitosan and the peptide. This characteristic signal is absent in the spectra of either unmodified chitosan or the free peptide, confirming that the observed triazole peak arises specifically from the CuAAC reaction (see Fig. 9D).

In addition to the triazole resonance, the incorporation of the $CRAMP_{18–35}$ peptide is further substantiated by the appearance of distinct aromatic proton signals corresponding to phenylalanine residues within the peptide sequence. Phenylalanine, present in two positions of $CRAMP_{18–35}$, contributes multiplet resonances in the δ = 7.08–7.30 ppm range, consistent with the aromatic region typical of phenyl groups. The detection of these peaks confirms the successful attachment of the peptide moiety, as chitosan itself lacks aromatic protons.

Moreover, the aliphatic region of the spectrum provides complementary evidence of peptide conjugation. Multiple resonances spanning δ = 0.79–4.48 ppm can be attributed to methylene and methine protons belonging to the aliphatic side chains of amino acids (such as leucine, isoleucine, valine, and alanine) as well as the peptide backbone. These broad signals reflect the chemical diversity of the peptide's aliphatic groups and further differentiate the conjugate spectrum from that of pristine chitosan, which typically displays resonances confined to carbohydrate ring protons.

Taken together, the triazole signal at δ = 7.94 ppm, the aromatic peaks of phenylalanine between δ = 7.08–7.30 ppm, and the aliphatic resonances distributed across δ = 0.79–4.48 ppm provide compelling spectral evidence for the successful conjugation of $CRAMP_{18–35}$ to chitosan through the CuAAC reaction (Fig. 9D). These results not only confirm the chemical integrity of the conjugation process but also highlight the sensitivity of NMR analysis in detecting both the newly formed linkage (triazole) and the peptide-derived structural features within the conjugate.

5.3 ^{1}H NMR Evidence of Ciprofloxacin Conjugation to Chitosan and Release Data

The successful conjugation of ciprofloxacin (CIP) to the chitosan backbone via click chemistry was clearly demonstrated by the ^{1}H NMR spectrum (Fig. 10A). The appearance of a distinct triazole proton signal at δ = 8.50 ppm confirmed the formation of the CuAAC-derived triazole linkage, while the aromatic resonances of the ciprofloxacin phenyl ring were evident in the δ = 7.30–8.30 ppm region. Overlapping

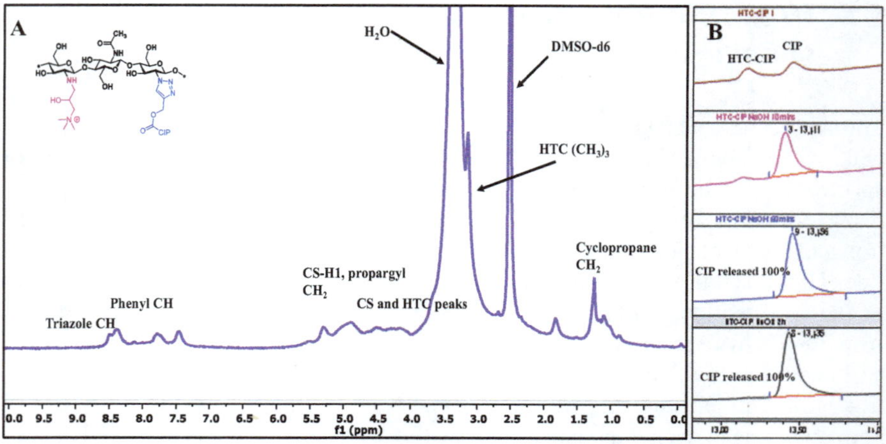

Fig. 10 Proton NMR for HTC-CIP (A) and HPLC profile for the CIP drug release by using NaOH (0.01 M) solution (B)

signals in the δ = 3.80–5.50 ppm range correspond to the combined contributions of chitosan H1 protons, the CIP-propargyl methylene group, and other HTC-related protons, reflecting successful covalent integration of the drug into the polymer matrix. Additional diagnostic peaks included the HTC trimethyl resonance at δ = 3.32 ppm and the ciprofloxacin cyclopropane signal at δ = 1.00 ppm, both supporting the presence of intact structural motifs from each component. Drug release studies, analyzed by HPLC (Fig. 10B), confirmed the release of ciprofloxacin from the HTC–CIP conjugate due to hydrolysis of the ester linkage in the presence of 0.01 M NaOH, indicating that the conjugate can undergo controlled release under mildly basic conditions. This combined NMR spectroscopy and HPLC chromatographic evidence provides strong confirmation of both successful synthesis and functional drug release behavior of the HTC–CIP conjugate, underscoring its potential as a pH-responsive antimicrobial delivery system.

6 Future Perspectives and Conclusions

The application of CuAAC click chemistry to develop chitosan drug conjugates represents a powerful and versatile platform for next-generation antimicrobial therapeutics. Future work should focus on optimizing substitution degrees on the chitosan backbone to balance solubility, drug-loading capacity, and biological activity. Tailoring the type of triazole substituent (quaternary ammonium, sulfonyl, or phosphonyl) could further modulate the physicochemical properties and enhance target-specific delivery, especially against intracellular pathogens such as *Staphylococcus aureus* and *Mycobacterium tuberculosis*. Advanced nanocarrier formulations of these conjugates, such as nanoparticles, hydrogels, or films, could improve pharmacokinetics and provide sustained release profiles, thereby reducing dosing frequency and minimizing systemic side effects. Moreover, integrating multi-drug conjugation strategies (e.g., combining ciprofloxacin with antimicrobial peptides, such as $CRAMP_{18-35}$) may offer synergistic antibacterial effects and help combat multidrug-resistant (MDR) strains. Future studies should also evaluate the in vivo performance of these conjugates, including biodistribution, intracellular uptake, and immunological responses, to accelerate their translation toward clinical applications.

In summary, chitotriazolan-based conjugates synthesized via CuAAC click chemistry provide a robust and modular strategy for the functionalization of chitosan with antimicrobial agents such as ciprofloxacin and $CRAMP_{18-35}$ peptides. The triazole linkage offers high stability, regioselectivity, and efficiency, while the chitosan backbone contributes biocompatibility, biodegradability, and enhanced solubility. Structural confirmation through FTIR and NMR spectroscopy consistently validates the success of these conjugations, while biological assays demonstrate strong antibacterial activity, particularly for cationic derivatives. Collectively, these findings highlight chitosan triazole conjugates as a promising class of bioactive polymers for antimicrobial drug delivery systems, with significant potential in addressing the challenges of intracellular infections and antibiotic resistance.

References

1. Kolb HC, Finn MG, Sharpless KB (2001) Click chemistry: diverse chemical function from a few good reactions. Angew Chem Int Ed 40(11):2004–2021
2. Tornøe CW, Christensen C, Meldal M (2002) Peptidotriazoles on solid phase: [1,2,3]-Triazoles by Regiospecific copper(I)-catalyzed 1,3-dipolar cycloadditions of terminal alkynes to Azides. J Org Chem 67(9):3057–3064
3. Zhao D et al (2018) Biomedical applications of chitosan and its derivative nanoparticles. Polymers (Basel) 10(4)
4. Másson M (2021) Antimicrobial properties of chitosan and its derivatives, in Chitosan for biomaterials III: structure-property relationships. In: Jayakumar R, Prabaharan M (eds). Springer International Publishing, Cham, pp 131–168
5. Ravi Kumar MNV (2000) A review of chitin and chitosan applications. React Funct Polym 46(1):1–27
6. Ilium L (1998) Chitosan and its use as a pharmaceutical excipient. Pharm Res 15(9):1326–1331
7. Jayakumar R et al (2011) Biomaterials based on chitin and chitosan in wound dressing applications. Biotechnol Adv 29(3):322–337
8. Rathinam S et al (2020) *Selective synthesis of N,N,*N-trimethylated chitosan derivatives at different degree of substitution and investigation of structure-activity relationship for activity against P. aeruginosa and MRSA. Int J Biol Macromol
9. Rathinam S et al (2024) Conjugation of CRAMP18–35 peptide to chitosan and hydroxypropyl chitosan via copper-catalyzed azide–alkyne cycloaddition and investigation of antibacterial activity. Int J Mol Sci 25(17):9440
10. Rabea EI et al (2003) Chitosan as antimicrobial agent: applications and mode of action. Biomacromolecules 4(6):1457–1465
11. Huisgen R (1960) Zehnjahresfeier des Fonds der Chemischen Industrie, pp 73–102
12. Meldal M, Tornøe CW (2008) Cu-catalyzed azide–alkyne cycloaddition. Chem Rev 108(8):2952–3015

13. Himo F et al (2005) Copper(I)-catalyzed synthesis of azoles. DFT study predicts unprecedented reactivity and intermediates. J Am Chem Soc 127(1):210–216
14. Pandiakumar AK, Sarma SP, Samuelson AG (2014) Mechanistic studies on the diazo transfer reaction. Tetrahedron Lett 55(18):2917–2920
15. Goddard-Borger ED, Stick RV (2007) An efficient, inexpensive, and shelf-stable diazotransfer reagent: imidazole-1-sulfonyl azide hydrochloride. Org Lett 9(19):3797–3800
16. Kulbokaite R et al (2009) N-PEG'ylation of chitosan via "click chemistry" reactions. React Funct Polym 69(10):771–778
17. Rathinam S et al (2021) Chitotriazolan (poly(β(1–4)-2-(1H-1,2,3-triazol-1-yl)-2-deoxy-d-glucose)) derivatives: synthesis, characterization, and evaluation of antibacterial activity. Carbohydr Polym 267:118162
18. Rathinam S et al (2020) The antibacterial structure-activity relationship for common chitosan derivatives. Int J Biol Macromol 165:1686–1693
19. Rathinam S et al (2023) Water-soluble chitotriazolans derived from cationic, neutral, and anionic common chitosan derivatives: synthesis, characterization, and antibacterial activity. Eur Polym J 196:112311
20. Rathinam S et al (2024) Water-Soluble Quaternary and Protonable Basic Chitotriazolans: Synthesis by Click Chemistry Conversion of Chitosan Azides and Investigation of Antibacterial Activity. J Funct Biomater 15(3):63
21. De Brucker K et al (2014) Derivatives of the mouse cathelicidin-related antimicrobial peptide (CRAMP) inhibit fungal and bacterial biofilm formation. Antimicrob Agents Chemother 58(9):5395–5404
22. Russell DG (2011) Mycobacterium tuberculosis and the intimate discourse of a chronic infection. Immunol Rev 240(1):252–268

Sankar Rathinam

Aathira Pradeep

Maonian Xu

Rangasamy Jayakumar

Már Másson

Veterinary Clinical Applications of Chitin and Chitosan

Yoshiharu Okamoto

Abstract Since 1985, we have developed several materials from chitin and chitosan for veterinary clinical use. Specifically, these include sponge type chitin, cotton type chitin, cotton type chitosan, finely powdered chitin, finely powdered chitosan, and chitin-impregnated nonwoven fabrics. These materials have been used to treat a variety of traumatic injuries and infectious diseases, resulting in extremely good healing. Recently, chitin nanofibers have also been developed, and similar effects have been confirmed. These substances have been shown to promote wound healing through the formation of abundant, vascularized granulation tissue, the inhibition of scar formation, neutrophil activation, complement activation, analgesic effects, and hemostatic effects. Furthermore, D-glucosamine, a monosaccharide in chitosan, has been shown to play a major role in the repair of articular cartilage. Additionally, chitin oligosaccharides and chitosan oligosaccharides have been shown to have antitumor effects when administered orally.

Keywords Wound healing acceleration · Articular cartilage repair effect · Antitumor effect

1 Introduction

Chitin and chitosan are polysaccharides composed of β(1–4)-linked N-acetyl-D-glucosamine (GlcNAc) and D-glucosamine (GlcN) units. They are naturally biosynthesized and are components of the exoskeletons of crustaceans and insects as well as fungal cell walls. These substances are biodegraded by enzymes such as chitinase and chitosanase, subsequently yielding oligosaccharides (chitin oligosaccharides: NACOS, chitosan oligosaccharides: COS) and monosaccharides (N-acetyl-D-glucosamine: GlcNAc, D-glucosamine: GlcN).

Y. Okamoto (✉)
Veterinary Advanced Medical Institute, WOLVES HAND Co., Ltd., Osaka, Japan
e-mail: y.okamoto@wolveshand.jp

S.-L. Wang, M. Anpo (eds.), *Chitin and Chitosan*, Springer Series on Polymer and Composite Materials, https://doi.org/10.1007/978-981-95-7457-5_6

Since 1985, we have developed several materials from chitin and chitosan for veterinary clinical use [1]. We will discuss these materials in detail. We have also demonstrated that oral administration of GlcN has an articular cartilage repair effect, and oral administration of NACOS and COS has an antitumor effect. These applications will also be introduced.

2 Application of Chitin and Chitosan as Wound Healing Materials in Veterinary Clinical Practice

In the 1990s, we commercialized and released chitin (Chitipack S, Chitipack P) and chitosan (Chitopack C) as wound dressings for animals. Additionally, we commercialized finely powdered chitosan (Chitofine) for the treatment of contaminated wounds and bovine mastitis (though these products are no longer commercially available). Alongside clinical studies of these substances, we have conducted research into their excellent physiological activities, particularly the mechanisms underlying their wound healing promotion effects. The clinical applications of each product are listed below.

2.1 Chitin-Derived Substances [2, 3]

2.1.1 Sponge Type Chitin (Chitipack S)

Application of Chitipack S to Severe Skin Defects in Dogs

Figure 1a shows a case of a dog with severe contusions involving the skin, subcutaneous tissue, tendon sheath, and bone due to a traffic accident. The skin defect was covered with Chitipack S (Fig. 1b). Antibiotics were administered systemically but not locally. Over time, tissue repair occurred, and the skin regenerated satisfactorily. Ultimately, the severe contusions healed with almost no functional impairment (Fig. 1c). Compared to cotton type chitosan (Chitopack C), Chitipack S is characterized by skin regeneration accompanied by hair growth.

Application as a Wound Cavity Filling Material

Figure 2a shows a case in which Chitipack S was used to fill a dead space following tumor removal, and the skin was sutured. One week later, good healing was observed (Fig. 2b). Table 1 shows the results of using Chitipack S to fill wound cavities after tumor removal in dogs. While some cases were pathologically benign, the majority were malignant. In particular, with malignant mammary gland tumors, local

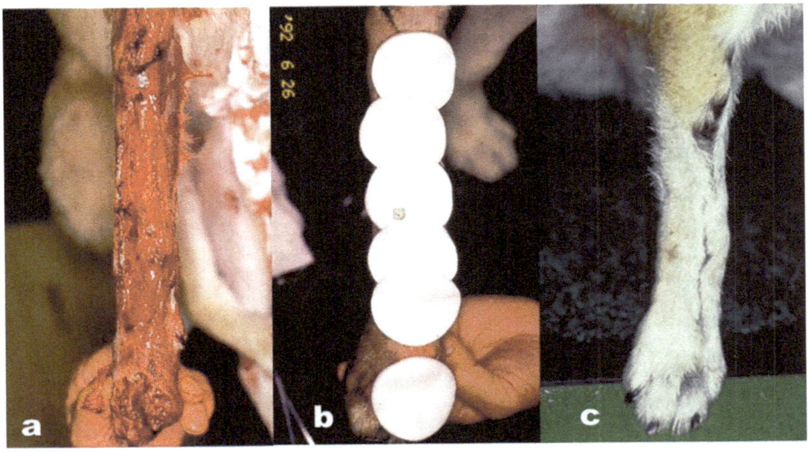

Fig. 1 (**a**) A case of a dog involved in a traffic accident. Severe contusions were observed in the skin, subcutaneous tissue, tendon sheaths, and bones. (**b**) The skin defect was covered with Chitipack S. (**c**) Ultimately, the severe laceration healed with little functional impairment

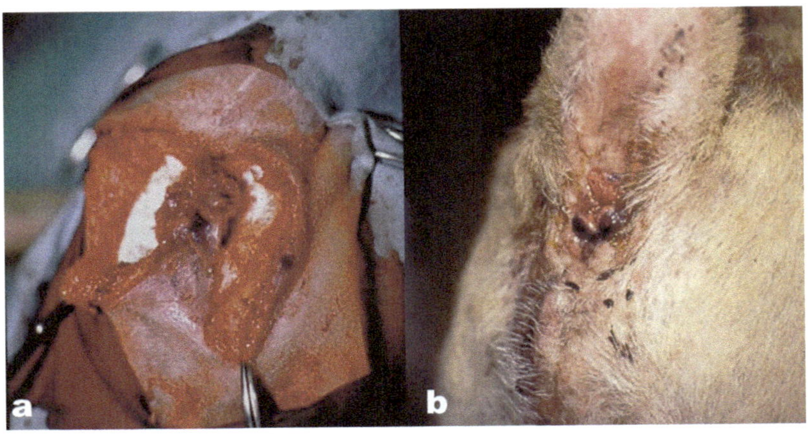

Fig. 2 (**a**) The dead space after tumor removal was filled with Chitipack S. (**b**) One week later, good healing was observed

recurrence is reported to occur in approximately 70% of cases, even after partial surgical removal. However, it has been suggested that filling the cavity with Chitipack S can suppress recurrence.

Table 1 Effect of Chitipack S filling the tumor resection cavity

No.	Species	Sex	Age (Y)	Site	Pathological diagnosis	Recurrence-free period
1	Mix	Female	10	Brest	Malignant MGT*	3 years
2	Shiba	Male	8	Lower jaw	Mast cell tumor	3 years
3	Yorkie	Female	8	Brest	Malignant MGT	2 years
4	Shiba	Male	10	Skin	Mast cell tumor	2 years
5	Mix	Male	10	Brest	Mast cell tumor	2 years
6	Shih Tzu	Female	12	Brest	Osteosarcoma	Recurrence after 1 month
7	Shih Tzu	Female	14	Brest	Osteosarcoma	1 year
8	Mix	Female	11	Brest	Lipoma	1 year
9	Mix	Male	8	Anus	Perianal adenoma	6 months
10	Shepard	Female	8	Brest	Malignant MGT	6 months
11	Yorkie	Female	8	Brest	Malignant MGT	6 months
12	Schnauzer	Female	8	Brest	Malignant MGT	6 months

*Mammary gland tumor

2.1.2 Chitin-Impregnated Nonwoven Fabric (Chitipack P)

In an experiment, we implanted Chitipack P subcutaneously in dogs, using a nonwoven fabric without chitin microparticles as the control. The condition of the nonwoven fabric was observed 18 days after implantation. In the control group, the nonwoven fabric was surrounded by mild granulation tissue, but the fabric itself had not become organized (Fig. 3a). In contrast, in the Chitipack P group, the fabric was completely organized (Fig. 3b). The histological findings for each nonwoven fabric are summarized in Table 2. It can be seen that the chitin- and chitosan-impregnated fabrics had more connective tissue infiltration than the control. These experimental results showed that nonwoven fabrics, which typically induce a foreign body reaction, organized without doing so. Based on this, we attempted their application in living organisms, as described below.

Application to Perineal Hernia in Dogs

Various techniques have been reported for perineal hernia repair in dogs. However, all have drawbacks, such as long surgical times and high recurrence rates. While silicone tubes have been used as fillers, they do not become integrated into tissue. Instead of silicone, a roll of Chitipack P was inserted into the hernia cavity (Fig. 4a). The caudal surface was sutured to the external anal sphincter, the obturator internus muscle in the ischial region, and the sacrotuberous ligament in three places to prevent caudal prolapse. The subcutaneous tissue was sutured in one layer, and the skin was closed to complete the surgery. The procedure took approximately 20 minutes.

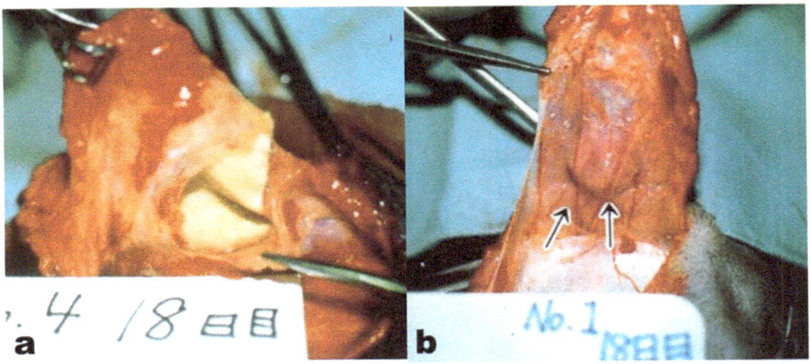

Fig. 3 (**a**) In the control group (non-chitin-impregnated nonwoven fabric), the nonwoven fabric itself had not become organized. (**b**) In the Chitipack P group (chitin-impregnated nonwoven fabric), the nonwoven fabric itself had completely become organized

Table 2 Summary of histological findings

Agent	Concentration (mg/ml)	Granulation tissue in implant	Granulation tissue surrounding implant	Inflammatory cells	Giant cells
Control	–	+	±	±	±
Chitin	0.1	±±	±	±	±
	1	++	±	+	+
	10	+++	+	+	++
	50	+	+++	+++	+
Chitosan	0.01	+	±	+	±
	0.1	++	±	+	+
	1	++	++	+	+
	10	+	+++	+++	±

±: slight, +: moderate, ++: severe, +++: more severe

Figure 4b shows the condition at the time of suture removal. The hernia healed without any particular complications.

Application to Bovine Umbilical Hernia

Umbilical hernias are common in calves. However, even when reduced using conventional methods, stress on the suture site often leads to recurrence. Using Chitipack P as a suture reinforcement material (Fig. 5a and b) resulted in almost no recurrence (Fig. 5c).

Chitipack P is a nonwoven fabric impregnated with chitin microparticles. Compared to chitosan, which will be discussed later, it has lower resistance to infection. Therefore, when using Chitipack P, careful attention must be paid to infection prevention.

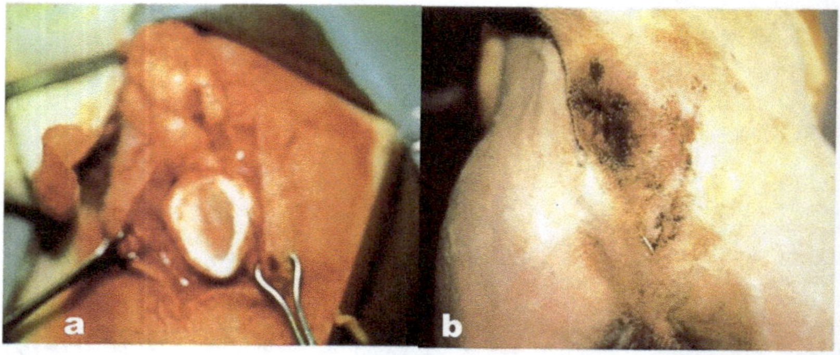

Fig. 4 (**a**) Instead of the silicone tube, a rolled up Chitipack P was inserted into the hernia cavity. (**b**) When the stitches were removed, the affected area was fine

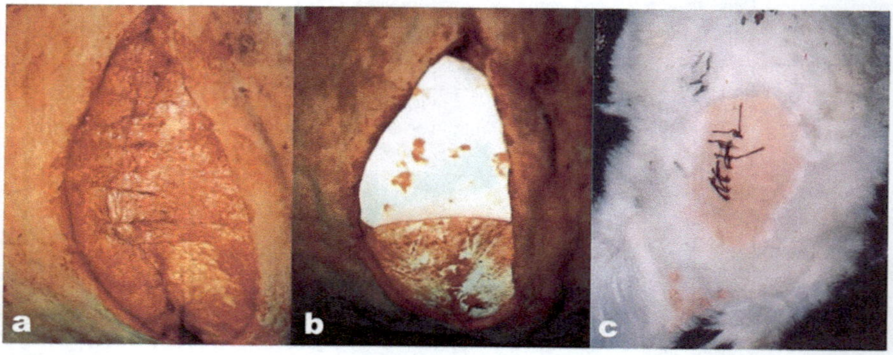

Fig. 5 (**a**) We have just repaired an umbilical hernia in a calf. (**b**) Chitipack P was used as a reinforcing material for sutures during reduction of an umbilical hernia in a calf. (**c**) One week later, no recurrence of the lesion was observed

2.2 *Chitosan-Derived Substances*

2.2.1 Cotton Type Chitosan (Chitopack C)

Application to Small Animals [4]

Application of Chitopack C to Severe Lacerations in a Cat

Figure 6a shows a cat that was involved in a traffic accident and visited the veterinary hospital a few days later. At the initial examination, severe subcutaneous tissue loss, bone contusions, and fractures were observed, with the wound surface dry. We recommended amputation to the owner, but the owner refused. Treatment with Chitopack C was attempted.

As shown in Fig. 6b, the affected area was covered with Chitopack C and bandaged (day 0). One week later (day 7), the bandage was removed (Fig. 6c). As seen

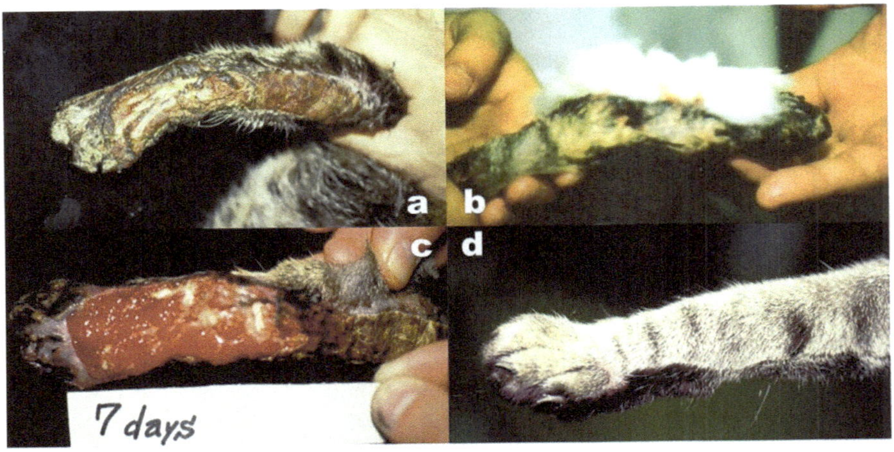

Fig. 6 (**a**) He was involved in a traffic accident and came to our hospital a few days later. The skin is severely damaged, and the metacarpals are exposed and fractured. (**b**) Chitopack C has been sprinkled over the wound surface, and a dressing applied. (**c**) On day 7, good granulation tissue formation is evident, and the bones are already embedded in the granulation tissue. (**d**) On day 70, the wound has completely healed, and there is virtually no functional deficit

in the photograph, granulation had proliferated to the point where the bone was no longer visible. The affected area was again covered with Chitopack C and bandaged. Another week later (day 13), granulation had proliferated even more vigorously than before. Treatment was then switched to a steroid and antibiotic ointment and continued. By day 19, the excess granulation had regressed, and the skin had begun to elongate. By day 70, the cat had healed, leaving only a slight scar (Fig. 6d).

Application of Chitopack C to a Dog Injured During Boar Hunting

Dogs used for wild boar hunting often suffer very severe injuries when fighting with wild boars. Most of these wounds are contaminated. Traditionally, antibiotics have been injected directly into the wound. To verify the effectiveness of Chitopack C, dogs injured during wild boar hunting were divided into two groups: one treated with the conventional method and the other with Chitopack C.

With the conventional method, the wound was cleaned, curettaged, antibiotics injected, sutured, and drained at the time of the initial consultation. In contrast, in the Chitopack C group, the wound was cleaned, packed with Chitopack C, sutured, and drained.

For follow-up treatment after the initial consultation, the conventional method involved daily drainage, cleansing, and antibiotic injection into the wound. In the Chitopack C group, drainage was performed when necessary, but cleansing and reapplication of Chitopack C were not performed.

Table 3 shows the cases and the number of treatments required for complete healing. In the five cases treated with the conventional method, there were three

Table 3 Results of each treatment for wounds

No.	Species	Sex	Age (Y)	Type of wound	Site	Number of treatments
Conventional method						
1	Kishu	Male	3	Contused wound	Shoulder	8
2	Kishu	Male	2	Lacerated wound	Chest	2
3	Terrier	Male	3	Lacerated wound	Chest–Abdomen	11
4	Dogo	Female	2	Contused wound	Chest	30
5	Dogo	Female	3	Lacerated wound	Chest	13
					Average	12.8
					SD	10.5
New method using chitosan						
6	Kishu	Female	3	Lacerated wound	Shoulder	1
7	Kishu	Male	2	Lacerated wound	Chest	3
8	Terrier	Female	3	Lacerated wound	Thigh	1
9	Dogo	Female	2	Contused wound	Thigh	1
10	Dogo	Female	6	Lacerated wound	Chest	1
11	Kishu	Male	3	Lacerated wound	Upper arm	1
					Average	1.3
					SD	0.8

lacerations and two contusions, while in the Chitopack C group, there were five lacerations and one contusion. With the conventional method, the number of treatments ranged from 2 to 30 (average: 12.8 ± 10.5), whereas in the Chitopack C group, five of the six cases received one treatment and the remaining case received three (average: 1.3 ± 0.8). It is clear that the Chitopack C group required fewer treatments.

We present a case of injury sustained during wild boar hunting. Fig. 7a shows an extensive subcutaneous contusion accompanied by muscle rupture. The wound cavity was heavily contaminated with hair, mud, and grass. The cavity was washed with saline, and after removing visible contamination, the subcutaneous tissue and skin were sutured. Fig. 7b shows the condition at suture removal, 18 days after the procedure. No treatment was administered during this time. The wound healed completely without suppuration.

Application to Farm Animals [5, 6]

Conventional treatments for infectious diseases in dairy cows, particularly hoof diseases, involve disinfection of the affected area and systemic antibiotic therapy. However, antibiotic residues in meat and milk pose serious food hygiene concerns. Chitosan, on the other hand, has already been approved by the Ministry of Health, Labor, and Welfare in Japan as a food additive and is widely used as a preservative and thickener. The use of chitosan as a wound treatment for farm animals—where

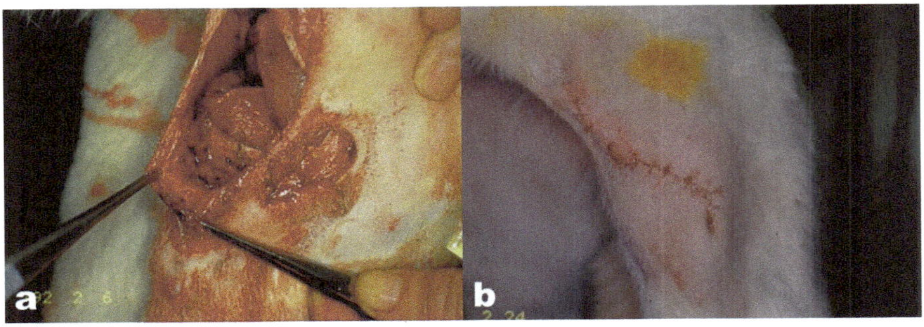

Fig. 7 (**a**) A case of injury sustained while hunting wild boar. Extensive subcutaneous tissue lacerations with muscle ruptures are observed. The wound cavity is significantly contaminated with hair, mud, and grass. (**b**) 18 days after the treatment, the wound had completely healed without suppuration

drug residues are a concern—has the potential to be extremely effective. It has also been used to treat equine hoof diseases with good results.

The authors applied Chitopack C and a chitosan suspension (a suspension of finely powdered chitosan in saline) to 141 cases of infectious hoof disease, 41 cases of interdigital phlegmon, 43 cases of abscess, and 17 cases of trauma in cattle. Good results were obtained in 238 of 245 cases (97%). Furthermore, treatment efficacy was comparable to that of conventional antibiotic-based treatments (Table 4). An example of application in cattle is shown below.

Application to Interdigital Phlegmon in Cattle

Interdigital phlegmon in calves is a suppurative disease. Usually, after washing the affected area, antibiotic ointment is applied. However, because the animal is constantly in contact with soil, complete healing can take time. In this case, in addition to interdigital phlegmon, there was erosion of the coronal area (Fig. 8a). The affected area was covered with Chitopack C (Fig. 8b) and bandaged. One month later, slight scarring was evident, but the condition had completely healed without suppuration (Fig. 8c). It was confirmed that granulation tissue could be produced with just one treatment, even under adverse conditions.

Application to Femoral Abscess in Cattle

Figure 9a shows a large abscess that developed in the thigh of a cow. Normally, this type of disease is left untreated in the field until the animal is euthanized. This is because, even if treated by incision and drainage, the abscess usually recurs, leading to euthanasia.

As a trial, an incision was made, the abscess cavity was washed with povidone-iodine solution, filled with Chitopack C, and the animal was observed (Fig. 9b).

Table 4 Comparison of conventional and chitosan treatments for septic pododermatitis and interdigital phlegmon

Disease	No.	Cure rate (%)	Average number of treatments[a]
Septic pododermatitis*			
Conventional treatment	93	95	2.4 ± 1.6[b]
Chitosan treatment	144	96	2.9 ± 2.8
Interdigital phlegmon			
Conventional treatment	12	100	2.3 ± 1.4
Chitosan treatment	41	100	2.8 ± 2.8

[a]For cured cases
[b]Mean ±SD
*Including sole ulcers

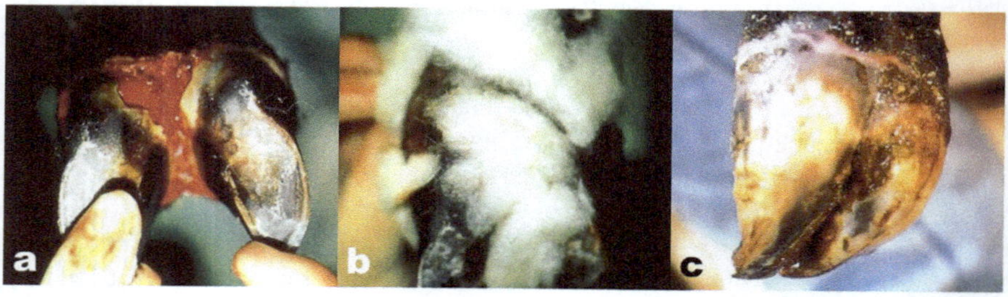

Fig. 8 (**a**) Cases with erosion of the coronal area in addition to interdigital rot. (**b**) Cover the affected area with Chitopack C and bandage it. (**c**) The affected area is healing well

After about 1 month, the affected area had completely healed (Fig. 9c). Furthermore, improvement in the animal's nutritional condition was also observed.

2.3 Other Chitin- and Chitosan-Derive Substances

2.3.1 Cotton Type Chitin

We present a case in which the subcutaneous tissue of the buttocks became infected, and the skin fell off from the buttocks to the tip of the tail. Treatment was carried out for an extended period using antibiotic ointments, but no skin regeneration was observed.

Treatment included applying cotton type chitin to the wound and spraying povidone-iodine solution over it. Figure 10a shows the image three days after treatment with cotton type chitin. Thin granulation tissue was already evident. Subsequent similar treatments led to the progression of epithelialization and hair growth (Fig. 10b). Figure 10c shows the condition one month after treatment. While some scarring remains, the skin has completely re-epithelialized and hair growth is evident.

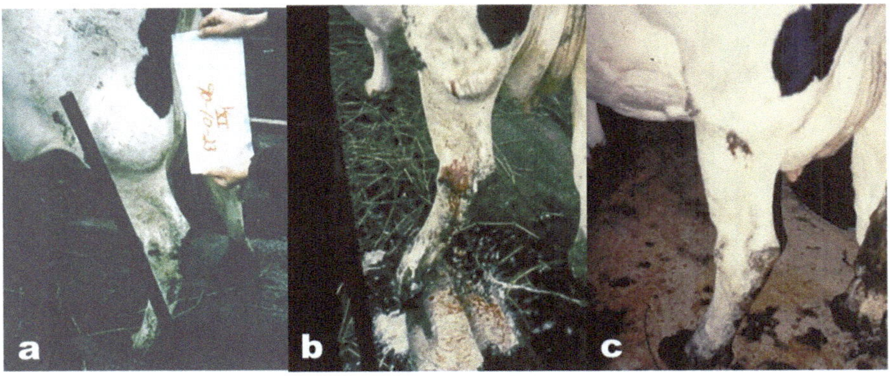

Fig. 9 (**a**) A large abscess in the thigh of a cow. (**b**) After incision and drainage, the wound cavity was washed with povidone-iodine solution and filled with Chitopack C. (**c**) After about a month, the lesion was completely healed

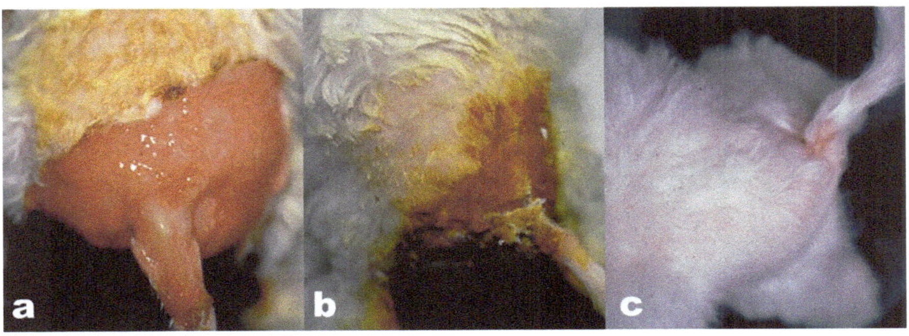

Fig. 10 (**a**) A case of subcutaneous suppuration in the buttocks and loss of skin from the buttocks to the tip of the tail. Cotton type chitin was applied to the wound surface, and povidone-iodine solution was sprayed on top of it. The photo was taken 3 days after treatment. Slight granulation tissue was already observed. (**b**) Subsequent similar treatments led to the progression of epithelialization of the skin and hair growth. (**c**) Condition 1 month after treatment. Although there are some scars, the skin has completely re-epithelialized and hair growth has been observed

2.3.2 Fine Powder Chitin

This cat was involved in a traffic accident and suffered complete skin loss from the hock to the extremity, as well as loss of the bottom of the extremity (Fig. 11a). Finely powdered chitin was sprinkled on the wound at intervals of several days (Fig. 11b).

Figure 11c shows the cat several days after treatment with finely powdered chitin. The skin had stretched considerably. Furthermore, cat litter was found adhering to the bottom of the extremity, indicating that the cat was walking and bearing weight on the affected limb. This suggests that chitin has an analgesic effect. The nails eventually fell off, but the skin had almost completely re-epithelialized (Fig. 11d).

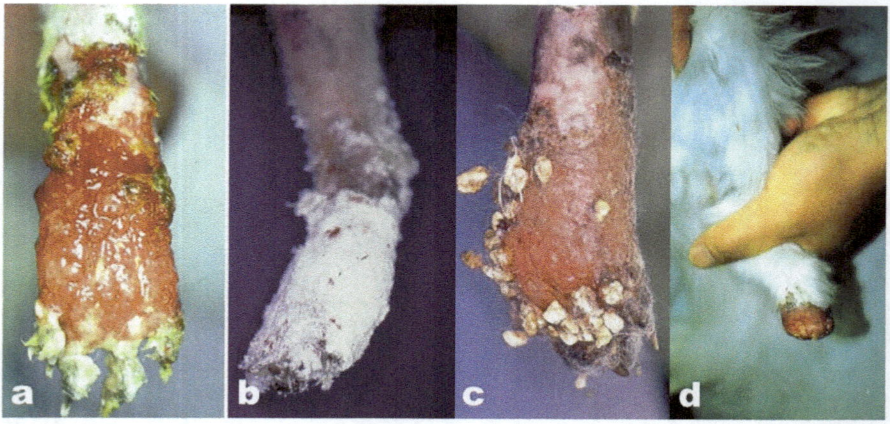

Fig. 11 (**a**) A case of complete loss of skin from the hock to the axil due to a traffic accident. (**b**) The injured section covered with flake type of chitin in fine powder form. (**c**) The injured leg 1 week after initiation of treatment. Granulation tissue regeneration is evident on the landing surface, and the adherence of litter sand to the surface indicates that the animal has regained weight bearing. (**d**) Eventually the nail fell off, but the skin re-epithelialized almost completely

2.3.3 Fine Powder Chitosan (Chitofine) [6]

This case involved empyema in a cat. X-ray images revealed characteristic pulmonary edema (Fig. 12a). Approximately 200 ml of purulent pleural fluid was collected by puncturing the chest wall (Fig. 12b). After lavaging the pleural cavity several times with saline, 8 ml of a Chitofine suspension (0.8 mg chitosan) was administered intrapleurally. Subsequent observations showed that the patient's condition improved day by day. X-ray examination 1 week after treatment revealed complete resolution of the pulmonary edema (Fig. 12c).

An attempt was also made to treat bovine mastitis using Chitofine in farm animals. Injecting 5 ml of Chitofine suspension (10 mg/ml) via the teat into the udders of three normal cows and four mastitis cows resulted in an increase in somatic cells (mostly neutrophils) in the milk (Table 5). Furthermore, the amount of lactoferrin in the milk also increased (Fig. 13).

Based on this basic research, Chitofine suspension was administered to cows with mastitis. The results are shown in Table 6. Four out of seven cases (57%) were cured. When treating bovine mastitis, the usual method involves injecting antibiotics into the teat. However, when antibiotics are administered into the udder, milk shipment is restricted for a certain period. Treatment with Chitofine eliminated this restriction and was found to be effective.

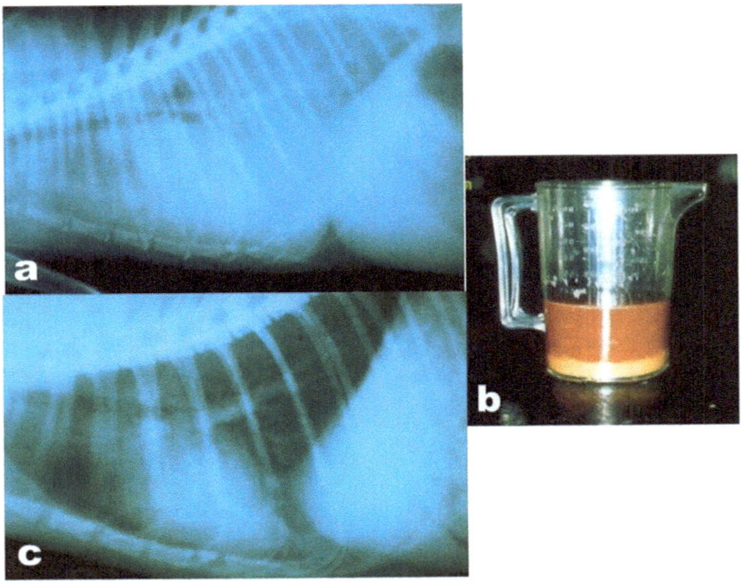

Fig. 12 (**a**) Cases of empyema in a cat. X-rays show characteristic pulmonary edema. (**b**) The chest wall was punctured and approximately 200 ml of purulent pleural fluid was collected. (**c**) X-ray taken 1 week after treatment. Pulmonary edema was completely cured

2.4 *New Chitin-Derived Substances [7–10]*

In 2009, Ifuku et al. succeeded in producing nanofibers from chitin. Experiments have confirmed that these nanofibers are as effective as conventional chitin products. The authors used chitin nanofibers in a case involving a severe skin defect in a cat (Fig. 14a) and obtained extremely good results (Fig. 14 b and c). This material has been commercially available since 2017 and is expected to become even more widely used in the future.

2.5 *Wound Healing Mechanism of Chitin and Chitosan*

So far, we have described the application of chitin and chitosan to wound healing. As a result, it has become clear that both chitin and chitosan strongly promote wound healing. The following points summarize the key characteristics of wound treatment using chitin and chitosan:

1) Formation of highly vascularized granulation tissue and inhibition of scar formation

Table 5 Changes in milk phagocytes following intramammary administration of chitosan

	0*	1*	3	7
Healthy cow				
1	600**	10300	400	100
2	300	16700	2700	900
3	700	5200	2400	900
Average	500	10700[a)]	1800	600
SD	200	5800	1300	500
Mastitis cow				
1	31600	26900	22800	19800
2	15200	66400	45300	37200
3	3300	34500	17500	7700
4	3500	9500	2300	1200
5	5600	32800	200	100
6	5200	32000	11000	16100
7	1700	17100	19800	26600
Average	9500	31300[a]	17000	15500
SD	10700	18000	15200	13600

*Days after chitosan administration (0 is before administration)
**Number of phagocytes per mm^3 of milk (values less than 100 were rounded to the nearest 100)
[a]Significant increase compared to day 0 ($p<0.05$)

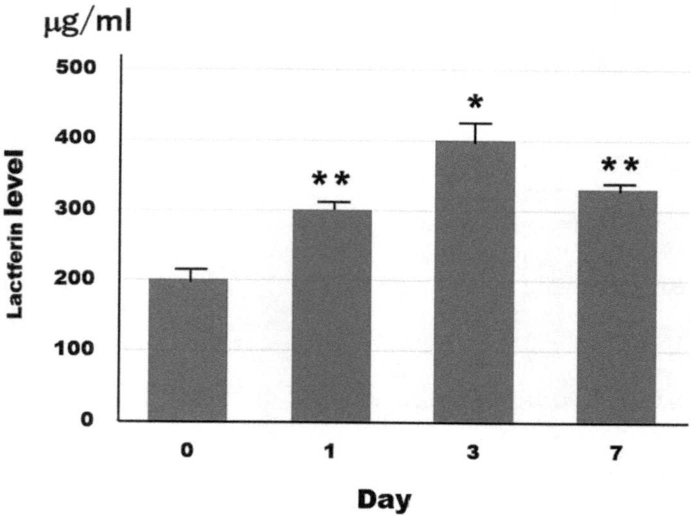

Fig. 13 Changes in lactoferrin content in milk following administrations of Chitofine suspension into the milk tank of healthy dairy cows. #: The value before chitosan administration was set as 0. *Significant increase compared to day 0 ($p < 0.05$). **Significant increase compared to day 0 ($p < 0.01$)

Table 6 Mastitis outcomes following intramammary administration of chitosan

No.	Age (Y)	PL test*	Causative bacteria	Outcome**
1	7	2+	*S. aureus*	Cure
2	5	2+	*S. aureus*	Invalid
3	5	2+	*S. aureus*	Recurrence
4	4	2+	*S. epidermis*	Cure
5	4	2+	*S. aureus*	Cure
6	4	2+	*S. epidermis*	Cure
7	4	2+	*S. aureus*	Invalid

*This is a simple diagnostic method for measuring the number of somatic cells in milk
**Cure: Healed within 10 days and did not recur for more than 1 month. Invalid: PL test did not become negative within 10 days. Recurrence: Recurrence within 1 month

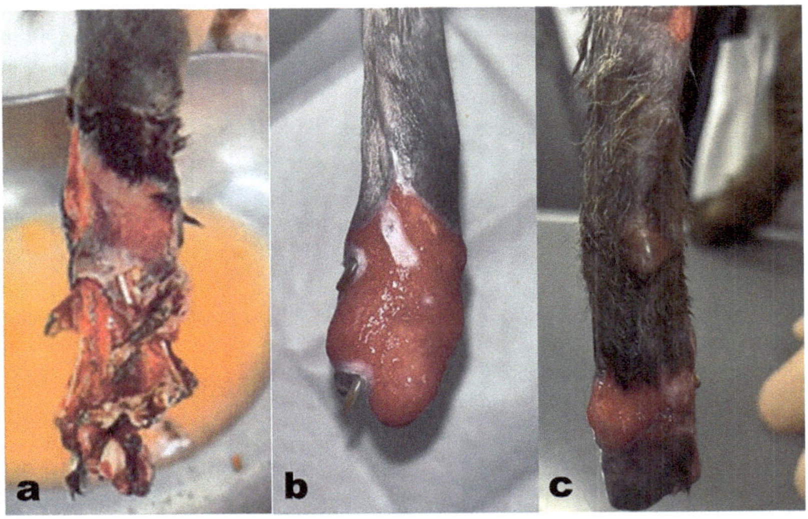

Fig. 14 (**a**, **b**) Severe skin defect in a cat. (**c**) On the 33rd day after treatment, granulation and epithelialization of the skin are evident. (**d**) On the 145th day, epithelialization has almost progressed

2) Dramatic effects on contaminated wounds (PMN activation, complement activation)
3) Analgesic effects
4) Hemostatic effects

The mechanisms of each therapeutic effect are described below.

2.5.1 Formation of Highly Vascularized Granulation Tissue and Inhibition of Scar Formation

(a) Tissue reactions to chitin and chitosan [11–13]

When chitin was implanted into the subcutaneous tissue of dogs, abundant vascularized granulation tissue formed within approximately 1 week. Early accumulation of polymorphonuclear leukocytes (PMNs) was also observed. This phenomenon was more pronounced with chitosan than with chitin. It was further confirmed that these substances eventually decompose and disappear in the body, with chitin biodegrading more rapidly than chitosan. Numerous multinucleated giant cells were observed within the granulation tissue, which disappeared within 2 weeks after administration. These experimental results confirmed the biological responses observed in clinical cases. Nakade et al. (1997) reported that chitin-induced matrix metalloproteinase (MMP) activity, particularly MMP2 and MMP9, in rat skin and subcutaneous tissue. MMP2 is produced by neutrophils and macrophages, while MMP9 is produced by macrophages, fibroblasts, and endothelial cells. Notably, MMP9 is involved in wound remodeling during the healing process.

(b) Effects of chitin/chitosan and their oligomers/monomers on fibroblasts and vascular endothelial cells [14]

The effects of chitin, chitosan, and their oligomers/monomers were evaluated on a mouse fibroblast cell line (3T6) and vascular endothelial cells (human umbilical vein endothelial cells: HUVEC) in vitro. Direct migration assays using the blind well chamber method revealed that the chemotactic activity of 3T6 cells was reduced by chitin, chitosan, and D-glucosamine (GlcN) (Fig. 15), indicating that these substances directly inhibit fibroblast migration.

In contrast, HUVEC migration activity was promoted by chitin, chitosan, and N-acetyl-D-glucosamine (GlcNAc), but suppressed by chitooligosaccharide (COS) and GlcN (Fig. 16). The effect of chitosan was dose-dependent. Latex beads—small, water-insoluble particles used as a physical control—also slightly promoted HUVEC migration, suggesting that HUVECs may be attracted to microparticles. However, the number of cells attracted by chitin and chitosan was significantly greater than that attracted by latex beads at the same concentration, suggesting that a non-physical mechanism is involved in the migration-promoting effect of chitin and chitosan.

Mori et al. (1997) studied cytokine release from fibroblasts stimulated by chitin, chitosan, and their constituent sugars. All tested sugars stimulated the production of interleukin-8 (IL-8). Among GlcN and GlcNAc, the latter exerted a stronger stimulatory effect. Similarly, the N-acetylchitooligosaccharide hexamer (NACOS6) had a more potent effect than chitooligosaccharide hexamer (COS6), but the polymeric form of chitosan showed the most pronounced effect.

All of these sugars stimulated fibroblasts to release IL-8, a key factor that promotes PMN migration. Chitin, chitosan, and their oligomers not only directly promote PMN migration but also enhance it indirectly by stimulating fibroblasts at the wound site.

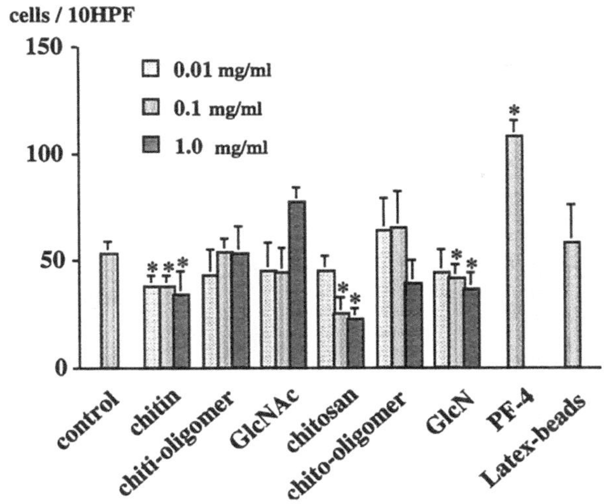

Fig. 15 Direct migration activity of 3T6 with a modified Boyden chamber method. Control: culture medium, GlcNAc: N-acetyl-D-glucosamine, GlcN: D-glucosamine, chitin-oligomer: mixture of GlcNAc1- GlcNAc6, chito-oligomer: mixture of GlcN1- GlcN6, PF-4: human recombinant platelet factor-4. Data are displayed as the mean ± SD for three or more experiments. $^{*}p < 0.05$ from negative control (culture medium)

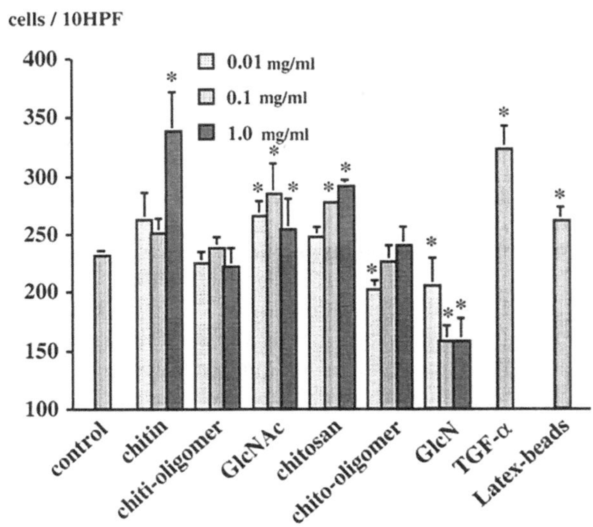

Fig. 16 Direct migration activity of HUVEC with a modified Boyden chamber method. Control: culture medium, GlcNAc: N-acetyl-D-glucosamine, GlcN: D- glucosamine, chitin-oligomer: mixture of GlcNAc1- GlcNAc6, chito-oligomer: mixture of GlcN1- GlcN6, TGFa: transforming growth factor a. Data are displayed as the mean ± SD for three or more experiments. $^{*}p < 0.05$ from negative control (culture medium)

Fibroblasts and vascular endothelial cells are central to the granulation tissue proliferation phase during wound healing. Fibroblasts are involved in collagen synthesis, essential for extracellular matrix formation during this phase. Furthermore, in the remodeling phase—the final stage of wound healing—fibroblasts release matrix metalloproteinases such as collagenase, which are involved in the breakdown of excess granulation tissue.

2.5.2 Dramatic Effects on Contaminated Wounds (PMN Activation, Complement Activation)

PMN Migration Ability [15]

The migration ability of PMNs was examined using the blind well chamber method. The results are shown in Table 7. When Hank's balanced salt solution (HBSS) was used as a control, the number of migrated PMNs was 52 ± 18. In contrast, the numbers of PMNs that migrated in response to chitin and chitosan were 457 ± 25 and 181 ± 50, respectively. Compared to the control, chitin exhibited approximately ten-fold higher migration activity, and chitosan approximately three-fold. Furthermore, in the presence of serum, chitosan's migration activity increased approximately two-fold.

Chemiluminescence of PMNs [16]

Table 8 shows results from a chemiluminescence reader used to assess the extent to which finely powdered chitin and chitosan were phagocytosed. Zymosan was used as a positive control. The chemiluminescence intensity per 1000 PMNs was 573.4 ± 22.3 for zymosan, compared to 156.0 ± 12.3 for chitin and 269.8 ± 27.3 for

Table 7 Effect of chitin and chitosan on canine polymorphonuclear cell migration (Checker Board Assay)

Migration-stimulating factor	Upper compartment (Factor concentration*)	Lower compartment (Factor concentration*)		
		0	0.1	1
Chitin	0	35 ± 8	178 ± 8	511 ± 20
	0.1	31 ± 7	85 ± 7	333 ± 15
	1	13 ± 4	65 ± 12	137 ± 25
Chitosan	0	40 ± 9	75 ± 8	205 ± 20
	0.1	25 ± 6	71 ± 4	202 ± 8
	1	2 ± 2	6 ± 2	58 ± 14

The migratory cell count (cells/mm^2) is shown as the mean ± standard deviation. *mg/mL

Table 8 Effect of chitin and chitosan on polymorphonuclear cell chemiluminescence*

Stimulating factor	Normal canine serum added	Inactivated canine serum added	No canine serum added
Zymosan	537.4 ± 22.3	215.0 ± 5.3	122.4 ± 12.9
Chitin	156.0 ± 12.3	2.4 ± 1.5	1.7 ± 0.3
Chitosan	269.8 ± 27.3	1.4 ± 0.7	1.5 ± 0.3

*The maximum chemiluminescence (LU) per 1000 polymorphonuclear leukocytes was measured using an LB-9505C Multi-biolumat (Berthold, Germany). The values are shown as the mean ± standard deviation

Table 9 Prostaglandin E_2 induction in a chitin and non-woven polyester fabric composite (Chitipack P) in canine subcutaneous tissue

Measurement substance	Prostaglandin E_2 concentration (ng/ml)
Normal canine serum	4.7 ± 0.5
Chitipack P	49.7 ± 12.1
Control*	9.5 ± 5.7

*Non-woven polyester fabric (Chitipack P base material)

chitosan. In other words, the chemiluminescence intensity of chitin and chitosan was 29.0% and 50.2%, respectively, of that of zymosan. These intensities were nearly eliminated in the absence of serum or in the presence of heat-inactivated serum. These findings suggest that chitin and chitosan activate PMN phagocytic activity through the involvement of heat-labile serum factors (likely complement). Furthermore, chitin's chemiluminescence intensity, even in the presence of serum, remained at only about 30% of zymosan's, indicating that chitin is not readily recognized by the body as a foreign substance.

PMN Production of Physiologically Active Substances [17].

Chitin implantation induces significant local exudation at the implantation site. Four days after implantation, exudate was collected under sterile conditions and prostaglandin E_2 (PGE_2) concentrations were measured. The PGE_2 concentration in the chitin-induced exudate was ten-fold higher than that in serum and five-fold higher than in the control (Table 9). The exudate precipitate contained numerous PMNs. PMNs were incubated with chitin or chitosan, and the concentrations of LTB_4, LTC_4, and PGE_2 in the culture supernatant were measured. These substances were found at elevated levels in the presence of chitin and chitosan (Table 10), with chitosan producing particularly high concentrations. PGE_2 increases vascular permeability, promoting exudation and PMN infiltration into tissues. Additionally, LTB_4 and PGE_2 stimulate PMN migration, serving as signals for PMN accumulation.

Table 10 Arachidonic acid synthesis and release from canine polymorphonuclear leukocytes due to stimulation by chitin and chitosan

	Substance produced**		
Stimulating substance*	LTB_4	LTC_4	PGE_2
Chitin	8300 ± 1500	64 ± 24	3700 ± 1000
Chitosan	12000 ± 2500	200 ± 32	5200 ± 3000
Latex beads	150 ± 20	73 ± 16	60 ± 15
Control	16 ± 5	19 ± 7	21 ± 11

*The control was Hank's buffer solution, the culture solution

**The values show the mean±standard deviation (pg/ml) for the concentration of the substances produced (leukotriene B_4 and C_4, and prostaglandin E_2) in the supernatant when canine polymorphonuclear leukocytes and the stimulating substance were mixed and incubated

Macrophage Activity [18]

The effects of chitin and partially deacetylated chitin on cytokine production in mouse peritoneal macrophages have been detailed by Nishimura et al. They reported that deacetylated chitin—especially 70% deacetylated—effectively promotes interleukin-1 (IL-1) and interferon (IFN) production. Tanigawa et al. (1992) also reported that chitin promotes IL-1 production from human blood monocytes. IL-1 promotes fibroblast proliferation and PMN activation but suppresses endothelial cell activation during wound healing. Shibata et al. (1997) reported that intravenous injection of chitin microparticles into mice activated alveolar macrophages. They also observed significant IFN-γ production from spleen-derived macrophages after stimulation with chitin—but not with chitosan or latex beads—in vitro. Ueno et al. (2001) reported that chitosan promoted the production of transforming growth factor beta 1 (TGF-β1) and platelet-derived growth factor (PDGF) from human monocyte-derived macrophages in vitro. TGF-β1 is a fibrogenic growth factor, and PDGF induces fibroblast proliferation and collagen synthesis.

The effects of chitin and its derivatives on macrophage migration were examined using the Boyden chamber method. As shown in Table 11, chitin and chitosan oligomers strongly attracted mouse peritoneal macrophages in a dose-dependent manner, whereas chitin and chitosan polymers did not. When chitin/chitosan was implanted subcutaneously, numerous multinucleated giant cells—presumably formed by macrophage fusion—were observed. This is a characteristic tissue reaction to chitin/chitosan. These results suggest that chitin/chitosan themselves do not directly attract macrophages, but that their biodegradation products do.

Complement Activity [19]

Subcutaneous administration of high doses (200 mg/kg) of chitosan to dogs resulted in hemorrhagic pneumonia, with 70% mortality within 1 week. Histological examination of the lungs revealed no vascular or lymph node infarction. Administration of 1 mg/kg of chitosan resulted in systemic PMN activation, demonstrated by

Table 11 Effect of chitin/chitosan and theiroligomers/monomers on peritoneal exudate macrophage migration

Agent	Concentration (mg/ml) 0.1	1.0
GlcNAc	5 ± 1	11 ± 4
NACOS	23 ± 1*	24 ± 7
Chitin	6 ± 1	2 ± 1*
GlcN	1 ± 1*	4 ± 4
COS	20 ± 6*	19 ± 2
Chitosan	4 ± 0*	1 ± 1*
Saline	6 ± 1	

The migratory cell count (cell/10 HPF) is displayed as the mean ± SD. *p:<0.05 vs. saline

Table 12 Complement activation by chitin, chitosan, and zymosan

Complement	Agent	Incubation time (min) 10	20	30	40
C3	Chitosan	63 ± 2**	60 ± 5**	59 ± 4**	58 ± 6**
	Chitin	83 ± 11*	77 ± 10**	73 ± 13**	63 ± 21**
	Zymosan	66 ± 3**	64 ± 4**	64 ± 4**	64 ± 4**
C4	Chitosan	100 ± 2	100 ± 2	100 ± 2	100 ± 2
	Chitin	100 ± 2	100 ± 2	100 ± 2	100 ± 2
	Zymosan	100 ± 2	100 ± 2	100 ± 2	100 ± 2
C5	Chitosan	56 ± 19**	23 ± 32**	23 ± 32**	23 ± 32**
	Chitin	73 ± 4**	45 ± 4**	45 ± 4**	45 ± 4**
	Zymosan	56 ± 19**	23 ± 32**	23 ± 32**	23 ± 32**

The concentration of each agent is 10 mg/ml. Data was expressed as the mean % and standard deviation in each agent

Individual data were converted to % relative to the initial complement level (pre-incubation). Statistical analysis was performed between each initial level and after incubation level (*p<0.05; **p<0.01)

sustained chemiluminescence responses to zymosan for approximately 1 week. These phenomena appear to result from the indirect, rather than direct, effects of chitosan on PMNs.

As described in the PMN section, fresh serum treated with chitin and chitosan promoted PMN phagocytosis, an effect not observed with heat-inactivated serum, suggesting complement activation. To investigate this, chitin and chitosan were tested for complement activation using single-radial immunodiffusion with fresh human serum. Chitin and chitosan reduced serum C3 and C5 levels but had no effect on C4 (Table 12), results consistent with those of zymosan. These findings indicate that chitin and chitosan activate the alternative complement pathway. Activation was rapid, as complement activity was detected after just 10 minutes of incubation. Upon activation, C5 is cleaved into C5a and C5b. Radioimmunoassay confirmed high concentrations of C5a in the culture supernatant.

3 Analgesic Effects [20]

In clinical trials, we have observed numerous cases in which the application of chitin and chitosan to wound sites demonstrated analgesic effects. To investigate this mechanism, we experimentally evaluated the analgesic effects of chitin and chitosan using the acetic acid-induced writhing test in mice, a method widely used in the development of analgesics.

Chitin and chitosan suspensions (average particle size: 3 μm) at concentrations of 1, 5, and 10 mg/ml were mixed with 0.5% acetic acid solution (referred to as chitin AC and chitosan AC, respectively) and administered intraperitoneally to mice. Both agents reduced the frequency of pain-induced abnormal behaviors (writhing), such as hind limb extension, abdominal rigidity, and abdominal twisting, in a dose-dependent manner. This effect was greater in mice administered chitosan AC compared to those administered chitin AC (Table 13).

In vitro studies showed that the addition of chitin or chitosan suspensions increased the pH of the acetic acid solution, with a greater effect observed for chitosan than for chitin (Fig. 17). Furthermore, the bradykinin concentration in the peritoneal lavage fluid of animals administered chitin AC was lower than that in animals administered chitosan AC. In vitro studies also showed that chitin particles absorbed more bradykinin than chitosan particles (Table 14).

Table 13 Effects of chitin and chitosan on number of writhing

Agents	Concentration (mg/ml)	Number of writhing
0.5 % Acetic acid	–	29.7 ± 1.5
Chitin	1	30.7 ± 1.2[a]
	5	24.0 ± 1.0*
	10	19.0 ± 1.0*
Chitosan	1	23.7 ± 3.5*
	5	19.7 ± 1.5*
	10	13.0 ± 1.0*

[a]Mean ± SD (n = 4)
*Significant difference compared to 0.5 % acetic acid ($p<0.05$)

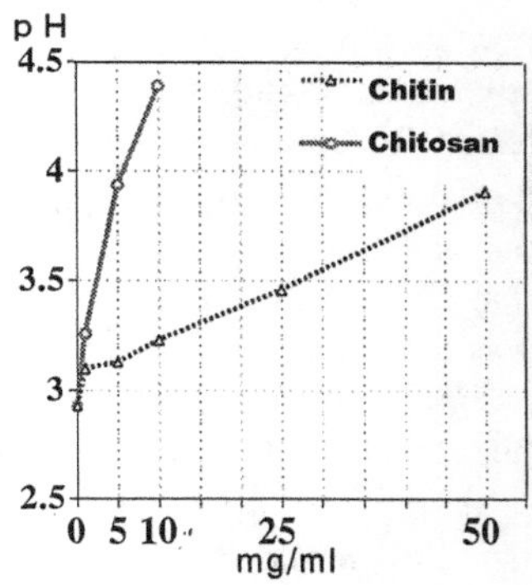

Fig. 17 The relationship between chitin/chitosan and proton level

Table 14 Absorption of bradykinin with chitin/chitosan in vitro

Agents	Concentration (mg/ml)	Bradykinin (ng/ml)	(%)
Saline	–	37.2[a]	(100)
Chitin	10	16.8	(45.2)
	100	16.8	(45.2)
Chitosan	10	30.0	(80.6)
	100	31.2	(83.9)

[a]Mean (n=2)

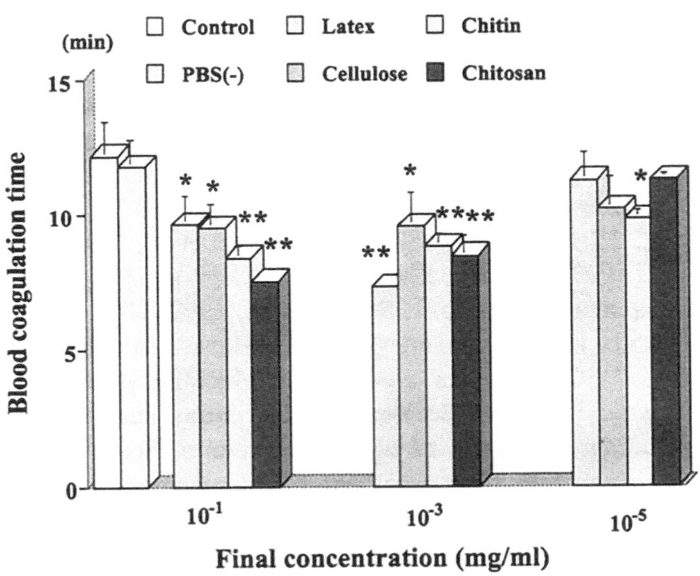

Fig. 18 Effect of chitin and chitosan on blood coagulation time. Chitin and chitosan with particle size of 2.8 μm were used. $^{*}p < 0.05$, $^{**}p < 0.01$ vs. control

These results suggest that the analgesic effect of chitin is exerted by adsorbing the pain-inducing substance bradykinin, while chitosan inhibits bradykinin production by adsorbing hydrogen ions and other substances that stimulate nerve endings.

4 Hemostatic Effect [21]

In the wound healing process, blood coagulation is the first event to occur at the time of injury. It forms a fibrin network, which serves as a scaffold for various cells. Blood coagulation is also accompanied by platelet aggregation, during which various physiologically active substances are released, promoting wound healing.

Experimental results showed that chitin and chitosan shortened blood coagulation time in a dose-dependent manner (Fig. 18). Chitin exhibited stronger

Table 15 Platelets aggregation in the presence of plasma

Final concentration of samples (mg/ml)	Platelet aggregation (%)		
	Latex	Chitin	Chitosan
0.01	0.3[a]	4.8[ae]	1.0[a]
0.03	0.7[a]	9.8[abe]	4.0[b]
0.1	19.7[b]	19.8[be]	9.9[c]
0.3	26.3[b]	46.2[c]	15.3[d]
1	23.7[b]	57.3[d]	20.3[e]
2	7.0[a]	14.3[e]	1.0[a]
PBS (-)	0[a]	0[a]	0[a]

We used chitin and chitosan with mean particles of 2.8 μm
*Different superscripts (a, b, c, d) in each column show significant difference ($p<0.05$)

aggregation ability than chitosan when used with platelets in the presence of plasma (PRP) (Table 15). Furthermore, the addition of chitin or chitosan to PRP promoted the release of platelet-derived PDGF-AB and TGF-β1.

5 The Regenerative Effect of Glucosamine (GlcN) on Articular Cartilage [22, 23]

Experimental studies using rabbits with knee articular cartilage injury, oral administration of GlcN, and observation of the healing process demonstrated that GlcN promotes repair of knee articular cartilage (Figs. 19a and b). Based on this data, we administered 1 g of GlcN and fish collagen peptide (FCP) orally for 3 months to 48 small animals with orthopedic diseases to examine whether symptoms improved. Lameness was classified into ten stages (Table 16) and scored. Figure 20a shows the lameness scores before administration; Figure 20b shows the scores 1 month after administration. Lameness improved in most cases.

Figure 20c shows the progression of hip dysplasia cases (17 cases). One month after the start of administration, the scores had decreased by approximately half. Similar evaluations were conducted in cases of spondylosis deformans (23 cases), and the results are shown in Fig. 20d. Similar effects were observed in spinal disorders as well as in other orthopedic conditions. These findings suggest that oral administration of GlcN and FCP not only improves articular cartilage damage but also has analgesic and neuroactivating effects.

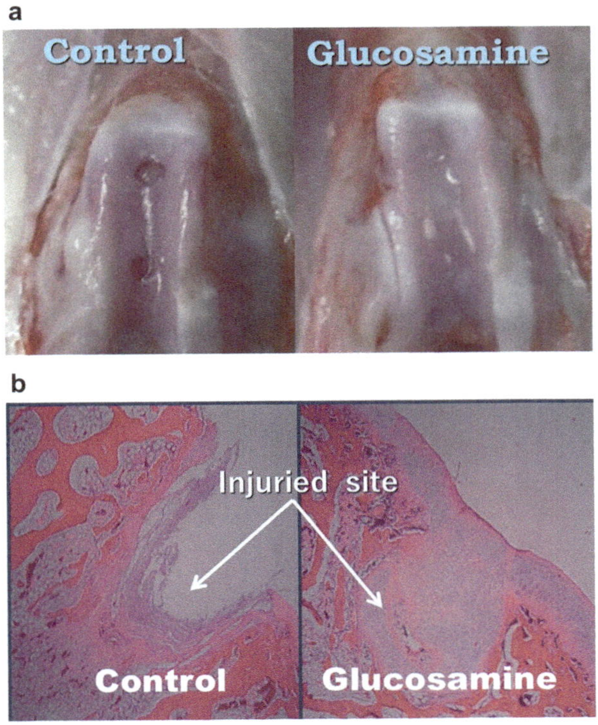

Fig. 19 (**a**) Macroscopic photograph of knee joint surface. In the control group, the articular cartilage had not yet been completely repaired macroscopically, whereas in the glucosamine group, it had been completely repaired macroscopically. (**b**) Microscopic photograph of knee joint surface. In the control group, obvious defects in the articular cartilage were still observed, whereas in the glucosamine group, the defects were completely repaired by chondrocytes

Table 16 Grading of lameness

Grade	Condition
1	No clinical symptoms
2	Owners sometimes notice strange gait
3	Occasional lameness
4	Lameness after running
5	Can run 10 m
6	Cannot run 10 m
7	Can walk 10 m
8	Can walk a few steps
9	Able to stand up on one's own
10	Inability to stand

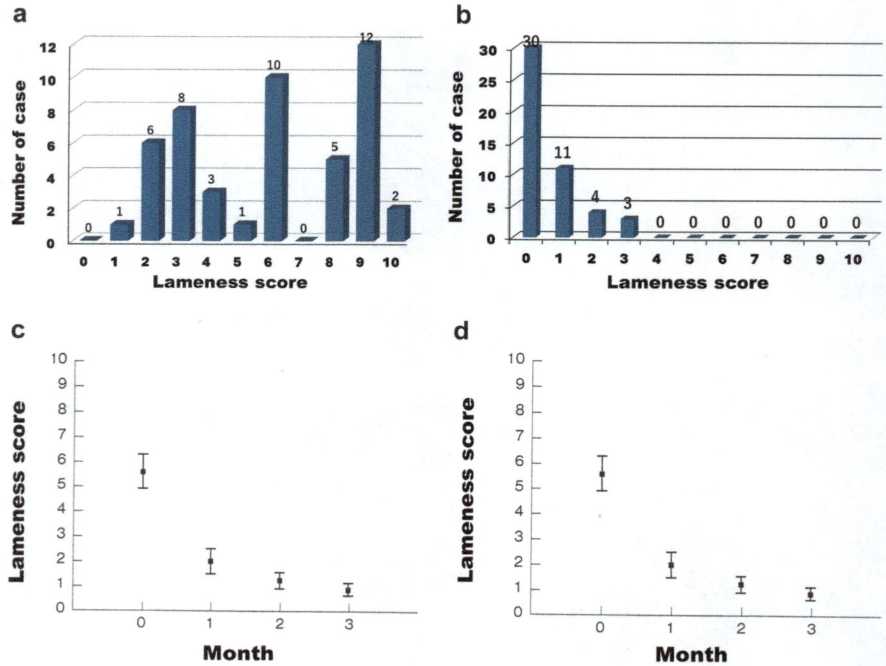

Fig. 20 (**a**) Lameness score before glucosamine administration. (**b**) Lameness score 1 month after glucosamine administration. (**c**) Changes in lameness score in hip dysplasia cases. (**d**) Changes in lameness scores in patients with spinal deformity

6 Antitumor Effects of Chitin Oligosaccharides (NACOS) and Chitosan Oligosaccharides (COS) [24]

In 1985, Suzuki et al. reported that chitin oligosaccharide hexamer (NACOS6) and chitosan oligosaccharide hexamer (COS6) had antitumor effects. However, no further studies were conducted to verify these results. In 2014, Masuda et al. orally administered NACOS and COS mixtures to tumor-bearing mice (Fig. 21a) and confirmed their antitumor effects (Fig. 21b). The mechanism of action was found to involve induction of IFN-γ and IL-12, as well as changes in the intestinal microbiota.

Based on these experimental results, NACOS or COS mixtures were orally administered to eight tumor-bearing dogs. Of these, four showed a marked or effective response, two showed no change, and two experienced disease progression (Table 17). Kan et al. also confirmed the effectiveness of NACOS mixture when administered orally to tumor-bearing humans.

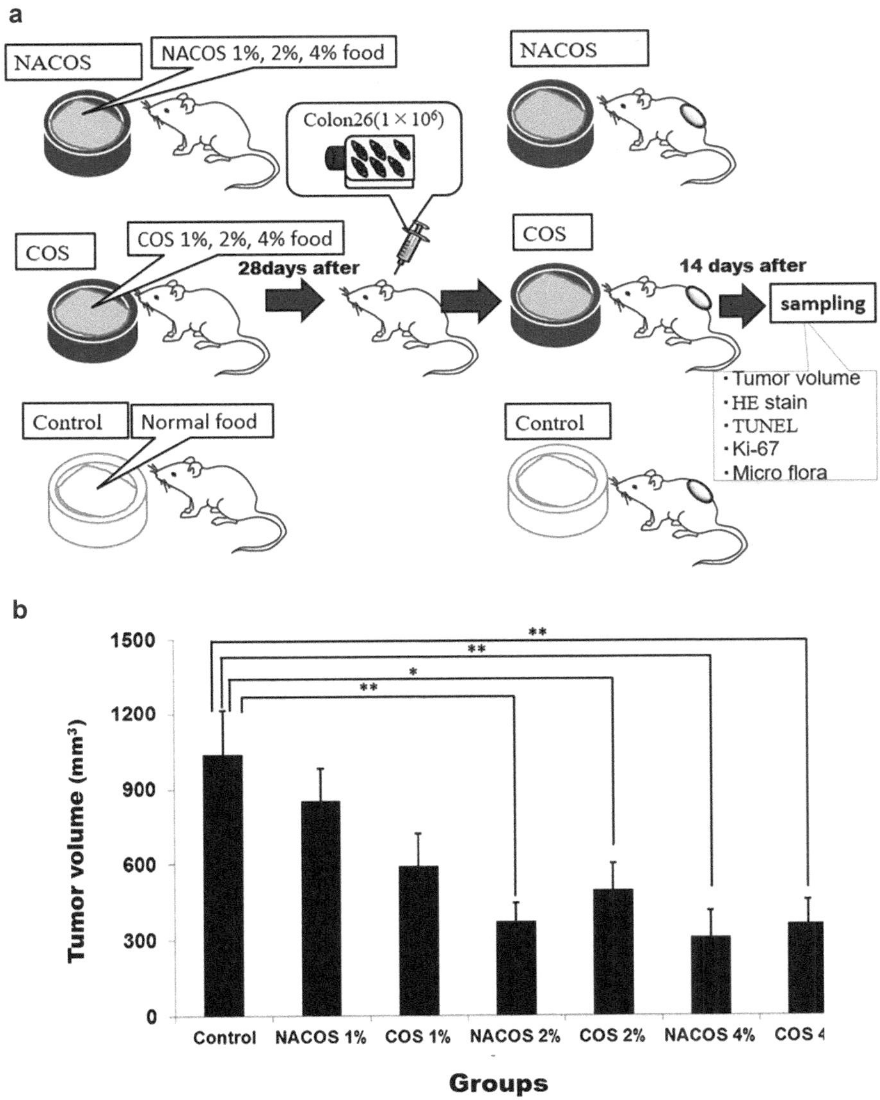

Fig. 21 (**a**) Experimental design for oral administration of NACOS/COS. (**b**) Tumor weight in each group. $^{*}p < 0.05$, $^{**}p < 0.01$ vs. control

Table 17 Effect of NACOS/COS on dogs with tumors

No.	Species	Diagnosis	Courses[a]	Result[b]
1	G. Retriever	Liposarcoma	1.5	CR
2	G. Russell Terrier	Mast cell tumor	4.5	CR
3	Mix	MGT[c]	6	PR
4	M. Poodle	Mast cell tumor	3	PR
5	Papillon	MGT	8.3	SD
6	M. Dachshunt	MGT	3	SD
7	L. Retriever	MGT	8.3	PD
8	Doberman	Osteosarcoma	3	PD

[a]1 course = 1 month
[b]CR: Complete disappear, PR: Decrease more than 20%, SD: Stable (change in -20~+20%), PD: Increase more than 20%
[c]Mammary gland tumor

7 Conclusion

This article introduced the veterinary clinical applications of chitin, chitosan, and their oligosaccharides and monosaccharides, and explained their mechanisms of action. In terms of wound healing, administration of chitin and chitosan directly promotes the production of physiologically active substances, blood coagulation (hemostasis), analgesic effects, and complement activation. Furthermore, upon degradation in vivo, their oligosaccharide and monosaccharide forms have been shown to stimulate fibroblasts and vascular endothelial cells, which are central players in wound healing, thereby promoting granulation tissue formation. However, many aspects of these substances remain unexplored, and further research is needed to elucidate more complex mechanisms.

In addition, D-glucosamine, a monosaccharide derived from chitosan, plays a significant role in articular cartilage repair. In recent years, glucosamine has been widely used as a supplement for arthritis, and our experimental results support its effectiveness in treating such conditions.

Furthermore, oral administration of chitin and chitosan oligosaccharides has demonstrated antitumor effects, raising hopes for the development of oral medications or supplements with such properties. However, many unknowns remain, and further studies are necessary.

References

1. Okamoto Y, Minami S, Matsuhashi A, Tokura S, Sashiwa H, Saimoto H, Shigemasa Y (1993) Developement of chitin and chitosan biomaterials. In: Yalpani M (ed) Carbohydrates and Carbohydrate Polymers. ATL Press, Mt. Prospect, pp 153–164
2. Minami S, Okamoto Y, Matsuhashi A, Tanioka S, Sashiwa H, Saimoto H, Shigemasa Y (1995) Wound management II: New approach with "Chitin-Chitosan". In: Proceedings of world veterinary congress, Org. Com. WVC, Tokyo, pp 275–282

3. Okamoto Y, Minami S, Matsuhashi A, Sashiwa H, Saimoto H, Shigemasa Y, Tanigawa T, Tanaka Y, Tokura S (1993) Application of polymeric *N*-acetyl-D-glucosamine (Chitin) to veterinary practice. J Vet Med Sci 55:743–747
4. Minami S, Okamoto Y, Tanioka S, Sashiwa S, Saimoto H, Matsuhashi A, Shigemasa Y (1993) Effects of chitosan on wound healing. In: Yalpani M (ed) Carbohydrates and Carbohydrate Polymers. ATL Press, Mt. Prospect, pp 141–152
5. Okamoto Y, Minami S, Matsuhashi A, Hamada K, Yanagihara G, Ohira J, Fukumoto Y, Murakami S (1996) Effect of chitosan on bovine wound healing. J Jpn Med Assoc 49:22–24. (in Japanese)
6. Minami S, Egawa T, Ohira J, Okamoto Y, Matsuhash A (1997) Effect of chitosan with intramammary administration on phagocytes in udder secretion. J Jpn Med Assoc 50:143–146. (in Japanese)
7. Ifuku S, Nogi M, Abe K, Yoshioka M, Morimoto M, Saimoto H, Yano H (2009) Preparation of chitin nanofibers with a uniform width as a-chitin from crab shells. Biomacromolecules 10:1584–1588
8. Azuma K, Ifuku S, Osaki T, Okamoto Y, Minami S (2014) Preparation and biomedical applications of chitin and chitosan nanofibers. J Biomed Nanotech 10:2891–2920
9. Izumi R, Komada S, Ochi K, Karasawa L, Osaki T, Murahata Y, Tsuka T, Imagawa T, Itoh N, Okamoto Y, Izawa H, Morimoto M, Saimoto H, Azuma K, Ifuku S (2015) Favorable effects of superficially deacetylated chitin nanofibrils on the wound healing process. Carbohydr Polym 123:461–467
10. Azuma K, Koizumi R, Izawa H, Morimoto M, Saimoto H, Osaki T, Ito N, Yamashita M, Tsuka T, Imagawa T, Okamoto Y, Inoue T, Ifuku S (2019) Hair growth-promoting activities of chitosan and surface-deacetylated chitin nanofibers. Int J Biol Macromol 126:11–17
11. Okamoto Y, Minami S, Matsuhashi A, Sashiwa H, Saimoto H, Shigemasa Y, Tanigawa T, Tanaka Y, Tokura S (1993) Polymeric N-acetyl-D-glucosamine (Chitin) induces histionic activation in dogs. J Vet Med Sci 55:739–742
12. Okamoto Y, Minami S, Matsuhashi A, Tanioka S, Shigemasa Y (1995) The fate of *N*-acetyl-D-glucosamine (Chitin) in canine subcutaneous tissues. Jpn J Biomaterials 13:112–116
13. Nakade T, Yokota H, Hori Y, Agata N, Ikeda T, Furusaki H, Yamada Y, Uchida Y, Yuasa A, Yamaguchi M, Taniyama H, Otomo K (1997) Induction of matix metalloproteinase (MMP) 2 and 9 in skin and subcutaneous tissue by implantation chitin in rats. In: Domard A, Roberts GAF, Varum KM (eds) Advances in chitin science Vol. II. Jacques Andre Publisher, Lyon, pp 733–737
14. Okamoto Y, Watanabe M, Miyatake K, Morimoto M, Shigemasa Y, Minami S (2002) Effects of chitin/chitosan and their oligomers/monomers on migration of fibroblasts and vascular endothelium. Biomaterials 23:1975–1979
15. Usami Y, Okamoto Y, Minami S, Matsuhashi A, Kumazawa NH, Tanioka S, Shigemasa Y (1994) Chitin and chitosan induce migration of bovine polymorphonuclear cells. J Vet Med Sci 56:761–762
16. Minami S, Masuda M, Suzuki H, Okamoto Y, Matsuhashi A, Kato K, Shigemasa Y (1997) Subcutaneous injected chitosan induces systemic activation in dogs. Carbohydr Polym 33:285–294
17. Minami S, Okamoto Y, Matsuhashi A, Sashiwa H, Saimoto H, Shigemasa Y, Tanigawa T, Suzuki T, Tanaka Y (1995) Polymeric *N*-acetyl-D-glucosamine (Chitin) induces prostaglandin E_2 in dogs. J Vet Med Sci 57:377–378
18. Okamoto Y, Inoue A, Miyatake K, Ogihara K, Shigemasa Y, Minami S (2003) Effects of chitin/chitosan and their oligomers/monomers on migrations of macrophages. Macromol Biosci 3:587–590
19. Suzuki Y, Okamoto Y, Morimoto M, Sashiwa H, Saimoto H, Tanioka S, Shigemasa Y, Minami S (1997) Influence of physio-chemical properties of chitin and chitosan on complement activation. Carbohydr Polym 42:307–310

20. Okamoto Y, Kawakami K, Miyatake K, Morimoto Y, Minami S (2002) Analgesic effects of chitin and chitosan. Carbohydr Polym 49:249–252
21. Okamoto Y, Yano R, Miyatake K, Imagawa T, Shigemasa Y, Minami S (2003) Effects of chitin and chitosan on blood coagulation. Carbohydr Polym 53:337–342
22. Tamai Y, Miyatake K, Okamoto Y, Takamori Y, Sakamoto H, Minami S (2003) Enhanced healing of cartilaginous injuries by glucosamine hydrochloride. Carbohydr Polym 48:369–378
23. Minami S, Hata M, Tamai Y, Hashida M, Takayama T, Yamamoto S, Okada K, Funatsu T, Okamura Y, Tsuka T, Okamoto Y (2008) Application of D-glucosamine hydrochloride and fish collage peptide on various orthopedic diseases and degenerative spondylopathy in dogs and cats. Glucosamine Res 4:37–40. (in Japanese)
24. Masuda S, Azuma K, Kurozumi S, Kiyose M, Osaki T, Tsuka T, Itoh N, Imagawa T, Minami S, Sato K, Okamoto Y (2014) Anti-tumor properties of orally administered glucosamine and N-acetyl-D-glucosamine oligomers in a mouse model. Carbohydr Polym 111:783–787

Yoshiharu Okamoto

Chitosan Nanoparticle Systems and Biocomposites for Nutraceutical Delivery

Jayachandran Venkatesan, Neema Kumari, Pandurang Appana Dalavi, Kranti Kiran Reddy Ealla, and Se-Kwon Kim

Abstract Nutraceuticals are often required to maintain a healthy body and metabolism. However, nutraceuticals are sometimes hydrophobic, poorly soluble, highly biodegradable, and have limited bioavailability. Nutraceutical nanodelivery systems can be used to improve stability, increase solubility, and improve bioavailability. Chitosan nanoparticles were used to deliver nutraceuticals. Chitosan-based systems also show promise in overcoming bioavailability barriers; however, safety, toxicity, and regulatory considerations remain essential. Current challenges highlight the need for scalable and standardized technologies. Overall, chitosan nanocarriers present significant potential for future nutraceutical innovation.

Keywords Chitosan nanoparticles · Biocompatibility · Delivery · Nutraceuticals

1 Introduction

Nutraceuticals have promising applications in maintaining health and preventing certain diseases. However, the problem exists its low solubility, bioavailability, stability, and targeted delivery limit its applications. Several methods have been used

J. Venkatesan (✉) · P. A. Dalavi
School of Dental Sciences and Technology, Department of Biomaterials and Transplant Sciences, Malla Reddy Vishwavidyapeeth, Hyderabad, Telangana, India

N. Kumari
Department of Microbiology, Malla Reddy Institute of Medical Sciences, Malla Reddy Vishwavidyapeeth, Hyderabad, India

K. K. R. Ealla
Department of Oral and Maxillofacial Pathology, Malla Reddy Institute of Dental Sciences, Malla Reddy Vishwavidyapeeth, Hyderabad, India

S.-K. Kim
Department of Marine Science and Convergence Engineering, Hanyang University, Ansan, Gyeonggi-do, South Korea

S.-L. Wang, M. Anpo (eds.), *Chitin and Chitosan*, Springer Series on Polymer and Composite Materials, https://doi.org/10.1007/978-981-95-7457-5_7

to overcome these limitations, including the formulation of different binders and other ingredients. Conventional nutraceutical delivery has disadvantages in terms of bioavailability, easy degradation in different atmospheres, and reaching the targeted place. Nanotechnology has recently gained attention for its potential to increase the surface area of nutraceuticals [1].

1.1 Chitosan: Structure, Sources, and Unique Properties

Chitosan is a natural cationic polysaccharide commonly derived from chitin (primarily from crustaceans, shells, and insects). Chitosan has been extensively used in several biological and biomedical applications (Fig. 1), including drug delivery systems, owing to its biocompatibility, biodegradability, and mucoadhesive properties. However, the poor solubility of chitosan at neutral pH restricts its use in real life applications. Hence, there is a need for chitosan modification in terms of derivatives or nanoformulation with chitosan [2]. Chitosan is known to have antifungal, antitumor, and antibacterial activities. In addition, chitosan has been extensively used for various other applications, including wastewater treatment, agriculture, fabric and textiles, cosmetics, nutraceutical application, and food processing applications [3].

2 Chemistry and Functional Properties of Chitosan

The bioavailability of nutraceuticals increases with nanoformulation, and the advantages of chitosan. Chitooligosaccharides and their nanoparticles include increased solubility and biocompatibility as well as improved absorption in the gastrointestinal system. The mucoadhesive properties of chitosan play a pivotal role in these properties [4].

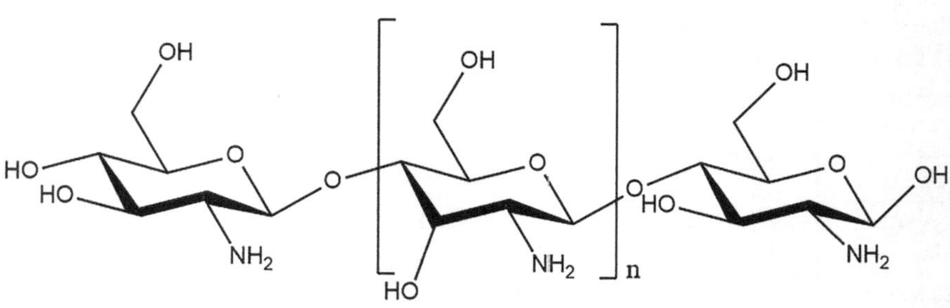

Fig. 1 Structure of chitosan

3 Methods of Chitosan Nanoparticle Preparation

3.1 *Ionic Gelation and Polyelectrolyte Complexation*

The simplest technique for creating chitosan nanoparticles is the ionic gelation process. Ionic gelation is often a one-shot synthesis that produces chitosan nanoparticles under mild synthetic conditions. In ionic gelation method, chitosan combined with sodium tripolyphosphate forms the chitosan nanoparticles. Chitosan solution and tripolyphosphate can be adjusted to get the uniform size of the nanoparticles from 100 nm to 1000 nm. A chitosan-tripolyphosphate nanoparticles containing cinnamaldehyde system were developed, applying the ionotropic gelation method [5] (Fig. 2).

3.2 *Emulsion and Reverse Micellar Techniques*

Emulsion method is extensively used to develop a chitosan nanoparticle. To develop chitosan nanoparticles, Brunel et al. (2008) applied reverse emulsion method. In the synthesis, along with the chitosan solution, Miglyol/Span 80 mixture was used. Besides this, the effect of the degree of acetylation of chitosan in the nanoparticle synthesis was studied [6].

3.3 *Spray Drying, Electro Spraying, and Antisolvent Precipitation*

Apart from emulsion and ion gelation, some more methods commonly used to synthesize chitosan nanoparticles are spray drying, electro spraying, and the antisolvent precipitation method. For instance, using spray drying method, Demir et al. (2013), synthesized chitosan nanoparticles. In the synthesis process, 10% chitosan and 20% chitosan was used [7].

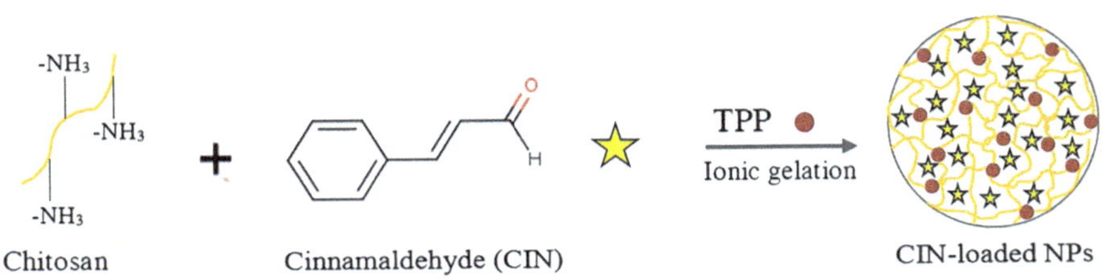

Fig. 2 Preparation of chitosan-loaded cinnamaldehyde nanoparticles. Adapted with permission from [5]

3.4 Advanced Green Synthesis and Biopolymer Cross-Linking Approaches

Producing chitosan nanoparticles using bioactive extracts via green synthesis approach is catching more attention in the biomedical field, because this approach can enhance biocompatibility and encourages biomedical applications. Cavalu et al. (2018) synthesized arabic gum coated chitosan nanoparticles containing propolis extract for biomedical applications. The ionotropic gelation method, followed by glutaraldehyde cross-linking, was used to produce the nanoparticles [8].

3.5 Critical Parameters Influencing Particle Size and Stability

Reported research articles disclose that there are several aspects that influence the particle size and stability of the chitosan nanoparticles. Some of the important parameters include the molecular weight of the chitosan, the chitosan-to-coating-agent ratio, the concentration of the chitosan, the degree of deacetylation of the chitosan, etc. Additionally, the ionic concentration of the solvent used and the type and concentration of the cross-linker used can affect the size and stability of the chitosan nanoparticles. Also, the synthesis method used to synthesize chitosan nanoparticles can influence the particle size and stability of the synthesized chitosan nanoparticles [9].

4 Encapsulation of Nutraceuticals in Chitosan Nanoparticles

Chitosan has been extensively used in nutraceutical and pharmaceutical drug delivery applications. Curcumin, Resveratrol, Quercetin, Catechins, Anthocyanins, Lutein, Astaxanthin, β-Carotene, vitamin, proteins and peptides, minerals, micronutrients were delivered by chitosan nanoparticles. Perishable food preservation and nutrient delivery are essential in the nutraceutical field. Various technologies have been extensively used for the preservation and delivery of nutraceuticals. Strawberries contain trans-resveratrol encapsulated in a chitosan matrix. Biocompatibility and biodegradability increased 150-fold and bioavailability by 3.5-fold after oral administration. The advantages of chitosan coating include preventing nutrient loss, inhibiting microbial growth, increasing nutraceutical values, and preserving strawberry quality [10].

4.1 Polyphenols

Kafirin, the prolamin protein from sorghum, containing carboxymethyl chitosan nanoparticles with curcumin prepared using acid induced aggregation method. The nutraceutical effect of curcumin is limited due to its poor solubility; hence, nanoparticles made up of kafirin-carboxymethyl chitosan increased curcumin cellular uptake. The average particle size of kafirin-carboxymethyl chitosan was 200.1 ± 3.4 nm, as measured by dynamic light scattering. The encapsulation and loading efficiency of curcumin in kafirin-carboxymethyl chitosan was 86.1 ± 2.1% and 6.1 ± 0.2%, respectively. Cellular uptake studies showed that curcumin in Caco-2 cells with formulated kafirin-carboxymethyl chitosan had higher uptake (>42%) than that of blank nanoparticles. The increased bioavailability of curcumin with nanoparticles can be further used in nutraceutical delivery applications [11].

Chlorogenic acid is an antioxidant that has been extensively studied for its free radical scavenging activity, which is associated with several diseases, including cancer. However, the poor solubility, stability, and low cellular permeability of chlorogenic acid hinder its use in nutraceutical applications. Homogeneous, stable, and biocompatible chlorogenic acid-chitosan nanoparticles were prepared using the ionic gelation method with tripolyphosphate. Physiochemical and morphological analyses were performed to confirm the formation of nanoparticles. The size and polydispersity index of the chlorogenic acid-chitosan nanoparticles were 134.44 ± 18.29 nm and 0.29 ± 0.03, respectively. The free radical scavenging activity of chlorogenic acid–chitosan nanoparticles IC50 5.2 ± 0.10 μM, which was higher than that of the blank nanoparticles IC50 6.4 ± 0.78 μM). Fluorescent microscopy and flow cytometry revealed that the nanoparticles accumulated in human renal cancer cells (786-O). The antiproliferative activity of the encapsulated chlorogenic acid-chitosan nanoparticles showed an IC_{50} of 16.20 μM. Hence, chitosan nanoparticles can deliver chlorogenic acid with enhanced accumulation, antioxidant, and antiproliferative properties [12].

4.2 Carotenoids and Pigments

Lutein-chitosan-alginate nanoparticles were developed using emulsification along with ionic gelation method. The particle size and zeta potential of the nanoparticles were 219 ± 14 nm and − 30.7 ± 0.6 mV, respectively. The encapsulation efficiency of the nanoparticles was 85.6 ± 1.7%. Nanoparticles initially released lutein in a burst, followed by sustained release under simulated gastrointestinal conditions. The bioaccessibility of the nanoparticles was 3.6 times higher than that of lutein encapsulated in the lipid matrix. The developed nanoparticles can potentially be used in nutraceuticals, functional foods, dietary supplements and therapeutics [13].

Astaxanthin is a hydrophobic xanthophyll carotenoid that has been widely studied for various biological and biomedical applications, including functional foods, nutraceuticals, and dietary supplements. However, the limited water solubility and low bioavailability of astaxanthin restrict its application in nutraceutical products. To overcome these issues, astaxanthin-loaded chitosan oligosaccharide/alginate nanoparticles were developed using ionotropic gelation and characterized for various applications. The particle size and zeta potential of the astaxanthin-loaded chitosan oligosaccharide/alginate nanoparticles were 264 ± 32 nm and −22.1 ± 1.3 mV, respectively, indicating their good stability. The encapsulation and loading capacities were 71.3 ± 2.2% and 6.9 ± 1.6%, respectively. The developed nanoparticles exhibited smooth surfaces and uniform size distribution. In addition, the nanoparticles were highly stable under exposure to UV light, heat, and acidic, alkaline, and simulated gastrointestinal fluid [14].

4.3 *Vitamins*

Carboxymethyl chitosan-soy protein isolate nanoparticles were prepared using a simple ionic gelation method. Vitamin D3 was incorporated with carboxymethyl chitosan-soy protein isolate nanoparticles and subsequently characterized with different analytical techniques. The particle sizes of the nanoparticles were 162 and 243 nm, respectively. The loading efficiency of vitamin d3 in the nanoparticles was 6.06%, with a higher encapsulation efficiency of 96.8%. Nanoparticles showed reduced the release of vitamin D3 in stimulated gastric fluid and increased release under stimulated intestinal conditions [15].

Vitamin B2 is important for health life and cellular metabolism. Its poor water solubility and light sensitivity limit its application as a nutraceutical. Chitosan-poly (lactic-co-glycolic acid) nanoparticles containing vitamin B2 were developed for food fortification and nutraceutical applications. The developed nanoparticles were 200 nm in size and exhibited increased stability compared to poly (lactic-co-glycolic acid) nanoparticles. Nanoparticles exhibited higher uptake in intestinal epithelial cells (Caco-2). Chitosan-poly (lactic-co-glycolic acid) nanoparticles containing vitamin B2 are potential candidates for nutraceutical, pharmaceutical, and food applications [16].

Vitamin B1, Thiamine, is a water-soluble vitamin and extensively used in food and nutraceuticals applications. Vitamins play important roles in several cellular functions and are critical for human health. Thiamine-incorporated chitosan nanoparticles were prepared using an ionic gelation method with sodium tripolyphosphate (TPP). The average size of the nanospheres was 504 nm, and in vitro release studies were performed. The results suggest sustained release from the nanoparticles for up to 30 days. The cellular uptake of thiamine-encapsulated chitosan was investigated in human cervical cancer cells and 3T3-L1 cells [17].

5 Functional and Biological Evaluation

5.1 *Antioxidant and Anti-inflammatory Activities*

Natural polyphenols exert significant antioxidant effects when administered at the appropriate dosage. However, their poor absorption leads to lower bioavailability. Therefore, in this study, two structurally different thiolated quaternary ammonium--chitosan conjugates were used to prepare nanoparticles by mixing a hyaluronic acid solution containing grape seed extracts. The developed nanoparticles were stable for 24 h, with an encapsulation efficiency of approximately 100%. Nanoparticles with a stronger positive surface charge were taken up by the cells more than nanoparticles with a less positive surface charge, indicating the role of the nanoparticle surface charge in their uptake. Furthermore, the resulting nanoparticles reduced oxidative stress in endothelial progenitor cells [18].

Chitosan-coated nanovesicles have been shown to increase drug absorption owing to their high mucoadhesive properties. When administered orally, it reaches the colon without degradation along the path, and once in the colon, it promotes controlled and persistent drug release due to partial degradation by the microbiota. Marinho et al. prepared chitosan- and nutriose-coated niosomes loaded with rosmarinic acid to evaluate their therapeutic potential against Ulcerative Colitis. The results indicated that RA-loaded niosomes could reduce inflammation and oxidative stress by modulating the NLRP3 inflammasome and inducing Nrf2/HO-1 signaling [19] (Fig. 3).

5.2 *Antimicrobial and Antifungal Effects*

Quercetin, a natural lipophilic polyphenol, is known to exert antimicrobial effects and can be used to disrupt biofilms. However, its low bioavailability makes its application difficult. Therefore, in a study, the authors developed quercetin-loaded chitosan nanoparticles, which successfully reduced multidrug-resistant *Escherichia coli* and *Staphylococcus aureus* isolates (colony-forming unit/ml). The authors also observed a reduction in biofilm formation in 83% of the bacterial samples by reducing exopolysaccharide production [20].

5.3 *Anticancer and Chemopreventive Properties*

In a study on colorectal cancer, chitosan-coated solid lipid nanoparticles also showed successful results by increasing the therapeutic efficacy of oregano and cinnamon by improving their stability and bioavailability. The results indicated that

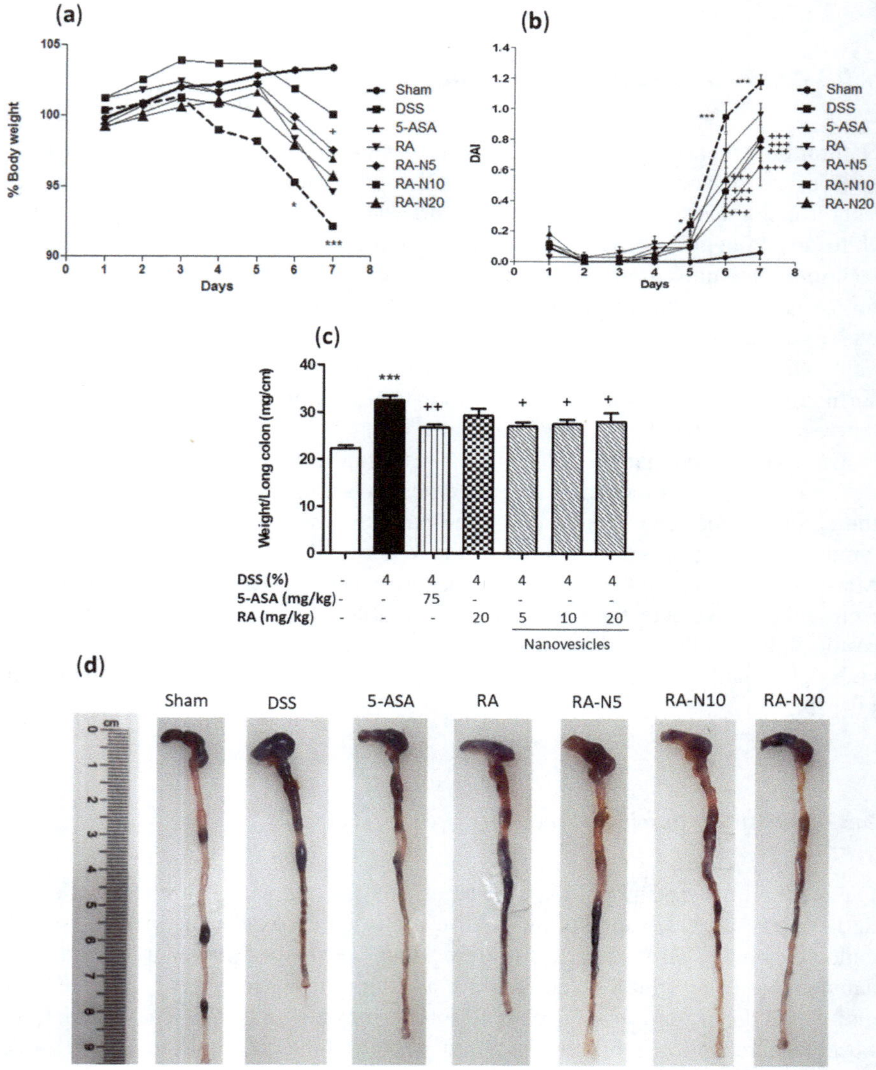

Fig. 3 (**a**) Biological activity of rosmarinic acid-loaded nanoparticles, (**b**) dose activity, (**c**) weight/length ratio of the colons, and (**d**) representative images of all colons. Adapted from [19]

chitosan-coated solid lipid nanoparticles significantly delayed the release of cinnamaldehyde and rosmarinic acid compared to cinnamon and oregano extracts, where cinnamaldehyde and rosmarinic acid dissolved in 6 h and 4 h, respectively [21].

5.4 Antidiabetic, Antiobesity, and Cardioprotective Effects

Chitosan has also been studied for its ability to reduce the cardiotoxicity of doxorubicin, a highly active antineoplastic agent. Chitosan nanoemulsions were prepared using different nutraceuticals with or without doxorubicin. The results indicated that the use of nanoemulsions resulted in reduced inflammation and protection against oxidative stress, thus reducing doxorubicin-induced cytotoxicity [22].

5.5 Probiotic and Gut Health Applications

Owing to its ability to adhere to negatively charged mucosal surfaces and cross the tight junctions between epithelial cells, chitosan has been explored for delivering nutraceuticals in the gastrointestinal tract. In one study, chitosan nanoparticles were encapsulated with active ingredients (hydrophobic or hydrophilic). Furthermore, to prevent the release of active ingredients in the simulated gastric fluid, the nanoparticles were coated with either soluble or resistant starch. These coatings resulted in the release of active ingredients in the small intestine or colon. The results indicated that by coating chitosan nanoparticles with soluble or resistant starch, the specific delivery of nutraceuticals to the small intestine or colon can be achieved [23].

6 Enhancement of Nutraceutical Bioavailability

6.1 Stability Under Gastrointestinal Conditions

Reported research discoveries shows that chitosan nanoparticles are highly stable under gastrointestinal environments. For instance, Amin et al. (2020), investigated the mucoadhesive qualities for the possible passage of protein antigen via the gastrointestinal tract (GIT) to possibly trigger a mucosal autoimmune reaction, as well as the stability of chitosan nanoparticles in simulated fluids. Findings reveal that the alginate-modified chitosan nanoparticles exhibited prolonged protein release characteristics and greater stability than polyethylene glycol (PEG) coated nanoparticles in simulated gastric and intestinal fluid. Protein ovalbumin release study from the nanoparticles were studied up to 96 h, the ovalbumin protein embedded in the alginate-coated chitosan nanoparticles did not exhibit any notable structural alterations. However, the alginate-coated nanoparticles demonstrated superior ex vivo mucosa mucoadhesion and mucin adherence compared to their PEG-coated counterparts. Chitosan-alginate nanoparticles can be used as a protein antigen carrier for oral vaccination administration to produce mucosal immunity [24].

6.2 Controlled and Targeted Release Mechanisms

A mucoadhesive polymer like a chitosan-based drug delivery system can be used to overcome the problem like a poor bioavailability in the gastrointestinal conditions. Nano-delivery methods based on chitosan-coated solid lipid nanoparticles (NC) were synthesized by the Ramalingam et al. (2016), to assist utilized curcumin into functional foods and nutraceuticals while also improving its long-term stability and oral bioavailability. The NC demonstrated extended physical stability at both room temperature and cool conditions. It has been demonstrated that following oral administration, NC increases curcumin's bioavailability more than curcumin solutions. Findings reveals that the fabricated nanoformulations can expand the use of bioactive curcumin in foods and nutraceutical applications [25].

6.3 Mucoadhesion and Permeation Enhancement

Chitosan nanoparticles are widely used in the food and nutraceutical industry due to its excellent biocompatibility, biodegradability, and mucoadhesion properties. To achieve higher mucoadhesion and permeation properties of the drug delivery system, chitosan was modified with thiol groups. By utilizing the rhodamine 123 as a marker for passive drug absorption, the permeation-promoting impact of the developed chitosan-4-thio-butylamidine blend (chitosan-TBA conjugate) when used in conjunction with the permeation facilitator glutathione (GSH) was assessed in Ussing chambers on freshly removed small intestinal mucosa from guinea pigs. Tensile tests and the rotating cylinder method were used to assess the mucoadhesive qualities of the chitosan-TBA combination adjusted to pH 3, 5, and 7. Medications comprising 10% cefadroxil, the model drug, 10% GSH, and 80% chitosan-TBA conjugate pH 3 in 100 mM phosphate buffer pH 6.8 at 37 degrees Celsius were utilized for release tests. The chitosan-TBA conjugate/GSH biocomposition had a three-fold greater permeation-boosting impact than untreated chitosan. Mucoadhesion experiments showed that the thiolated chitosan's mucoadhesive qualities increase with decreasing pH. Release experiments revealed that GSH and cefadroxil were released continuously over a number of hours. In order to increase the therapeutic effectiveness of certain medications that are poorly absorbed from the gastrointestinal tract, developed drug delivery system can be the best choice [26].

6.4 Synergistic Co-Encapsulation Strategies

Co-encapsulation of the bioactive components is the best option to obtain outstanding bioavailability of the chitosan-based delivery system in gastrointestinal conditions. In order to create soybean protein isolate-curcumin-chitosan quaternary ammonium salt nanoparticles (SPI-CUR-CQAS NPs), Weng et al. (2024) used acid-treatment of SPI and chitosan quaternary ammonium salt (CQAS) to build colloidal transporters for curcumin (CUR). The synthesized nanoparticles are spherical in shape and have diameter less than 200 nm, and possess exceedingly positively charged. The SPI CUR-CQAS NPs at pH 1.5 has excellent encapsulation efficiency, and loading capacity. Furthermore, in vitro digestion study reveals that, the developed system shows sustainable release of the CUR in the stomach, small intestine, and colon environments. Furthermore, compared with unbound CUR (16%), the packed CUR's ultimate bioavailability (65%) was significantly better. As a result, the SPI-CQAS NPs created in this work can be utilized as a successful CUR delivery technique in nutritional foods [27]. Another finding involved adding piperine to the outer shell and curcumin to the zein-hyaluronic acid nanoparticles. Additionally, the surface was covered with high molecular weight chitosan. Experimental study reveals that, in simulated gastrointestinal circumstances, this system exhibits regulated release of piperine and curcumin. Therefore, the developed new formulation has potential applications in the food and nutraceutical sectors [28].

6.5 Cellular Uptake and Pharmacokinetic Studies

Borges et al. (2006) chitosan nanoparticles were coated with the alginate and release of the ovalbumin protein from the system were examined at the different pH conditions, and compared with the non-coated chitosan nanoparticles. In vitro cell interaction study done on the mice spleen cells reveals the biocompatibility of the developed system. Finally, experiments performed on particle ingestion in rat Peyer's patches. It was shown that covering the nanoparticles with sodium alginate improved their stability at pH 6.8 and 7.4 at 37 °C in addition to preventing the rapid emission seen with uncoated particles. After 3.5 h of incubation in a low ionic strength buffer, the release was less than 5% at neutral pH. Chitosan, alginate and fabricated nanoparticles show excellent cell viability. Further study done on the confocal laser scanning microscope reveals that the rat Peyer's patches can uptake alginate-coated nanoparticles, making them appropriate carriers for intestinal mucosal immunization [29].

7 Conclusion

Chitosan-based nutraceutical delivery systems offer enhanced stability, protection, and bioavailability compared with conventional systems. Chitosan nanoparticles can be prepared using various method including ionic gelation method, emulsion and reverse micellar techniques, spray drying, electro spraying, and antisolvent precipitation. Among these techniques, ionic gelation is widely used for the preparation of chitosan nanoparticles. This method is simple, easily convertible to bulk production, and feasible for industrial production. Various nutraceuticals have been used, including polyphenols, carotenoids, pigments, vitamins, proteins, peptides, minerals, micronutrients, plant extracts, and natural antioxidants. These nutraceuticals combined with chitosan nanoparticles have several biological applications, including antioxidant, anti-inflammatory, antimicrobial, antifungal, anticancer, and chemopreventive effects, and have been studied for their antidiabetic, antiobesity, and cardioprotective effects. Despite their promising functional and biological performances, challenges in safety validation, regulatory approval, and large-scale production persist. Continued research and technological refinement are essential to fully realize the potential of chitosan in next-generation nutraceutical formulations.

References

1. Wang S, Su R, Nie S, Sun M, Zhang J, Wu D et al (2014) Application of nanotechnology in improving bioavailability and bioactivity of diet-derived phytochemicals. J Nutr Biochem 25(4):363–376. https://doi.org/10.1016/j.jnutbio.2013.10.002
2. Chen L, Subirade M (2005) Chitosan/β-lactoglobulin core-shell nanoparticles as nutraceutical carriers. Biomaterials 26(30):6041–6053. https://doi.org/10.1016/j.biomaterials.2005.03.011
3. Nguyen TH, Kwak HS, Kim SM (2013) Physicochemical and biofunctional properties of crab chitosan nanoparticles. J Nanosci Nanotechnol 13(8):5296–5304. https://doi.org/10.1166/jnn.2013.7485
4. Hassan UA, Hussein MZ, Masarudin MJ (2017) Synthesis of a nanoparticle system for the enhanced accumulation of fluorescently-labelled amino acids encapsulated in monodispersed chitosan nanoparticle system. Malays Appl Biol 46(1):171–175
5. Joghataei M, Hosseini SF, Arab-Tehrany E (2019) Cinnamaldehyde loaded chitosan/tripolyphosphate nanoassemblies: fabrication, characterization, and in vitro evaluation of antioxidant activity. J Food Process Preserv 43(7):e13972
6. Brunel F, Véron L, David L, Domard A, Delair T (2008) A novel synthesis of chitosan nanoparticles in reverse emulsion. Langmuir 24(20):11370–11377
7. Demir GM, Degim IT (2013) Preparation of chitosan nanoparticles by nano spray drying technology. Fabad J Pharm Sci 38(3):127
8. Cavalu S, Bisboaca S, Mates IM, Pasca PM, Laslo V, Costea T et al (2018) Novel formulation based on chitosan-Arabic gum nanoparticles entrapping propolis extract. Rev Chim 69:3756–3760
9. Rampino A, Borgogna M, Blasi P, Bellich B, Cesàro A (2013) Chitosan nanoparticles: preparation, size evolution and stability. Int J Pharm 455(1–2):219–228
10. Hu Q, Zhou F, Ly NK, Ordyna J, Peterson T, Fan Z et al (2023) Development of multifunctional nanoencapsulated trans-resveratrol/chitosan nutraceutical edible coating for strawberry preservation. ACS Nano 17(9):8586–8597. https://doi.org/10.1021/acsnano.3c01094
11. Xiao J, Nian S, Huang Q (2015) Assembly of kafirin/carboxymethyl chitosan nanoparticles to enhance the cellular uptake of curcumin. Food Hydrocoll 51:166–175. https://doi.org/10.1016/j.foodhyd.2015.05.012
12. Rajan RK, Hussein MZ, Fakurazi S, Yusoff K, Masarudin MJ (2019) Increased ROS scavenging and antioxidant efficiency of chlorogenic acid compound delivered via a chitosan

nanoparticulate system for efficient in vitro visualization and accumulation in human renal adenocarcinoma cells. Int J Mol Sci 20(19). https://doi.org/10.3390/ijms20194667
13. Sorasitthiyanukarn FN, Singcharoen K, Kulpreechanan N (2024) Enhancing lutein delivery through chitosan/alginate nanoparticles: a study on improved stability, bioaccessibility and biological activity. Res J Pharm Technol 17(10):4771–4779. https://doi.org/10.52711/0974-360X.2024.00735
14. Sorasitthiyanukarn FN, Muangnoi C, Pornchai Rojsitthisak P, Rojsitthisak P (2022) Chitosan oligosaccharide/alginate nanoparticles as an effective carrier for astaxanthin with improving stability, in vitro oral bioaccessibility, and bioavailability. Food Hydrocoll 124. https://doi.org/10.1016/j.foodhyd.2021.107246
15. Teng Z, Luo Y, Wang Q (2013) Carboxymethyl chitosan-soy protein complex nanoparticles for the encapsulation and controlled release of vitamin D3. Food Chem 141(1):524–532. https://doi.org/10.1016/j.foodchem.2013.03.043
16. Sathiensathaporn S, Solé-Porta A, Baowan D, Pissuwan D, Wongtrakoongate P, Roig A et al (2025) Nanoencapsulation of vitamin B2 using chitosan-modified poly(lactic-co-glycolic acid) nanoparticles: synthesis, characterization, and in vitro studies on simulated gastrointestinal stability and delivery. J Food Sci 90(1). https://doi.org/10.1111/1750-3841.17631
17. Kaur H, Kaur M, Rajesh P, Kaur H (2023) Formulation of thiamine-laden chitosan nanosphere for nutraceutical application. Curr Bioact Compd 19(8):61–68. https://doi.org/10.2174/1573407219666230331100611
18. Felice F, Zambito Y, Belardinelli E, D'Onofrio C, Fabiano A, Balbarini A et al (2013) Delivery of natural polyphenols by polymeric nanoparticles improves the resistance of endothelial progenitor cells to oxidative stress. Eur J Pharm Sci 50(3–4):393–399. https://doi.org/10.1016/j.ejps.2013.08.008
19. Marinho S, Illanes M, Ávila-Román J, Motilva V, Talero E (2021) Anti-inflammatory effects of rosmarinic acid-loaded nanovesicles in acute colitis through modulation of NLRP3 inflammasome. Biomolecules 11(2):1–17. https://doi.org/10.3390/biom11020162
20. Messias de Souza G, Gervasoni LF, Rosa RDS, de Souza Iacia MVM, Nai GA, Pereira VC et al (2022) Quercetin-loaded chitosan nanoparticles as an alternative for controlling bacterial adhesion to urethral catheter. Int J Urol 29(10):1228–1234. https://doi.org/10.1111/iju.14958
21. Kamel KM, Khalil IA, Rateb ME, Elgendy H, El-Hawary S (2017) Chitosan-coated cinnamon/oregano-loaded solid lipid nanoparticles to augment 5-fluorouracil cytotoxicity for colorectal cancer: extract standardization, nanoparticle optimization, and cytotoxicity evaluation. J Agric Food Chem 65(36):7966–7981. https://doi.org/10.1021/acs.jafc.7b03093
22. Quagliariello V, Vecchione R, Coppola C, Di Cicco C, de Capua A, Piscopo G et al (2018) Cardioprotective effects of nanoemulsions loaded with anti-inflammatory nutraceuticals against doxorubicin-induced cardiotoxicity. Nutrients 10(9). https://doi.org/10.3390/nu10091304
23. Sampathkumar K, Loo SCJ (2018) Targeted gastrointestinal delivery of nutraceuticals with polysaccharide-based coatings. Macromol Biosci 18(4). https://doi.org/10.1002/mabi.201700363
24. Amin MK, Boateng JS (2022) Enhancing stability and mucoadhesive properties of chitosan nanoparticles by surface modification with sodium alginate and polyethylene glycol for potential oral mucosa vaccine delivery. Mar Drugs 20(3):156
25. Ramalingam P, Yoo SW, Ko YT (2016) Nanodelivery systems based on mucoadhesive polymer coated solid lipid nanoparticles to improve the oral intake of food curcumin. Food Res Int 84:113–119
26. Bernkop-Schnürch A, Guggi D, Pinter Y (2004) Thiolated chitosans: development and in vitro evaluation of a mucoadhesive, permeation enhancing oral drug delivery system. J Control Release 94(1):177–186
27. Weng S, Su H, McClements DJ, Wang Y, Li T, Fu Y (2024) Delivery of curcumin using colloidal complexes formed from ammonium chitosan and acid treated soybean protein. Food Biophys 19(3):676–687

28. Chen S, McClements DJ, Jian L, Han Y, Dai L, Mao L et al (2019) Core–shell biopolymer nanoparticles for co-delivery of curcumin and piperine: sequential electrostatic deposition of hyaluronic acid and chitosan shells on the zein core. ACS Appl Mater Interfaces 11(41):38103–38115
29. Borges O, Cordeiro-da-Silva A, Romeijn SG, Amidi M, de Sousa A, Borchard G et al (2006) Uptake studies in rat Peyer's patches, cytotoxicity and release studies of alginate coated chitosan nanoparticles for mucosal vaccination. J Control Release 114(3):348–358

Jayachandran Venkatesan

Neema Kumari

Pandurang Appana Dalavi

Kranti Kiran Reddy Ealla

Se-Kwon Kim

Chitosan-Based Biomedical Hydrogels

Lihua Li, Lu Lu, Changren Zhou, Yifan Zhang, Zexiang Zheng, Xinshuo Liu, and Junxiao Xiang

Abstract Chitosan-based hydrogels represent a cornerstone class of biomaterials in the field of biomedical engineering, distinguished by their unique biocompatibility, biodegradability, inherent antimicrobial activity, and extensive chemical tunability. This chapter systematically reviews the fabrication, key properties, and diverse biomedical applications of chitosan-based hydrogels, while also addressing current challenges and future research directions. In terms of fabrication, two primary crosslinking strategies of chemical and physical crosslinking are detailed, highlighting their respective mechanisms, advantages, and limitations. The key properties of these hydrogels, including pH/temperature-responsive swelling behavior, tunable mechanical stiffness, self-adhesion, controlled biodegradation, biocompatibility, inherent antimicrobial activity, and efficient drug loading/release capabilities, are elaborated, emphasizing how these features are tailored to meet biomedical requirements. Furthermore, this chapter comprehensively showcases their applications in critical biomedical scenarios, including wound healing dressings, trigger-responsive drug delivery systems, targeted therapy, tissue engineering, and other emerging areas such as biosensing and functional contact lenses. Finally, the current challenges with limited mechanical strength, batch-to-batch variability, and potential immunogenicity are discussed, and future perspectives are proposed. This chapter aims to provide a comprehensive overview of chitosan-based hydrogels, offering insights for researchers and facilitating their further translation into clinical practice.

Keywords Chitosan · Hydrogel · Wound healing · Drug delivery · Tissue engineering

L. Li (✉) · L. Lu · C. Zhou · Y. Zhang · Z. Zheng · X. Liu · J. Xiang
Engineering Research Center of Artificial Organs and Materials, Ministry of Education; Department of Materials Science and Engineering, Jinan University, Guangzhou, China
e-mail: tlihuali@jnu.edu.cn; tlulu@jnu.edu.cn; tcrz9@jnu.edu.cn

S.-L. Wang, M. Anpo (eds.), *Chitin and Chitosan*, Springer Series on Polymer and Composite Materials, https://doi.org/10.1007/978-981-95-7457-5_8

1 Introduction

Hydrogels, defined as three-dimensional hydrophilic polymer networks capable of absorbing and retaining large volumes of water or biological fluids, have emerged as indispensable materials in biomedical research and applications. Their structural similarity to the native extracellular matrix (ECM) of tissues, combined with tunable physicochemical properties (e.g., porosity, swelling degree, and degradation rate), enable them to create a biomimetic microenvironment that supports cell adhesion, proliferation, and differentiation. This unique characteristic makes hydrogels ideal for a wide range of biomedical uses, including wound dressings, drug delivery carriers, tissue engineering scaffolds, biosensors, and implantable devices. Among the diverse polymers used for hydrogel fabrication, chitosan stands out as a premier natural polymer candidate.

Chitosan possesses several inherent advantages that make it particularly suitable for biomedical hydrogel development. Its excellent biocompatibility ensures minimal adverse reactions when in contact with biological tissues or cells, as its degradation products (low-molecular-weight chitosan oligosaccharides) are non-toxic and can be metabolized by the human body [1]. In addition, its biodegradability, mediated by enzymes such as lysozyme and chitosanase, allows for controlled degradation that can be synchronized with tissue regeneration or drug release processes, eliminating the need for secondary surgical removal [2]. Chitosan exhibits intrinsic antimicrobial activity, primarily attributed to its protonated amino groups ($-NH_3^+$) under acidic conditions, which interact with the anionic surfaces of bacterial cell membranes, disrupting membrane integrity and inhibiting bacterial growth, which a critical feature for applications like wound dressings and antimicrobial coatings. Additionally, the abundant amino ($-NH_2$) and hydroxyl (-OH) groups on the chitosan backbone provide versatile sites for chemical modification and crosslinking, enabling precise tuning of hydrogel properties to match specific application requirements [3].

Chitosan-based hydrogels have established themselves as a versatile and promising platform in biomedicine. Their primary applications span drug delivery, where they function as intelligent carriers for controlled and targeted release of therapeutics (e.g., pH-responsive cancer therapy, glucose-sensitive insulin delivery) [2]; tissue engineering, serving as scaffolds that mimic the extracellular matrix to support the regeneration of bone, cartilage, and skin; and wound healing, where they act as advanced dressings that provide hemostasis, prevent infection, and promote moist healing. Beyond these core uses, they are also being developed for biosensing by transducing biological signals into measurable outputs and for ophthalmic applications like therapeutic contact lenses. The ability to engineer these hydrogels through molecular design and composite formation to meet specific mechanical, biological, and kinetic requirements underscores their significant potential for advancing regenerative medicine and precision therapy.

2 Fabrication of Chitosan Hydrogels

Chitosan-based hydrogels are formed through two primary mechanisms: (1) Chemical crosslinking: Involves covalent bonds formed between chitosan chains using crosslinking agents. These gels are more stable and permanent but require careful selection of non-toxic crosslinkers for biomedical use. (2) Physical crosslinking: Involves secondary forces like ionic interactions, hydrogen bonding, crystallite formation and hydrophobic interactions. These gels are often reversible and can be sensitive to environmental changes. The choice of methods depends on the desired gelation time, mechanical strength, stability, and ultimate biological application. The trend is moving toward mild, cytompatible crosslinking methods and the use of natural crosslinkers to maximize the innate biocompatibility of chitosan for advanced biomedical uses [4, 5] (Fig. 1).

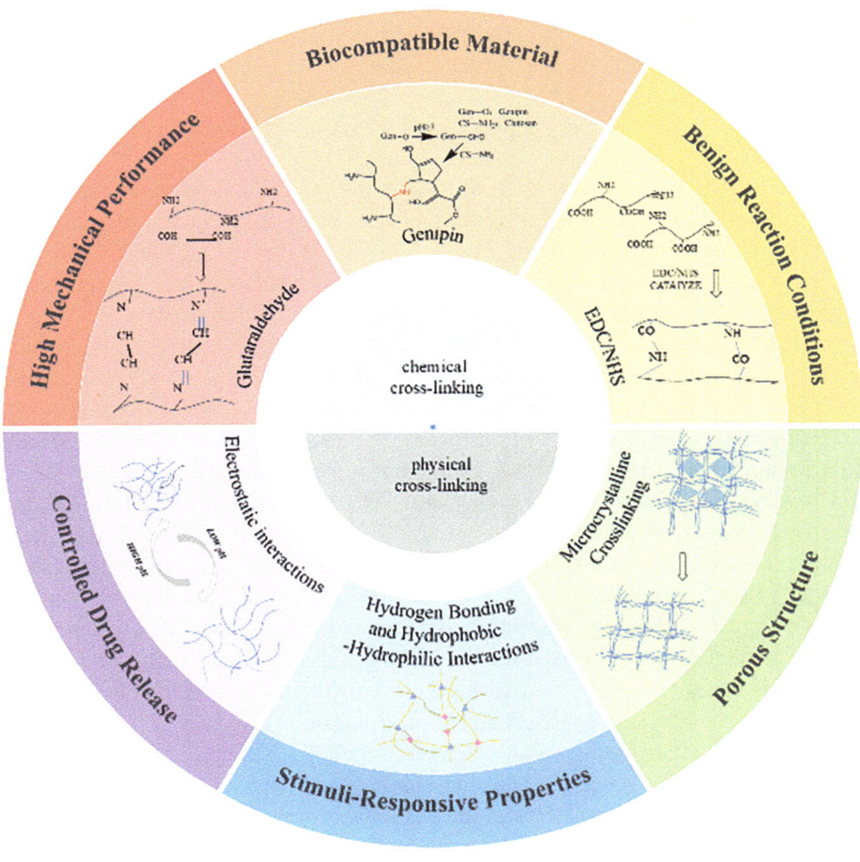

Fig. 1 Classification of Crosslinking Modalities in Chitosan-Based Hydrogels

2.1 Chemical Crosslinking

Common chemical crosslinking methods include free radical polymerization, condensation reactions and addition reactions. The amino group at the C2 position and the hydroxyl groups at the C3 and C6 positions of the sugar are often utilized in free radical polymerization and are commonly crosslinked using agents such as glutaraldehyde, genipin and carbodiimide hydrochloride (EDC)/N-hydroxysuccinimide (NHS).

The chemical crosslinking of glutaraldehyde and chitosan is a classic, highly efficient and widely applied covalent crosslinking strategy, where the aldehyde groups (-CHO) undergo a characteristic Schiff base reaction with the abundant free primary amino groups ($-NH_2$) on the chitosan molecular chains. This reaction constructs a stable and dense three-dimensional covalent network structure between chitosan molecules [6]. However, this technique also has an aspect that requires careful consideration: glutaraldehyde possesses certain cytotoxicity. Residual free glutaraldehyde or incompletely stabilized imine bonds may adversely affect surrounding cells and tissues. Therefore, in practical applications, researchers minimize toxicity by optimizing the crosslinker concentration, reaction time, and conditions, and by implementing thorough post-reaction washing steps, such as treatment with glycine or sodium borohydride to reduce and stabilize the imine bonds.

Due to its excellent biocompatibility and low cytotoxicity, genipin is widely regarded as an ideal substitute for toxic synthetic crosslinking agents such as glutaraldehyde, for preparing chitosan hydrogels. It crosslinks chitosan in two fast, sequential steps: (1) the glucoside ring opens to expose an aldehyde that reacts with a chitosan $–NH_2$, and (2) the same aldehyde condenses with a second $–NH_2$, creating an intermolecular –N=CH– (Schiff) bridge [7]. The resulting network locks genipin's antioxidant and estrogen-mimetic moieties directly into the hydrogel. Meanwhile, the resulting crosslinked network markedly improves the mechanical strength and structural stability of chitosan gels, enabling them to better withstand the mechanical demands of physiological environments.

The synergistic catalytic action of EDC/NHS for the chemical crosslinking of chitosan is a highly efficient and biocompatible covalent crosslinking strategy in the field of biomaterials. Its core principle lies in the use of EDC as a zero-length crosslinker [8], which efficiently activates the carboxyl groups on the chitosan molecular chains (typically from its derivatives or blends). These activated groups then undergo amidation reactions with the abundant amino groups on chitosan itself to form stable covalent amide bonds, thereby constructing a three-dimensional network structure with excellent mechanical properties and high stability. Since chitosan itself only contains amino and hydroxyl groups, it is usually necessary to blend it with other carboxyl-containing biomacromolecules (such as hyaluronic acid, collagen, sodium alginate, etc.) or use carboxyl-modified chitosan derivatives (e.g., carboxymethyl chitosan) to provide reaction sites. EDC/NHS crosslinking is fast, clean, and safe: NHS protects activated acids from water, so nearly every –COOH couples to an –NH_2, giving a tight, uniform net. Notably, the catalysts leave no toxic residue after a quick wash. The dense mesh boosts strength and elasticity, slows acid-driven dissolution, and permits precise, long-term control of degradation, which is ideal for implants and sustained drug release.

2.2 *Physical Crosslinking*

Physical crosslinking is a simple process where the hydrogels polymer chains are held together either by electrostatic force, Van der Waals force, protein interaction, hydrophobic force, ionic bond, dipole-dipole, hydrogen bond, hydrophilic interaction or entanglement. These weak and temporary bonds are reversible and altered by temperature or pH change. In comparison to chemical crosslinking approaches, physical crosslinking strategies impart chitosan-based hydrogels with enhanced biosafety, reversible and dynamic network characteristics, and stimuli-responsive behavior, thereby establishing them as more favorable candidates for the development of intelligent and biocompatible materials tailored for advanced biomedical applications.

The formation and properties of chitosan hydrogels via electrostatic interactions are profoundly influenced by several key factors. Foremost among these is pH, which plays a central role. The amino groups on chitosan molecular chains must be protonated and positively charged (-NH_3^+) under acidic conditions to effectively interact with negatively charged polyanions (such as carboxylate ions, -COO^-). At low pH, the polymer size increases, resulting in a loose network structure that facilitates drug diffusion and release. In contrast, at high pH, the crosslinked network contracts, leading to increased crosslinking density and reduced size, which generally enhances the elastic modulus and strength of the hydrogel. Sébastien Ladet et al. [9] utilized pH as a regulatory handle to precisely modulate electrostatic interactions, thereby directing the formation of physically crosslinked chitosan hydrogels. Without relying on chemical crosslinking agents, this approach enables the induction of a sol-gel transition solely through pH adjustment. Based on this mechanism, they successfully engineered complex hydrogels with a multi-membrane "onion-like" architecture. Secondly, the ratio of positive to negative charges must be close to the isoelectric point, although this does not necessarily correspond to the point of maximum turbidity. Thirdly, the ionic environment strongly interferes with the process. Salt ions (e.g. Na^+, Cl^-) can screen the electrostatic attraction between polymer chains, reducing gel strength but potentially accelerating the self-healing process. It is particularly noteworthy that differences in the hydration capacity of counterions (e.g., weakly hydrated Cl^- vs strongly hydrated Ac^-) affect their binding to polycations and participate in dynamic electrostatic competition, which is essential for the gel's self-repair functionality. Finally, external conditions such as temperature and mechanical force cannot be overlooked. Elevated temperatures, for example above 60 °C, may disrupt synergistic interactions like hydrogen bonding, while applied strain influences the mechanical response of the material, exhibiting elastic and self-healing behavior under small deformations, and potentially transitioning to plastic behavior under large deformations. These factors are interconnected and collectively determine the structure, stability, and functional performance of chitosan hydrogels formed through electrostatic interactions.

Electrostatic interaction relies on the attraction between polyelectrolytes. Its core advantages lie in its mild and efficient gelation process and dynamic responsiveness (pH/ion-triggered swelling or self-healing), making it suitable for injectable drug carriers or hemostatic materials. However, this method has significant limitations. Under normal physiological conditions (with pH about 7.4 and high ionic strength), charge shielding leads to gel disintegration. The inferior mechanical strength of purely electrostatic crosslinking, along with its restricted applicability under acidic conditions, poses a major obstacle to its long-term in vivo implementation.

Hydrogen bonding constructs hydrogels through intermolecular hydrogen bond networks, offering outstanding biosafety. Moreover, they are less affected by ions, providing better stability in physiological environments. The dense hydroxyl groups (-OH) and amino groups (-NH_2) on the chitosan backbone act as hydrogen bond donors, forming a three-dimensional hydrogen bond network with water molecules, oxygen atoms on adjacent chains such as C=O in residual acetyl groups, or externally added polymers such as -OH in polyvinyl alcohol. This type of interaction exhibits reversibility and environmental responsiveness.

Hydrophobic interactions in chitosan originate from two types of structures in its molecules. (1) Natural hydrophobic groups, i.e., residual acetyl groups (-$COCH_3$) when the degree of deacetylation (DD) is less than 100%; and (2) Artificially modified groups, hydrophobic units such as alkyl chains (C6-C18), phenyl groups, or cholesterol introduced through chemical grafting. These hydrophobic domains are dispersed in low-temperature aqueous solutions due to hydration. Once the temperature exceeds the lower critical solution temperature (LCST), the hydration layer collapses, and the hydrophobic chains aggregate with each other to form micelle-like crosslinking points, triggering a sol-gel transition. In most practical systems, hydrogen bonding and hydrophobic interactions do not exist in isolation. Instead, they are deeply intertwined to form a hierarchical network structure, enabling breakthroughs in performance. Their intelligent responsiveness facilitates adaptation to wound microenvironments. Nevertheless, this approach suffers from slow gelation kinetics, insufficient wet-state strength due to water molecules competing for hydrogen bonds leading to swelling collapse, and requires precise adjustment of the deacetylation degree to balance hydrogen bonding and hydrophobic interactions (high DD enhances hydrogen bonding, low DD strengthens hydrophobic effects), resulting in relatively complex processing. Such gels are more suitable for long-term wound dressings or bioinks.

Chitosan-based hydrogel formation via microcrystalline crosslinking relies on creating microcrystalline domains as physical junctions. This process begins with an elevation in pH, which reduces intermolecular repulsions and promotes chain association. Subsequent conformational reorganization, driven by hydrogen bonding, then yields nanoscale microcrystalline regions. These microdomains critically serve as physical crosslinks, consolidating the surrounding polymer chains through an extensive and robust network of multivalent hydrogen bonds, which ultimately gives rise to a continuous three-dimensional hydrogel architecture. Microcrystalline domains serve not only as critical crosslinking points that define the topological architecture of the hydrogel network but also directly contribute to its ultimate mechanical properties and structural stability. Furthermore, incorporating nanoparticles into the hydrogel precursor solution, followed by repeated freeze-thaw cycles, promotes the organization of polymer chains into a systematic crystalline structure nucleated around the nanoparticles. This method significantly enhances the porosity of the hydrogel. In the context of wound dressing applications, such a structural configuration facilitates the absorption of tissue exudate and supports skin regeneration [10]. However, microcrystalline crosslinking has inherent limitations. Its gelation is irreversible, leading to an absence of self-healing and reversible responsiveness, which restricts applications like smart drug delivery. Furthermore, the stochastic nucleation of microcrystalline domains causes structural heterogeneity and divergent mechanical properties. Finally, the process's high sensitivity to material parameters and processing conditions challenges its controllability and reproducibility.

3 Properties of Chitosan Hydrogels

Chitosan hydrogels integrate cationic, biocompatible, biodegradable, and bioactive features with facile tunability, enabling programmable performance for drug delivery, tissue engineering, wound care, and biosensors.

3.1 Swelling Property

Chitosan hydrogels swell due to the thermodynamic affinity of the polymer network for water, balanced by the elastic retractive forces of the crosslinked polymer chains. The primary driver is the presence of hydrophilic functional groups (-NH_2 and -OH) on the chitosan backbone. These groups form hydrogen bonds with water molecules, causing the network to absorb water and expand.

Their swelling is governed by a competition between hydrophilic solvation of –NH_2/–OH groups and network elasticity; protonation at pH < 6.3 generates intrachain repulsion that expands pores, whereas deprotonation above pH 6.3 collapses the matrix. Cross-link density is therefore used to calibrate swelling: dense networks (low swelling) provide the 0.2–10 MPa modulus required for cartilage or bone scaffolds [11], while loose networks (high swelling) mimic soft ECM, absorb exudate and serve as atraumatic dressings.

3.2 Mechanical Property

The mechanical properties of chitosan hydrogels are a key factor in their suitability for any given application, especially in the biomedical field, where they often have to withstand physiological loads and pressures while matching the viscoelastic properties of different human tissues.

Mechanical matching is further refined by double-network architectures, directional porosity or interpenetrating phases that replicate ligament (65–447 MPa), skin or adipose viscoelasticity; stress relaxation times ($10–10^3$ s) and initial moduli (1–30 kPa) direct stem-cell lineage toward fat or bone [12, 13]. When cells interact with the gel, stress relaxation occurs. When the gel has a high degree of crosslinking and a dense network, the stress relaxation time is short and the modulus is high. Conversely, when the gel has a low degree of crosslinking, the opposite is true.

With the development of flexible electronics such as wearable sensors, flexible electrodes, chitosan hydrogels have emerged as potential flexible substrate materials due to their combination of flexibility, conductivity, which can be combined with conductive materials like graphene and carbon nanotubes. This type of hydrogel usually maintains the structural stability of the hydrogel network through electrostatic interactions and hydrogen bond interactions between chitosan and other substances. The protonation of amino groups can also endow the hydrogel with good electrical conductivity. Moreover, the hydrogel has a highly dense network with a relatively low average pore size, ensuring that it can maintain high elasticity.

3.3 Self-Adhesion Properties

Self-adhesion arises from hydrogen bonding, electrostatic attraction and a pH-triggered "molecular stitching" mechanism in which chitosan diffuses into tissue, entangles with host matrices and forms an interfacial layer; composite IPNs with polyacrylamide or alginate augment wet shear strength and tear resistance.

Chitosan hydrogels exhibit unique advantages in the biomedical field due to their stable adhesion in moist environments and excellent biocompatibility. For instance, in wound care, traditional dressings tend to fall off easily due to frequent movement or wound exudate, whereas chitosan hydrogels, leveraging their self-adhesive property, can closely adhere to skin or mucosal wounds, forming a closed or semipermeable barrier. This barrier not only isolates the wound from external bacterial contamination but also retains wound moisture to promote healing. For surgical incisions on fragile organs such as the liver and spleen, as well as minor bleeding points in laparoscopic surgery, traditional suturing is prone to causing tissue tearing. In contrast, chitosan hydrogels can function as biological sealants. Their self-adhesive property enables them to quickly adhere to the moist surface of organs, achieving hemostasis through physical occlusion and the promotion of blood coagulation, while simultaneously sealing the incisions to prevent body fluid leakage. In summary, the self-adhesive property of chitosan hydrogels breaks the application limitations of traditional biomaterials in moist and dynamic biological environments, making them a better choice in scenarios such as wound care and drug delivery. Their core value lies in "on-demand adhesion"—they can not only stably attach to target tissues but also gently detach or degrade when needed.

3.4 *Biodegradation*

Chitosan hydrogels owe their broad biomedical applicability largely to inherent biodegradability: their degradation is mild, products are safe, and they interact synergistically with biological environments.

Enzymatic degradation is the primary in vitro/in vivo pathway, with enzyme specificity dictating outcomes. Lysozyme recognizes "deacetylated-acetylated unit" sequences in chitosan (DD 70%–85%) and cleaves β-(1,4) glycosidic bonds at a mild rate. Chitosanase (prevalent in intestines/tumor microenvironments) acts on high-DD chitosan (DD >85%), directly breaking glucosamine-unit glycosidic bonds without sequence recognition, and degrades far faster than lysozyme. In vivo, weak acids in body fluids also hydrolyze chitosan glycosidic bonds, synergizing with enzymes to accelerate degradation, and the body's metabolism quickly clears degradation-derived oligosaccharides/monosaccharides to avoid accumulation.

Chitosan hydrogel degradation time is a dynamic range, shaped by material properties and biological context. The range is varying by application: 3–7 days for superficial wound dressings (synchronized with 5–7 days granulation tissue growth), 1–4 weeks for tumor-site drug carriers (accelerated in acidic microenvironments, releasing 80% drug in 1–2 weeks and fully degrading post-tumor shrinkage), and 2—5 days for traumatic hemostasis (degrading post-hemostasis to prevent thrombosis/infection from long-term retention).

3.5 *Biocompatibility and Cytotoxicity*

The biocompatibility of chitosan-based hydrogels is defined by two core traits: low cytotoxicity and weak immunogenicity, which underpin their extensive use in biomedical contexts and distinguish them as safe, tissue-compatible biomaterials.

Chitosan's inherent nature as a natural polysaccharide ensures minimal cytotoxicity, with neither the hydrogel itself nor its degradation products disrupting cell function or inducing abnormal physiological responses. In vivo, chitosan hydrogels are degraded by lysozyme and chitosanase into low-molecular-weight oligosaccharides and glucosamine monosaccharides: oligosaccharides are directly absorbed by cells via membrane transporters, while glucosamine integrates into human glucose metabolism, ultimately breaking down into CO_2 and H_2O without toxic residues. Beyond non-toxicity, the hydrogel also fosters a cell-friendly microenvironment:

under physiological pH, free $-NH_2$ groups on chitosan protonate into positively charged $-NH_3^+$, which form electrostatic bonds with negatively charged ECM adhesion proteins such as fibronectin and laminin. This interaction acts as a bridge to promote cell adhesion, growth, and maintenance of native functions.

Chitosan hydrogels elicit only mild, controllable immune responses post-implantation or contact, avoiding pathological damage or strong immune rejection, for example antibody production or massive immune cell aggregation. This is attributed to two key factors: (1) as a naturally abundant polysaccharide, unlike synthetic polymers, chitosan is not recognized as a foreign threat by the immune system, and (2) its molecular structure shares similarities with the body's endogenous glycosaminoglycans, further reducing the likelihood of specific immune attacks.

3.6 Antimicrobial Activity

Chitosan hydrogels inherit chitosan's intrinsic broad-spectrum antimicrobial activity-rooted in its molecular amino groups, making them invaluable in biomedicine and healthcare. Unlike traditional materials, they avoid harsh synthetic agents or antibiotic resistance risks, aligning with green, sustainable biomaterial trends. Its practical applications can be serving as wound dressings for burns, abrasions, surgical incisions, and chronic non-healing ulcers. It is also used to form protective films on oral mucosa via chitosan-based mouthwashes/oral gels, isolating bacteria while directly inhibiting pathogens like *Streptococcus mutans* and *Porphyromonas gingivalis* to treat gingivitis and periodontitis.

Chitosan's antibacterial action relies on multi-target interactions with bacteria, driven by protonation of its amino groups in acidic environments: cell membrane disruption, intracellular interference and nutrient Deprivation. Three core factors including DD, chemical modification, and molecular weight-determine chitosan's antimicrobial efficacy by regulating its charge density, solubility, and interaction with bacteria. Activity is maximized at medium MW (100–1000 kDa) and DD 70–90%, while quaternization or polyamine grafting further lowers the minimum inhibitory concentration (MIC) or minimum bactericidal concentration (MBC) [14].

3.7 Drug Loading and Release Capabilities

Chitosan hydrogels support diverse drug loading strategies compatible with hydrophilic/hydrophobic, small-molecule/macromolecular drugs, while preserving drug activity. Physical adsorption relies on non-covalent interactions (e.g., electrostatic forces between protonated chitosan amino groups and negatively charged drugs like nucleic acids) for simple, low-impact loading. Chemical bonding uses covalent links between chitosan's hydroxyl/amino groups and drug functional groups (aldehyde, carboxyl) for stable loading, with release triggered by bond cleavage. Physical entrapment embeds drugs in the hydrogel's porous network via co-gelation with chitosan precursors, requiring no drug modification. For hydrophobic drugs, residual acetyl groups in chitosan or grafted hydrophobic moieties enable binding via hydrophobic interactions.

Drug release from chitosan hydrogels combines passive and active modes. Passive release includes diffusion-dominated release and degradation-triggered release. By diffusion-driven release, drug molecules diffuse through the porous network, with rate tuned by crosslinking degree, drug size, and swelling. And by degradation-triggered release, enzymatic hydrolysis by lysozyme or colonic bacteria breaks down the hydrogel, releasing drugs-ideal for colon targeting. More critically, stimuli-responsive release enables precise delivery via internal pH/enzymes or external temperature/light triggers. pH responsiveness exploits lesion-tissue pH differences to alter hydrogel swelling or bond stability, while temperature responsiveness, via thermosensitive polymers like PNIPAM, uses temperature-induced network shrinkage/swelling to control release, and light responsiveness, with photosensitive groups like azobenzene, uses specific wavelengths to trigger bond cleavage or isomerization for spatiotemporal release.

4 Biomedical Applications of Chitosan Hydrogels

4.1 Wound Healing Dressings

Chitosan-based hydrogels are exceptionally well-suited for wound healing dressings, functioning as an ideal bioactive matrix that actively promotes the entire repair process. Their inherent cationic nature enables rapid hemostasis by electrostatically aggregating red blood cells and activating platelets, initiating primary hemostasis. Furthermore, the hydrogel's moist environment and porous structure facilitate exudate absorption, maintain optimal hydration, and allow for gas exchange, while its intrinsic antimicrobial properties help prevent infection by disrupting bacterial membranes. Beyond these passive benefits, advanced formulations can be engineered with enhanced properties: self-healing capabilities to maintain integrity, stimuli-responsiveness (e.g., to pH or enzymes) for on-demand drug release (e.g., antibiotics, growth factors), and incorporation of bioactive agents to actively accelerate angiogenesis and tissue regeneration. This combination of passive protection and active therapeutic intervention makes chitosan-based hydrogels a superior and versatile platform for managing a wide range of wounds, from acute burns to chronic ulcers. Some studies have used a quaternary system of "quaternized chitosan (cationic) + phosphorylated chitosan (simulating clotting factors) + tannic acid (crosslinking and auxiliary hemostatic agent) + poly-ε-lysine (bioadhesive)" to prepare a bifunctional chitosan composite hydrogel with excellent hemostatic activity, achieving a coagulation time of 225 s [15].

The use of chitosan dressings can reduce the risk of wound infection, thereby decreasing the need for systemic antibiotics and avoiding the development of drug resistance. It can also synergistically enhance effects with other antibacterial components such as silver ions and lysozyme, making it particularly suitable for the anti-infective treatment of chronic infected wounds. Simultaneously, by controlling bacterial load, chitosan can effectively mitigate excessive and persistent inflammatory responses, creating a relatively clean microenvironment for the wound and facilitating its progression into the proliferation phase. The excellent antibacterial properties of chitosan originate from the protonated amino groups binding to the anionic groups on the bacterial cell membrane surface, forming a biofilm and disrupting membrane permeability, thereby preventing nutrients from entering the cell and ultimately leading to bacterial lysis. One study reported that chitosan was conjugated with L-arginine to form chitosan-arginine, which was then crosslinked with benzaldehyde-functionalized polyethylene glycol via dynamic Schiff base covalent bonds to form a hydrogel matrix with a porous structure (as shown in Fig. 2) [16]. Another study constructed a chitosan/antimonene nanosheet composite hydrogel. Chitosan non-selectively captures Gram-positive/negative bacteria through van der Waals forces, electrostatic forces, etc., while exerting its intrinsic antibacterial properties. Under NIR laser irradiation, the antimonene nanosheets convert light energy into localized hyperthermia, destroying the cell membranes and proteins of residual bacteria, achieving highly efficient sterilization.

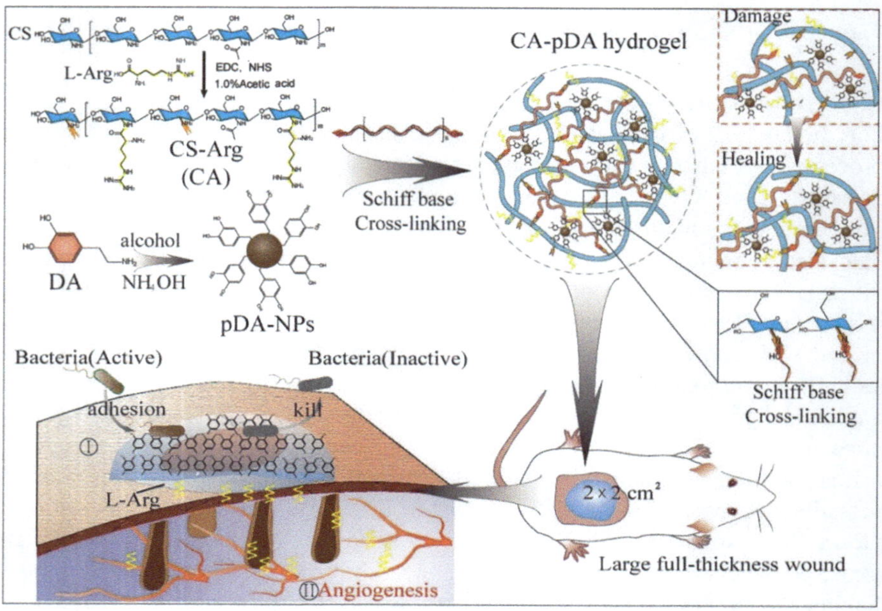

Fig. 2 Schematic illustration of the design and application of self-healing CA-pDA composite hydrogel for accelerated large skin full-thickness defect repair and the underlying mechanism of hydrogel dressing for enhanced wound repair [16].Reprinted with permission

4.2 *Stimuli-Responsive Drug Delivery System*

pH sensitive chitosan hydrogels demonstrate potential application in the acidic conditions of inflammation or targeted tumor drug delivery. A pH-responsive composite hydrogel was prepared by grafting acrylamide onto maleic acid-crosslinked chitosan and alginate [17], which exhibited a maximum swelling rate of 1246% under alkaline conditions, demonstrating pronounced pH sensitivity. By grafting gallic acid (GA) onto chitosan, a self-crosslinked CS-GA hydrogel exhibited significant swelling behavior under acidic conditions, with a swelling rate as high as 5072%, while showing only 144% swelling in weakly alkaline environments [18]. Another pH-responsive HC@PTM composite hydrogel was prepared by covalently crosslinking aldehyde-modified hyaluronic acid with succinic acid-modified chitosan and incorporating reactive oxygen species (ROS)-responsive drug-loaded micelles. The pH responsiveness of this hydrogel is jointly regulated by the pH sensitivity of Schiff base bonds and the protonation/deprotonation states of chitosan amine and hyaluronic acid carboxyl groups. Furthermore, this hydrogel combines excellent biocompatibility with ROS scavenging capacity, downregulating inflammatory cytokine expression and promoting M2 macrophage polarization. In a rheumatoid arthritis mouse model, it effectively alleviates joint swelling and repairs cartilage damage.

The temperature sensitivity of hydrogels manifests as intermolecular interactions that vary with temperature, which offers simple preparation conditions, avoids harmful components like organic solvents, and finds extensive applications in drug delivery and tissue engineering. For instance, dendritic chitosan was designed by modifying chitosan with a triple-folded dendritic polyethylene glycol, which exists as a liquid at room temperature, facilitating mixing with nanomicelles loaded with Tripterygium wilfordii homoflavone for interlamellar corneal injection. Upon contact with physiological temperature post-injection, it undergoes rapid thermo-induced aggregation, completing a sol-gel transition.

Chitosan-based hydrogels can be chemo-enzymatically functionalized with peptide or glycosidic linkers that act as substrate mimics for disease-overexpressed enzymes (e.g., matrix metalloproteinases, cathepsins, and hyaluronidase). Upon exposure to elevated enzymatic activity at the pathological locus, the selective cleavage of these linkers induces either (1) topological scission of the polymer chains, (2) dissolution of critical crosslinks, or (3) a sol–gel phase transition, thereby generating an on-demand release profile. A gut enzymse-responsive hydrogel was designed by modifying carboxymethyl chitosan with methacrylic anhydride, forming a three-dimensional network via UV crosslinking to encapsulate hydrophobic imatinib and the permeation enhancer sodium deoxycholate [19]. This achieves synergistic effects of precise controlled drug release and enhanced intestinal permeation. This hydrogel remains stable and non-degradable in the stomach, preventing premature drug release. Upon entering the intestine, intestinal lysozyme hydrolyzes the β-glycosidic bonds of carboxymethyl chitosan, triggering gradual hydrogel degradation and achieving complete drug release within 48 h. This addresses the challenges of traditional formulations, such as colonic precipitation and inconsistent release. The hydrogel exhibits strong enzyme responsiveness, degrading specifically in the intestine with excellent targeting.

Glucose-responsive drug-releasing chitosan hydrogels demonstrate significant application potential in diabetes treatment, blood glucose regulation, and interventions for related diseases. To develop an intelligent drug delivery system targeting periodontitis, researchers grafted the photosensitive compound methacrylic anhydride onto chitosan molecular chains [20], which formed a hydrogel membrane after crosslinking under UV irradiation. Subsequently, glucose oxidase (GOx) was immobilized on the membrane surface to endow it with glucose responsiveness, while simultaneously loading metronidazole, a drug for periodontitis treatment. Conventional therapies struggle to address random fluctuations in blood glucose: under hyperglycemic conditions, standard antibiotic doses are often insufficient and prone to inducing drug resistance and gastrointestinal side effects. In contrast, the hydrogel system can catalyze the oxidation of surrounding glucose into gluconic acid via GOx. This provides a novel smart local drug delivery system for the clinical treatment of diabetic periodontitis.

4.3 Targeted Therapy

Chitosan-based hydrogels have evolved into a versatile platform for precision pharmacotherapy. By integrating orthogonal chemical modification, stimulus-responsive motifs and hybrid nanomaterials, their targeting index, drug-payload capacity and therapeutic index have been simultaneously amplified, permitting disease-specific delivery across multiple pathologies.

The most fundamental targeting mechanism exploited in oncology is the Enhanced Permeability and Retention (EPR) effect. Tumor vasculature is often leaky, and lymphatic drainage is impaired, allowing nanoscale hydrogels or hydrogel nanoparticles (nanogels) to extravasate and accumulate preferentially within the tumor tissue. By carefully controlling the size (typically 10–200 nm) and surface chemistry of chitosan hydrogel particles, researchers can leverage this passive targeting pathway.

Active targeting involves decorating the surface of chitosan hydrogels with ligands that have high affinity for receptors overexpressed on target cells. This approach enhances cellular uptake and specificity beyond what passive accumulation can achieve. Common ligands include,

Folic acid: For cancer cells over-expressing folate receptors.
Hyaluronic acid: For targeting the CD44 receptor, prevalent on cancer stem cells.
Peptides: Such as RGD (Arg-Gly-Asp) sequences for targeting integrins on angiogenic endothelial cells and tumor cells.
Antibodies or aptamers: Offering high specificity for unique cell surface markers.

For instance, a chitosan hydrogel nanoparticle functionalized with folic acid will not only accumulate in a tumor via the EPR effect but will also be preferentially internalized by folate receptor-positive cancer cells, drastically improving therapeutic precision. Incorporating iron oxide nanoparticles (e.g., Fe_3O_4) allows the hydrogel to be guided to a specific site using an external magnetic field, adding a physical targeting dimension. Once localized, drug release can be triggered by the same magnetic field (via hyperthermia) or by the local TME.

Above mentioned stimuli-responsive chitosan hydrogels (X.4.2) undergo a dramatic structural change—such as swelling, dissociation, or sol-gel transition—in response to specific pathological stimuli, leading to triggered drug release. The most advanced platforms combine multiple targeting strategies to overcome biological barriers. A single hydrogel can be designed to be both pH- and redox-sensitive, providing a two-step release mechanism. Furthermore, biomimetic strategies, such as coating hydrogel nanoparticles with cell membranes (e.g., from red blood cells for prolonged circulation or from cancer cells for homologous targeting), are gaining traction. These "camouflaged" hydrogels evade the immune system and exhibit superior targeting capabilities. For example, chitosan and polypyrrole form a Schiff base–crosslinked nanogel loaded with docetaxel (DTX) and siRANK, then coated with PC-3 prostate cancer cell membrane (CM) [21]. Surface proteins on the CM bind homologous receptors on prostate cancer cells, enabling tumor-targeted accumulation. In the acidic tumor microenvironment, Schiff base hydrolysis swells the network, accelerating DTX release. As the gel degrades, siRANK enters cancer cells, downregulates RANK, blocks the RANK/RANKL pathway, and inhibits bone metastasis.

4.4 *Tissue Engineering*

Chitosanbased hydrogel are pivotal in tissue engineering for mimicking the ECM to guide cell behavior. Their customizable properties, enabled by an easily modified chemical structure, support applications in bone, cartilage, and skin regeneration. Key mechanisms include providing mechanical and osteoinductive cues in bone repair via hydroxyapatite composites, activating cartilage-specific signaling through structural similarity to glycosaminoglycans, and leveraging inherent antibacterial action for wound healing.

Functionalization modifications that introduce cell-specific recognition sites are a key strategy for enhancing the targeted biological activity of chitosan. In a distinct application for bone regeneration, chitosan was functionalized with bisphosphonates to form a dual-crosslinked hydrogel with poly(acrylamide). This network combines covalent bonds for stability with dynamic metal coordination bonds with Mg^{2+} ions, enabling sustained ion delivery. The material exhibits robust mechanical properties, including a tensile strength of 0.62 MPa and extensibility of 973%, alongside favorable creep recovery. This platform synergistically promotes the proliferation and differentiation of mouse embryonic osteoblast precursor cells in vitro via the BMP2/Wnt pathway. In a rat calvarial defect model, the coordinated delivery of Mg^{2+} ions and BMP2 significantly accelerated bone regeneration [22] (Fig. 3).

The application of chitosan in load-bearing bone regeneration is constrained by a significant mechanical mismatch with native cortical bone. Advanced composite designs address this by synergizing mechanical reinforcement with bioactivity. The primary strategy involves incorporating rigid inorganic phases. While nanohydroxyapatite (nHA) provides mechanical augmentation and inherent osteoconductivity, further sophistication is achieved by doping with therapeutic ions. For example, strontium-doped bioglass nanoparticles within a chitosan-gelatin matrix not only improve compressive strength but also actively stimulate bone regeneration by depositing a cell-supportive apatite layer. Beyond static reinforcement, smart release systems are critical for ions like Mg^{2+}, whose efficacy depends on sustained availability. The encapsulation of $MgO/MgCO_3$ within PLA microspheres dispersed in a chitosan hydrogel creates a dual-phase barrier that successfully modulates the release kinetics, preventing a cytotoxic burst and providing a long-term signal that concurrently promotes osteogenesis and angiogenesis.

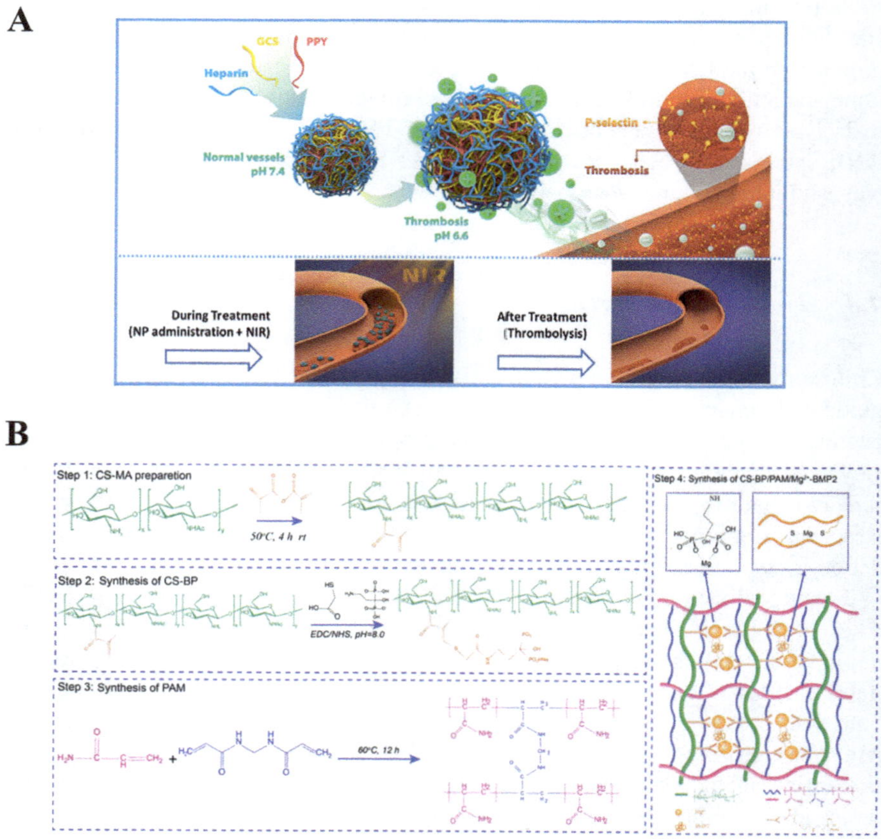

Fig. 3 Biochemical bionics of chitosan materials [22]. Reprinted with permission

Injectable chitosan hydrogels are ideal carriers for articular cartilage repair, offering minimally invasive implantation and in situ gelation. A critical challenge lies in balancing rapid thermosensitive gelation with enhanced mechanical strength. This was addressed by formulating a chitosan solution with weak alkalis like sodium bicarbonate and β-glycerophosphate, resulting in a hydrogel with robust mechanical properties and a fast gelation rate. To further improve its bioactivity, chondroitin sulfate was incorporated into the system. The resulting composite gel exhibits favorable thermosensitivity and injectability, with a pH and osmotic pressure close to physiological levels, demonstrating high biocompatibility and making it a suitable platform for cell-based therapy of cartilage defects. Another design encapsulates sinomenine in chitosan microspheres within photo-crosslinked gelatin methacryloyl (GelMA) hydrogel. This system provides sustained drug release, enhances chondrocyte autophagy, and effectively delays osteoarthritis progression in vivo. A system combines water-soluble chitosan, hyaluronic acid, and silanized hydroxypropyl methylcellulose (Si-HPMC) to form a tunable injectable hydrogel. Optimized Si-HPMC content improves swelling and rheology while promoting chondrocyte proliferation and extracellular matrix deposition, achieving approximately 79.5% bone regeneration in 21 days [23].

Chitosan-based hydrogels are pivotal in skin tissue engineering (STE) due to their abilities to accelerate wound closure, promote re-epithelialization, facilitate neovascularization, and modulate the inflammatory microenvironment. Their tunable porosity, mechanical stiffness, and degradation kinetics further enable customization to match distinct wound phenotypes (acute burns, chronic diabetic ulcers or full-thickness defects), addressing critical limitations of conventional dressings (poor moisture retention, lack of biological activity or scarring). Physically cross-linked chitosan-TPP hydrogels, for instance, leverage ionic gelation to form porous networks that retain moisture and exert antimicrobial effects via cationic chitosan, reducing *Staphylococcus aureus* colonization and boosting re-epithelialization in rat burn models. Meanwhile, photocross-linkable chitosan such as methacryloyl (CSMA) hydrogels enable 3D bioprinting of patient-specific skin equivalents. Moving beyond basic wound closure, a key frontier is the regeneration of skin appendages. A bilayer scaffold features a dense chitosan-gelatin membrane to restore the epidermal barrier, underpinned a porous chitosan-hyaluronic acid hydrogel seeded with hair follicle dermal papilla cells. This construct not only supports dermal regeneration but also creates a niche that activates signaling pathways (e.g., Wnt/β-catenin), successfully inducing de novo hair follicle genesis in preclinical models, a significant step toward truly functional skin restoration.

4.5 Other Applications

In the field of biosensing, chitosan hydrogels serve as critical transduction matrices that convert biological recognition events into measurable signals. They are primarily utilized in two strategic approaches: as stimuli-responsive materials whose volumetric changes (swelling or deswelling) in response to pH, humidity, or ionic strength provide a direct readout for environmental monitoring (e.g., bacterial growth via pH shifts); and as immobilization scaffolds for bioreceptors such as enzymes and antibodies, facilitating highly specific detection of biomarkers with significant sensitivity and selectivity. The integration of nanomaterials, like magnetic nanoparticles, further enhances signal amplification, enabling versatile optical or electrical outputs. For example, immobilizing glucose oxidase (GOx) within chitosan gels for noninvasive blood glucose monitoring, and employing antigen–antibody crosslinked hydrogels to detect inflammatory biomarkers, both demonstrating high sensitivity and selectivity. The future trajectory points toward their incorporation into flexible and implantable devices for real-time health monitoring, leveraging their compatibility with microelectronics to create next-generation wearable diagnostics.

Simultaneously, in contact lens technology, chitosan hydrogels address fundamental limitations of conventional materials, such as low oxygen permeability and surface hydrophobicity. By forming interpenetrating polymer networks (IPNs) with standard lens polymers (e.g., pHEMA) or through surface functionalization techniques, chitosan imparts essential biological functions without compromising optical transparency. This integration yields lenses with inherent antimicrobial properties, improved surface wettability, and the capacity for sustained drug release, transforming passive vision correction devices into active therapeutic platforms. For instance, lenses incorporating quaternized chitosan and anti-inflammatory agents have demonstrated efficacy in treating bacterial keratitis by simultaneously clearing pathogens and mitigating inflammation, thereby promoting corneal healing [24], as illustrated in Fig. 4. Another route is to functionalize the lens surface. For example, polydopamine-assisted co-deposition can be used to form a coating of chitosan or other biopolymers on silicone hydrogel lenses, thereby introducing hydrophilic

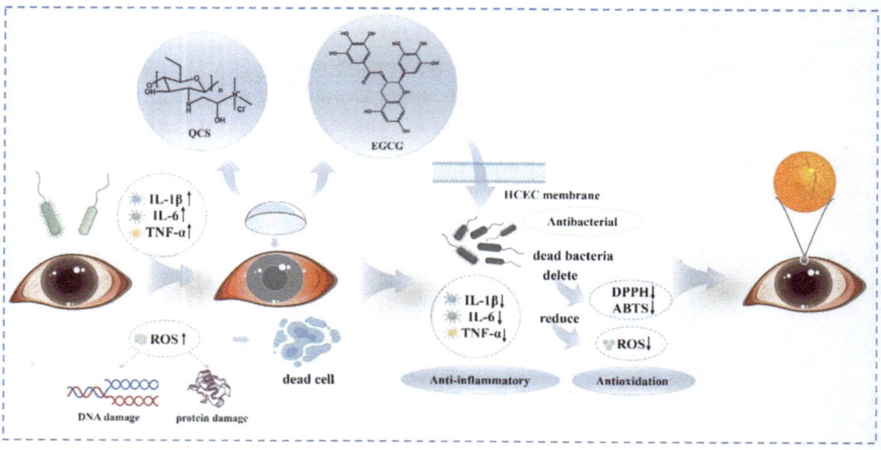

Fig. 4 Hydrogel contact lenses modified with quaternized chitosan and epigallocatechin gallate (EGCG) exhibit enhanced antibacterial activity against bacterial keratitis [24]. Reprinted with permission

functional groups and antimicrobial constituents. In addition, studies have proposed employing chitosan and its nanocomposites as lens coatings to prevent Candida colonization and biofilm development that cause recalcitrant keratitis, while also exerting pro-healing effects.

5 Challenges and Future Perspectives

Chitosan-based hydrogels have become a cornerstone of biomedical research and translation, owing to their inherent biocompatibility, biodegradability, and chemical tunability-properties that align with diverse applications ranging from drug delivery and tissue engineering to biosensing. However, their clinical utility and widespread adoption are still hindered by several critical challenges, while emerging research directions promise to address these limitations and unlock new biomedical potentials.

A primary obstacle is the limited mechanical strength of many chitosan hydrogels, particularly those crosslinked via weak physical interactions. This often renders them unsuitable for load-bearing applications such as bone and cartilage repair, and may lead to structural failure under physiological conditions. Secondly, the natural origin of chitosan results in significant batch-to-batch variability in parameters like molecular weight and DD. This variability poses a major hurdle to manufacturing reproducibility, quality control, and obtaining regulatory approval. Finally, while generally considered biocompatible, there are concerns regarding the potential immunogenicity of chitosan, especially with high molecular weight or poorly purified samples, which could trigger undesirable inflammatory responses.

To address these challenges, future efforts should focus on several key directions. The development of novel crosslinking strategies is paramount. Exploring high-density ionic crosslinking, enzyme-mediated coupling, or dual-network hydrogels that combine covalent and reversible bonds can significantly enhance mechanical robustness and stability. Secondly, the incorporation of bioactive molecules such as growth factors, antimicrobial peptides, or genetic material, can transform passive chitosan scaffolds into active therapeutic systems that precisely guide cell behavior and enhance regenerative outcomes. Furthermore, the fabrication of multifunctional hydrogels that integrate diagnostic capabilities (sensing pH or biomarkers, etc.) with therapeutic action, for example, on-demand drug release, represents the frontier of smart biomaterial design. Lastly, there is ample room for the exploration of new biomedical applications, including but not limited to immunomodulation, cancer immunotherapy, neural interfaces, and organ-on-a-chip technologies.

In conclusion, while challenges related to mechanics, reproducibility, and biocompatibility remain, the future of chitosan hydrogels is bright. By embracing innovative crosslinking chemistry, incorporating sophisticated bioactivity, and designing multifunctional systems, researchers can unlock the full potential of this versatile biopolymer, paving the way for its successful translation into advanced clinical therapies.

References

1. Kantak MN, Bharate SS (2022) Analysis of clinical trials on biomaterial and therapeutic applications of chitosan: a review. Carbohydr Polym 278:118999
2. Budiarso IJ, Rini NDW, Tsalsabila A, Birowosuto MD, Wibowo A (2023) Chitosan-based smart biomaterials for biomedical applications: progress and perspectives. ACS Biomater Sci Eng 9:3084–3115
3. Li R, Lin Z, Zhang Q, Zhang Y, Lyu Y, Li X, Zhou C, Wu G, Ao N, Li L (2020) Injectable and in-situ formable thiolated chitosan coated liposomal hydrogels as curcumin carriers for prevention of in vivo breast cancer recurrence. Carbohydr Polym 230:1–11
4. Zhu M, Luo B, Hu X, Liu H, Zhou C, Lu L (2022) Antibacterial peptide encapsulation and sustained release from chitosan-based delivery system. Eur Polym J 181:111640
5. Huang L, Chu Y, Zhang L, Liu X, Hao W, Chen Y, Dai J (2021) High strength pure chitosan hydrogels via double crosslinking strategy. Biomed Mater 16:045048
6. Svang-Ariyaskul A, Huang R, Douglas P, Pal R, Feng X, Chen P, Liu L (2006) Blended chitosan and polyvinyl alcohol membranes for the pervaporation dehydration of isopropanol. J Membr Sci 280:815–823
7. Chellathurai M, Chung L, Hilles A, Sofian Z, Singha S, Ghosal K, Mahmood S (2024) Pharmaceutical chitosan hydrogels: a review on its design and applications. Int J Biol Macromol 280:135775
8. Khor E (1997) Methods for the treatment of collagenous tissues for bioprostheses. Biomaterials 18:95–105
9. Ladet S, David L, Domard A (2008) Multi-membrane hydrogels. Nature 452:76–79
10. Bingren T, Shiyao H, Yu T, Jiayue L (2020) Chemical and physical chitosan hydrogels as prospective carriers for drug delivery: a review. J Mater Chem B 8:10050–10064
11. Zhou L, Guo P, D'Este M, Tong W, Xu J, Yao H, Stoddart M, van Osch G, Ho K, Li Z, Qin L (2022) Functionalized hydrogels for articular cartilage tissue engineering. Engineering 13:71–90
12. Zhu H, Wang C, Xin C, Dai Z, Yang Y, Li Y, Wang M, Zhang Y, Yue S, Cai R, Qian K (2025) Directional anneal-casting strategy enables mechanically enhanced biomimetic hydrogel for ligament reconstruction. Compos Part B Eng 292:112068
13. Fu L, Li L, Bian Q, Xue B, Jin J, Li J, Cao Y, Jiang Q, Li H (2023) Cartilage-like protein hydrogels engineered via entanglement. Nature 618:7966
14. Jia Z, shen D, Xu W (2001) Synthesis and antibacterial activities of quaternary ammonium salt of chitosan. Carbohydr Res 333(1):1–6
15. Singh G, Nayal A, Malhotra S, Koul V (2020) Dual functionalized chitosan based composite hydrogel for haemostatic efficacy and adhesive property. Carbohydr Polym 247:116757
16. Ling Z, Chen Z, Deng J, Wang Y, Yuan B, Yang X, Lin H, Cao J, Zhu X, Zhang X (2021) A novel self-healing polydopamine-functionalized chitosan-arginine hydrogel with enhanced angiogenic and antibacterial activities for accelerating skin wound healing. Chem Eng J 420:130302
17. Medha S (2024) Sethi, RSM optimized chitosan based composite hydrogel for sustained drug delivery applications. Mater Today Commun 38
18. Park G, Gwak M, Choi Y, Park W (2023) pH-sensitive gallol-rich chitosan hydrogel beads for on-off controlled drug delivery. Inter J Bio Macromol 240:12346
19. Wang C, Sun M, Fan Z, Du J (2022) Intestine enzyme-responsive polysaccharide-based hydrogel to open epithelial tight junctions for oral delivery of imatinib against colon cancer. Chin J Polym Sci 40:1154–1164
20. Liu J, Xiao Y, Wang X, Huang L, Chen Y, Bao C (2019) Glucose-sensitive delivery of metronidazole by using a photo-crosslinked chitosan hydrogel film to inhibit Porphyromonas gingivalis proliferation. Inter J Bio Macromol 122:19–28
21. Yu Q, Gao Y, Dai W, Li D, Zhang L, Hameed M, Guo R, Liu M, Shi X, Cao X (2024) Cell membrane-camouflaged chitosan-polypyrrole nanogels co-deliver drug and gene for targeted chemotherapy and bone metastasis inhibition of prostate cancer. Adv Heal Mater 13:2400114

22. Xiong A, He Y, Gao L, Li G, Liu S, Weng J, Wang D, Zeng H (2021) The fabrication of a highly efficient hydrogel based on a functionalized double network loaded with magnesium ion and BMP2 for bone defect synergistic treatment. Mater Sci Eng C Mater Biol Appl 128:112347
23. Hu M, Yang J, Xu J (2021) Structural and biological investigation of chitosan/hyaluronic acid with silanized-hydroxypropyl methylcellulose as an injectable reinforced interpenetrating network hydrogel for cartilage tissue engineering. Drug Deliv 28:607–619
24. Lin S, ZhengG Y, Xue P, Cong Y, He X, Zhang C, Yuan C (2025) Hydrogel contact lenses modified by quaternary chitosan and epigallocatechin gallate with enhancement of antimicrobial for bacterial keratitis. Int J Biol Macromol 306:141787

Lihua Li

Lu Lu

Changren Zhou

Yifan Zhang

Zexiang Zheng

Xinshuo Liu

Junxiao Xiang

Chitosan Hydrogels in Tissue Engineering and Regenerative Medicine

Sona Sunil and Jayakumar Rangasamy

Abstract Chitosan hydrogels are biomaterials that are revolutionising the field of tissue engineering and regenerative medicine with their unique properties, making them stand out among other classic polymeric hydrogels. Chitosan hydrogels can be fabricated using different methods. These hydrogels encompass distinctive physico-chemical attributes that can be tailored for various applications in drug delivery, wound healing, tissue regeneration, etc. This book chapter reviews the preparation techniques of chitosan hydrogels and their applications in regenerative medicine, including drug delivery, wound healing, and haemostasis. It also concentrates the utility of chitosan hydrogels in adipose, bone, cartilage, ligament, and tendon tissue engineering. This book chapter has outlined various research findings on using chitosan hydrogels for distinct fields of tissue engineering and regenerative medicine.

Keywords Chitosan hydrogel · Drug delivery · Wound healing · Haemostasis · Adipose · Bone · Cartilage · Ligament · Tendon · Tissue engineering · Regenerative medicine

1 Introduction

Chitosan (CS) is a natural cationic biopolymer that is obtained via the chitin deacetylation. Chitin is extracted from the shells of crustaceans, shrimps, crabs, lobsters, etc., and is chemically processed to obtain CS [1–5]. CS is a linear polysaccharide, comprising both deacetylated and acetylated units, β-(1,4)-linked D-glucosamine and N-acetyl-D-glucosamine, respectively. The degree of deacetylation (usually

S. Sunil · J. Rangasamy (✉)
Polymeric Biomaterials Lab, School of Nanosciences and Molecular Medicine, Amrita Vishwa Vidyapeetham, Kochi, India
e-mail: rjayakumar@aims.amrita.edu

S.-L. Wang, M. Anpo (eds.), *Chitin and Chitosan*, Springer Series on Polymer and Composite Materials, https://doi.org/10.1007/978-981-95-7457-5_9

>50%) also influences the properties of CS. The presence of hydroxyl and amino groups (Fig. 1) influences the activity of CS and enables it to participate in various chemical interactions, such as the Schiff base reactions. These existing functional groups in CS play a major role in introducing properties such as its biodegradability, biocompatibility, and non-immunogenicity, which makes it an ideal prospect for bone tissue applications. It also has inherent antimicrobial activity, anti-inflammatory activity, mucoadhesive property, enabling its extensive use in biomedical applications [1, 2, 5].

Hydrogels are water-absorbing materials formed by the crosslinking of the same or different polymers containing hydrophilic functional groups. The presence of amine (-NH_2) and hydroxyl (-OH) groups in the CS backbone enables it to participate in the process of crosslinking and thereby form a stable gel [2]. CS hydrogels come under the class of naturally formed hydrogels, as CS is classified as a natural polymer. These natural hydrogels are gaining wide attention over hydrogels derived from synthetic polymers, due to their biocompatible nature and ability to produce non-toxic by-products upon degradation [2]. The application of CS hydrogels, when compared to other physical forms such as membranes, scaffolds, sponges, etc., has acquired interest in the field of regenerative medicine and tissue engineering due to various advantages. CS hydrogels are well known for their ease of preparation and its dimensions range from micro to even nanometres (Fig. 2). Since these hydrogels possess an inherent hydrophilic nature, they have an immense swelling ability or in other words have a high water-absorbing capability, which makes it useful in carrying drugs and small molecules, making them impactful in drug delivery applications (Fig. 2) [1–3]. The injectable property is one of the main highlights of the CS hydrogel that makes it distinct from other forms of CS. It gives an added advantage of injecting hydrogel into the localised area of application and allows for moldability with respect to the defect site (Fig. 2) [1–5]. The in situ gelling property of CS hydrogels makes them fall under the category of smart hydrogels, which enables the hydrogel to undergo a sol-gel transition 'in situ' or at the local site, resulting in better tissue regeneration (Fig. 2) [3]. This makes it useful in drug delivery and implant associated applications. Thus, the idea and application of 'smart' or 'intelligent' CS hydrogels is upcoming and being extensively used in the field of tissue engineering and regenerative medicine. These hydrogels will be able to recognise and respond

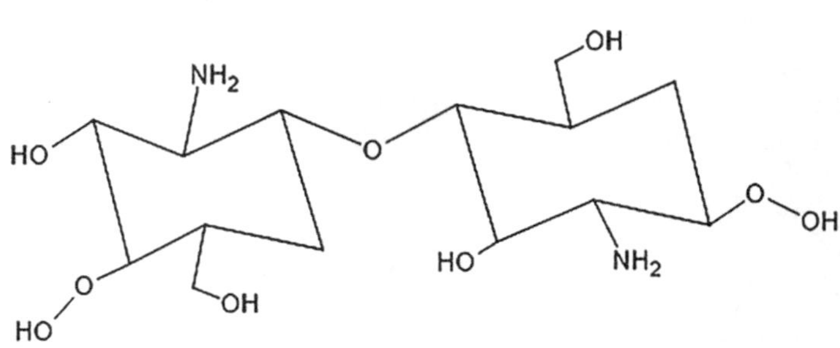

Fig. 1 Structure of CS

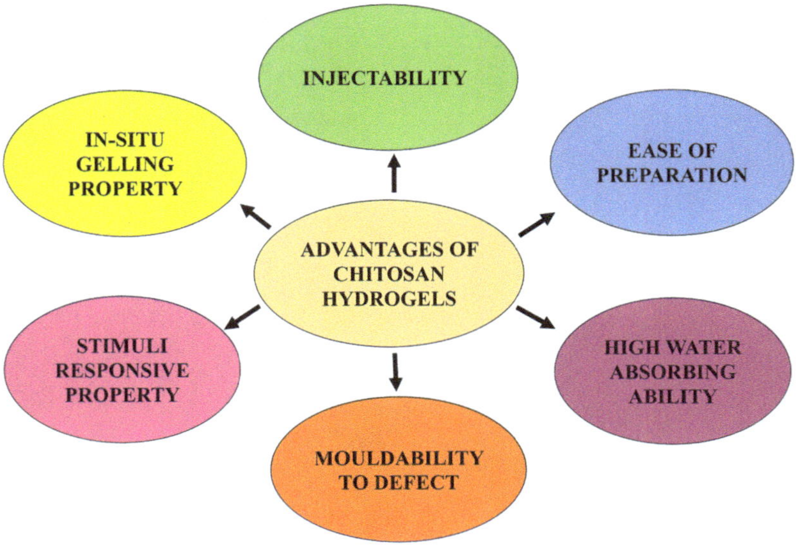

Fig. 2 Advantages of CS hydrogels

to different external stimuli such as light, pH, temperature, magnetic field, etc. [3]. Researchers tend to focus on developing derivatives of CS through various chemical modifications to bring out the utility of this biomaterial for a wide variety of applications. This book chapter aims to highlight the current advancements of CS hydrogels, their preparation and utility in various fields of tissue engineering and regenerative medicine.

2 Fabrication Techniques of CS Hydrogel

CS hydrogels are generally prepared by various methods, including crosslinking, via polymerisation, by electrospinning, etc. The synthesis protocols for CS hydrogels have been studied in detail by researchers, and several methods have hence been developed to prepare them. It was found that the simplest way to form a CS hydrogel was by exposing the acidic CS solution to a higher alkaline pH, either by the addition of ammonia or sodium hydroxide [5]. One of the major methods involved in the synthesis of CS hydrogels is the crosslinking technique. They are categorised as physically and chemically cross-linked CS gels [1, 2, 5].

The physically cross-linked gels involve temporary or reversible interactions such as hydrophobic, ionic or host-guest interactions (Fig. 3). They make use of small anionic molecules, polyelectrolytes, or host-guest interactions via using CS derivatives for the physical union between the CS chains. The use of small anionic molecules, especially those that contain phosphate groups such as tripolyphosphate,

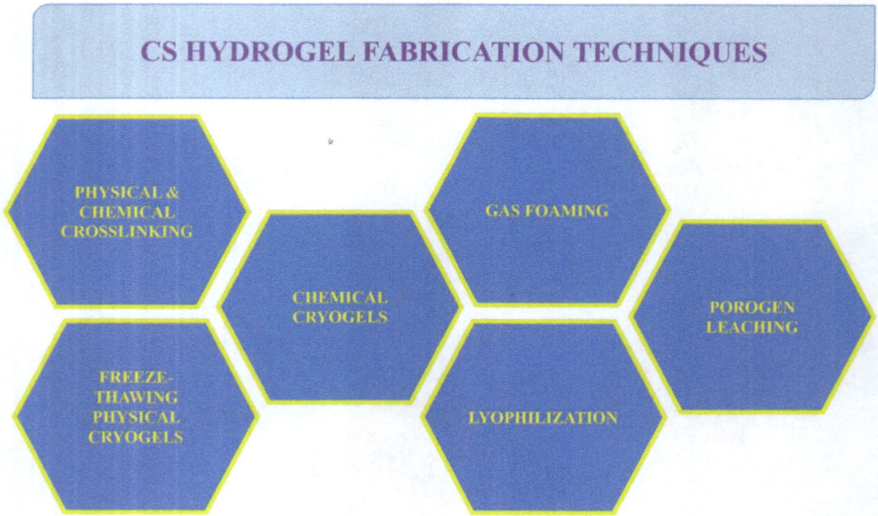

Fig. 3 CS hydrogels fabrication techniques

citrate and β-glycerophosphate, etc. resulted in the formation of physically cross-linked CS hydrogels. The main drawback of such a preparation method is the non-uniform crosslinking that occurs between the CS chains [2–4].

Chemically cross-linked gels are reported to be have superior mechanical strength when compared to the physically cross-linked counterparts, mainly due to the covalent interaction between the CS and the crosslinking agents, or via a functionalized monomer or polymer (Fig. 3). Chemical cross-linkers such as glutaraldehyde (GA) are known to provide enhanced mechanical properties, but the associated toxicity is a major challenge in the preparation of GA-CS hydrogels. On the other hand, genipin is a naturally found alternative to GA, exhibiting comparable mechanical properties and offering a low toxicity profile [2, 5]. The Schiff base reaction is a common method involved in synthesising chemically linked CS hydrogels, comprising an interplay between the amino groups of CS and the aldehyde group-containing candidates such as oxidised alginate and oxidised glucose. It is quick, stimulates rapid gelation, and drastically improves the mechanical properties. There are various other methods for the fabrication of chemically cross-linked hydrogels, such as Michael addition, formation of five interpenetrating networks, etc. [4].

Research has been conducted to find distinct strategies in developing porous CS hydrogels, such as freeze-thawing to generate physical cryogels, chemical cryogels, lyophilisation, gas foaming, and porogen leaching [1, 3, 5]. Physical cryogels are synthesised by utilising a concentrated polymeric aqueous solution, which is subjected to a freeze-thawing process (Fig. 3). These cryogels impart enhanced interconnected porous structure along with high swelling ability, making them ideal constructs for biomedical applications. Chemical cryogels utilise the process of free radical polymerisation, which involves cross-linkers and initiators at freezing temperatures, where the ice crystal serves as the template (Fig. 3). Macro-porous hydrogels can be produced using the lyophilization method (Fig. 3). It involves the preparation of polymeric solutions with an aqueous base, and later allowed for freezing, and then subjected to solvent sublimation under vacuum. Super-porous 3D matrices can be developed using foaming agents like $NaHCO_3$ or NH_4HCO_3 in the gas foaming method. Porogen leaching is a strategy that involves porogens, salts or inorganic particles in a free radical polymerisation reaction, to create porous hydrogels (Fig. 3) [5].

The previously described approaches have been majorly adopted to fabricate CS hydrogels for its implementation in the sector of regenerative medicine and tissue engineering.

3 Application of CS Hydrogel in Regenerative Medicine

3.1 Haemostasis

The need to control excess blood loss that occurs during trauma and injury is a serious concern that needs to be dealt with in the field of regenerative medicine. The intrinsic hemostatic ability of the body might not suffice during major bleeding or blood loss situations. CS is well studied for its hemostatic ability, wherein the positively charged NH_2 groups of CS interact with the negatively charged membrane of erythrocytes, and this leads to a plug formation at the haemorrhage site. The CS hydrogel is known to integrate with the tissue at the applied location, due to its shear-thinning and mucoadhesive abilities [4, 5].

Countless studies have investigated improving the natural hemostatic capability of CS by forming composite materials. A study explored the hemostatic ability of an injectable CS-nano-bioglass (CS-nBG) composite hydrogel. nBG, being a bioceramic, is known to release silica, phosphorus, and calcium ions, which can interfere with the bleeding process. To examine the hemostatic capability of the gel system, it was studied in an in vivo rat liver and femoral artery injury model and the blood clotting was observed to be within 54 ± 3 s and 185 ± 9 s, respectively. The blood clotting efficiency of CS-nBG gel was higher when compared to the control CS gel. It was accounted for by the combined effect of the intrinsic hemostatic property of CS and the release of silica, phosphorus, and calcium ions from the system, which activated coagulation factor XII, initiated the extrinsic pathway of coagulation cascade and activation of coagulation factors, respectively (Fig. 4) [6].

A similar study was conducted wherein they incorporated whitlockite nanoparticles (nWH) in CS hydrogel (CS-nWH) to check its effectiveness as a hemostatic agent in vitro and in vivo. nWH is also a bioceramic material, usually found in the natural bone composition and is being heavily researched for bone tissue engineering applications. nWH is well known to promote the release of calcium, magnesium, and phosphate ions. The hemostatic efficiency of the prepared biocompatible

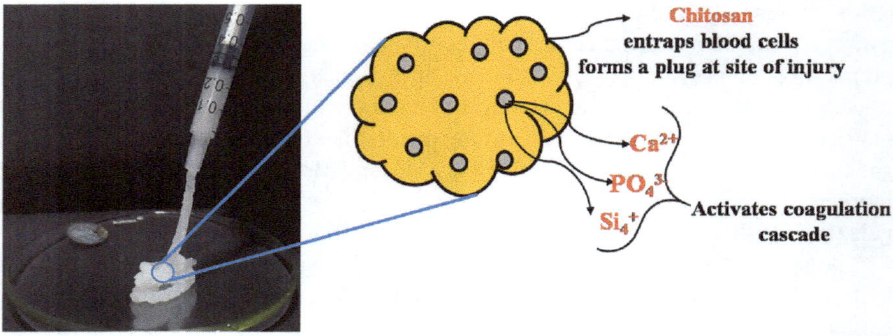

Fig. 4 Schematic diagram showing the possible mechanism of action of CS-nBG hydrogel in blood. Reproduced with permission from Elsevier [6]

and injectable gel system was studied in an in vivo rat liver and femoral artery injury model. The blood clotting time of the CS-nWH system was observed to be 62 ± 3 s and 229 ± 9 s, respectively. It was found that the CS-nWH hydrogel showed a higher blood clotting efficacy than the control CS gel and commercially available hemostatic agent Floseal®. The gel system shows an enhanced blood clotting property, by virtue of the interaction of CS with the negatively charged red blood cells and the release of calcium, magnesium, and phosphate ions, which can further activate the coagulation cascade, leading to rapid blood clotting [7].

In an another study, in which they integrated a potassium aluminum sulfate (PA)- vasoconstrictor and calcium chloride (Ca)- coagulation activator into CS hydrogel (CS-PA-Ca) (Fig. 5). Both in vitro and in vivo studies of the gel system were performed in rat liver and femoral artery haemorrhage model and found the blood clotting time to be 20 ± 10 s, 105 ± 31 s, respectively. The rapid hemostatic effect of the CS-PA-Ca gel system is justified by the interaction of positively charged amine groups of CS that interact with the negatively charged erythrocyte membrane. The PA will precipitate proteins and lead to platelet aggregation, in addition to the vasoconstriction of the blood vessels at site of injury. Ca stimulates the coagulation factors and triggers the activation of instrinsic, extrinsic, and common coagulation pathways. According to the haemostatic time and mass loss of already established CS-based haemostatic hydrogels (Table 1), the CS-PA-Ca gel system exhibited the highest blood clotting efficiency when compared to the previously reported CS-nWH, CS-nBG hydrogels and other CS-based haemostatic hydrogels established in literature [8]. From the previous studies, the blood clotting time of CS-nWH and CS-nBG hydrogels studied in the rat liver and femoral artery haemorrhage model showed a fast and enhanced blood clotting when compared to the commercially available Floseal and Fibrin sealant [8]. These studies indicated that the addition of inorganic additives to chitosan hydrogel significantly reduced the blood clotting time by acting through all haemostasis mechanisms.

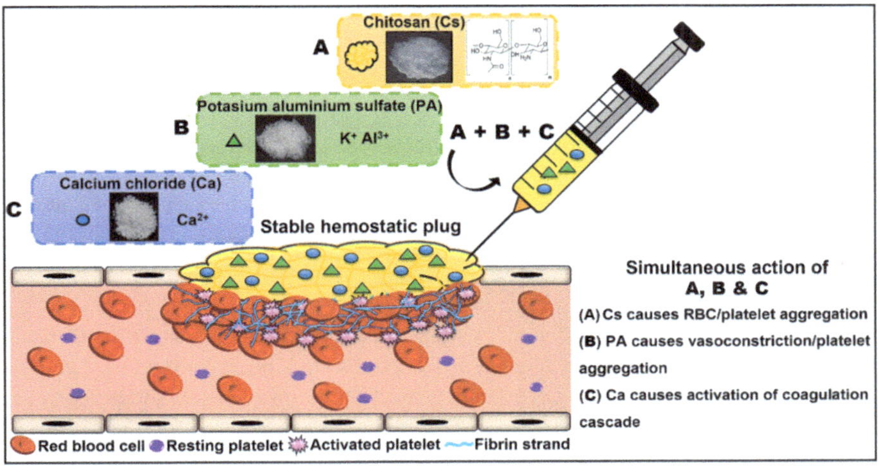

Fig. 5 Schematic illustration of CS-PA-Ca hydrogel mechanism of action in haemostasis. Reproduced with permission from Elsevier [8]

Table 1 Tabular description of the haemostatic time and mass of blood loss for CS-based haemostatic hydrogels reported in the literature. Reproduced with permission from Elsevier [8]

Hydrogel system	In vivo injury model	Haemostatic time	Mass of blood loss
Chitosan-nano-whitlockite hydrogel	Rat liver haemorrhage model	62 ± 3 s	478 ± 8 mg
	Rat femoral artery hemorrage model	229 ± 9 s	247 ± 10 mg
Chitosan-nano-bioglass hydrogel	Rat liver haemorrhage model	54 ± 3 s	350 ± 8 mg
	Rat femoral artery haemorrage model	185 ± 9 s	197 ± 8 mg
Carboxymethyl chitosan-gelatin--oxidised alginate hydrogel	Rat liver haemorrhage model	62 s	0.036 g
Hydrophobically modified chitosan/oxidised- dextran hydrogel	Rat liver haemorrhage model	Observation made after 120 s	15 mg
Catechol hydroxybutyl chitosan hydrogel	Rat liver haemorrhage model	30 s	30.7 ± 11.4 mg
Quaternized chitosan-phosphorylated chitosan-poly-Ɛ lysine-tannic acid hydrogel	Rat liver laceration model	Observation made after 120 s	0.19 g
O-nitrobenzyl alcohol-carboxymethyl chitosan/ carboxymethyl chitosan hydrogel	Rat liver injury model	20.67 ± 2.52 s	150.30 ± 11.98 mg

3.2 Drug Delivery

Drug delivery is a vast field employed in regenerative medicine, which aims to deliver mainly drugs, small molecules, or even genetic material to precise locations within the body in a controlled manner. It is a growing field where extensive research is being done on controlling the rate of the drug released and improving localised delivery of the drug molecule, and limiting off-target release to reduce side effects. Various subsectors of drug delivery have been developed, and they specialise in delivering a certain type of drug, such as antibacterial drug delivery, anti--inflammatory drug delivery, anti-cancer drug delivery, antifungal drug delivery, peptide delivery and gene delivery. Among these, antibacterial, antifungal, anti--inflammatory, peptide delivery, are mainly concentrated in the domain of regenerative medicine. CS hydrogels have been proven to be an ideal candidate for drug delivery due to its non-toxicity, biodegradability, injectability, and shear-thinning properties [4, 9].

Research has demonstrated the utility of an injectable CS gel system, where nanocurcumin (nC) and arginine (R) are loaded, to prevent hypoxia induced endothelial damage (Fig. 5). It was noted that at low or acidic pH, a higher release of nC and R was detected from the gel system, which could be replicated in a low pH ischemic site in the body. In vitro tube formation studies validated the angiogenic potential of the prepared CS-nC/R gel system. Western plot analysis was employed for validating the expression levels of endothelial nitric oxide synthase (eNOS) and phosphorylated eNOS (serine 1117). In cultured endothelial cells, subjected to hypoxic conditions, an increased phosphorylation of eNOS at serine 1177 was witnessed for the CS-nC/R gel system, when compared to the control hydrogels. According to literature, normoxia conditions express phosphorylated eNOS in the presence of an arginine substrate, which generates nitric oxide that helps to maintain the homeostasis of the endothelial cells (Fig. 6). In hypoxic conditions, due to the depleted pool of arginine, the production of eNOS system is affected, leading to endothelial dysfunction and disrupted homeostasis (Fig. 6). In such hypoxic conditions, the prepared CS-nC/R system was proven to substitute for the depleted L-arginine, which thereby phosphorylates eNOS to generate nitric oxide and the nanocurcumin released from the gel aids in tackling the excess free radicals in the hypoxic environment, thereby proving its antioxidant function (Fig. 6). Hence, the final CS-nC/R hydrogel system encompassed better antioxidant and angiogenic property than the control CS gel [10].

Another group studied the effect of harnessing the properties of CS-hyaluronic acid (CS-HA) hydrogel to deliver deferoxamine (DFO) loaded PLGA nanoparticles for elevated angiogenic potential. From Fig. 7, it can be understood that below the neutral pH, the CS-HA mixture is not able to form a stable hydrogel, and instead forms flocculates, whereas above the neutral pH, CS-HA mixture generates an irregular hydrogel, encompassing thread-like structures. A stable coacervate and injectable CS-HA hydrogel is produced at a pH range of 7 to 8, where the interactions between CS and HA are stable enough to stimulate coacervate gel formation

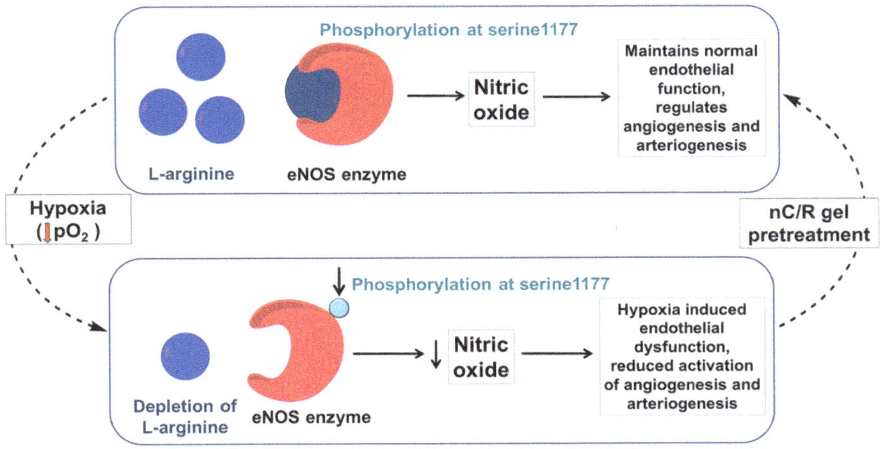

Fig. 6 Depicting the mechanism of action of the nanocurcumin and arginine peptide-loaded CS hydrogel system in treating hypoxic conditions. Reproduced with permission from Elsevier [10]

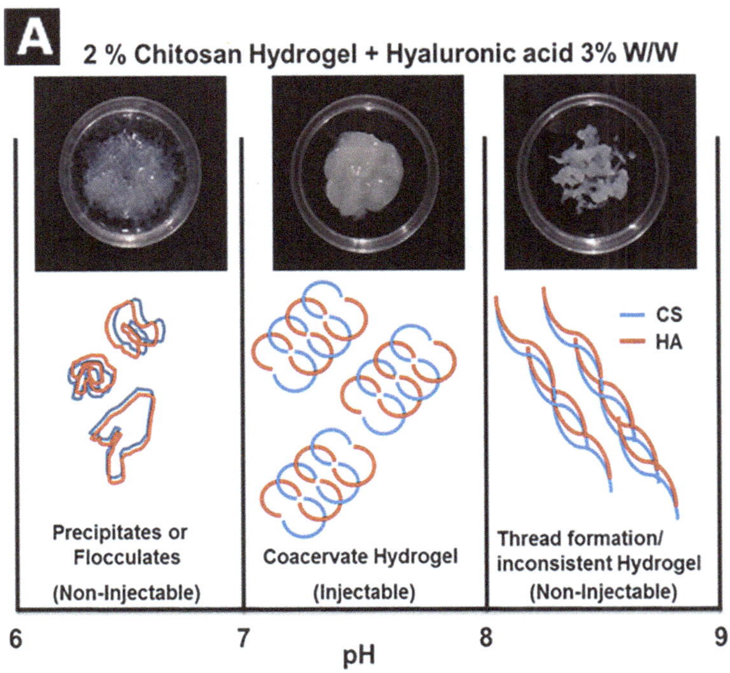

Fig. 7 The coacervate CS-HA hydrogel formation and its pH dependency. Reproduced with permission from Elsevier [11]

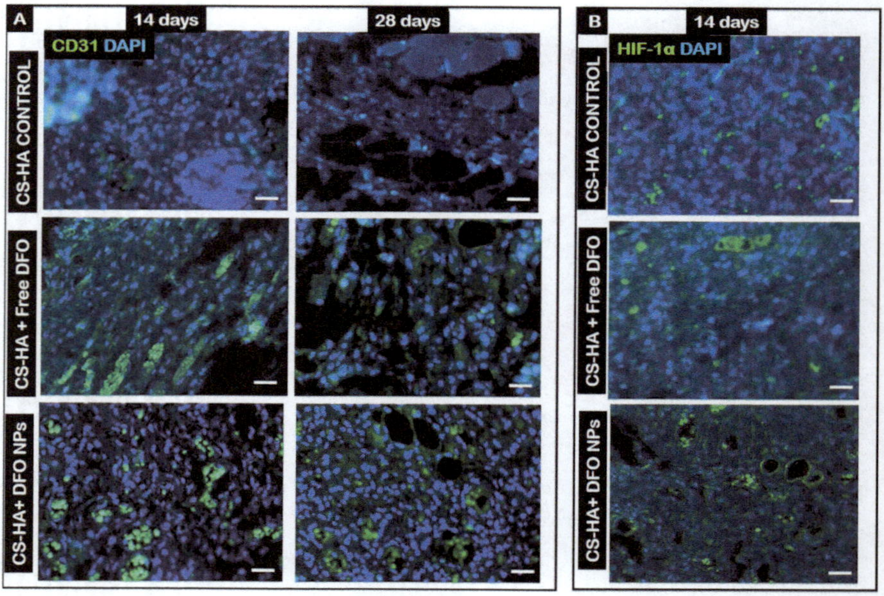

Fig. 8 Immunohistochemistry staining images of CS-HA control, CS-HA + Free DFO, and CS-HA + DFO NPs. (A) Staining of CD 31 marker on 14 and 28 days, (B) Staining of HIF-1 on 14th day (Scale bar = 25 m). Reproduced with permission from Elsevier [11]

(Fig. 7). The biocompatible hydrogel allowed for the controlled release of the DFO from the PLGA nanoparticles. On conducting in vivo angiogenic experiments in mice for a period of 2–4 weeks, the immunohistochemistry results obtained highlighted the increased expression of the endothelial marker CD-31 (CD-31: green and Nuclei: DAPI Blue) on days 14 and 28 when compared to control gels (Fig. 8A), suggesting the sustained release of DFO from CS-HA + DFO NPs gel system. From Fig. 8B, it can be observed that HIF-1α expression is significantly higher in the final gel system, when compared to the control systems. This suggests that the sustained release of DFO from the prepared gel system triggers a hypoxic environment, maintaining the HIF-1α factor, which can boost the expression of angiogenic factors such as VEGF, ANG2, confirming the enhanced neo-vascularisation of the gel system when compared to the control system. Hence, the prepared gel system can be used for therapeutic angiogenesis applications [11]. These reports suggest that the use of CS hydrogels as efficient drug carriers and can be utilised for applications in antibacterial, ant-inflammatory, antifungal, peptide and gene delivery.

3.3 Wound Healing

Wound healing is an integral part of regenerative medicine applications, and typically refers to the treatment of wounds, which mainly include infectious wounds, chronic wounds, burn wounds, diabetic ulcerous wounds, etc. Persistent infection in the wound area leads to prolonged or delayed healing. The infections might be caused by gram-positive bacteria, gram-negative bacteria, or even polymicrobial populations, including fungal agents. The main aim of wound healing applications is to deplete and destroy the underlying infectious microbes and promote or initiate the healing of the wound. CS hydrogels turn out to be excellent prospects for wound healing, due to their inherent antibacterial, biocompatible and moisture-retaining capacity in the wound site [4, 12].

A study revealed the effectiveness of an injectable, CS, and Platelet Rich Plasma (PRP) composite hydrogel, into which tigecycline (tg) nanoparticles were entrapped. This hydrogel system (tg-CSNPs-CS-PRP) was developed mainly to treat gram--positive bacteria, such as *Staphylococcus aureus*. The hydrogel was found to be biocompatible, hemocompatible, and injectable and demonstrated a sustained release of tg nanoparticles from the gel system. The addition of PRP to the gel, in turn, leads to higher cell proliferation and migration. In vitro and Ex vivo antibacterial studies proved the efficacy of tg-CSNPs-CS-PRP gel system, which showed an enhanced antibacterial activity against *S. aureus*. The control CS-PRP gel failed to show anti-staphylococcal activity, indicating that PRP did not contribute to the antibacterial activity; instead, it was solely due to sustained release of tg. In vivo Fly survival assay of this gel system was conducted in *Drosophila melanogaster*, by infecting SApCgfp via intrathoracic infection, and were provided with CS-PRP, tg--CS-PRP, and tg-CSNPs-CS-PRP gels (Fig. 9A). The gels systems containing tg demonstrated increased survivability of the infected flies, indicating an enhanced *S. aureus* inhibition. The tg-CS-PRP (~90%) displayed improved survivability of flies than the tg-CSNPs-CS-PRP system (~80%), due to the higher level of drug release, which might be accounted for by the burst release of tg from the tg-CS-PRP gel system (Fig. 9B). This indicates that the final system provides a controlled and sustained release of tg, indicating its applications for *S. aureus* related infections. Figure 9C represents the bacterial load quantification within the flies, which indicates the lowest bacterial presence in tg-containing gel systems. Fluorescence microscopy images were captured to highlight the bacterial contents (green) within the flies (Fig. 9C). The tg-containing gel systems, including the tg-CSNPs-CS-PRP system, show decreased levels of fluorescence when compared to the infected control. Hence, the CS-PRP gel system loaded with tigecycline nanoparticles serves to be an effective prospect for both the treatment of gram-positive *S. aureus* related infections and promotes healing of the wound [12].

The occurrence of multi-drug-resistant (MDR) infections is rising day by day, and they are resistant to the conventional antibiotics and drugs available, posing a serious threat to society. A group of researchers studied the preparation and application of colistimethate sodium-CS hydrogel system (CS-CSM), for the localised treatment of multidrug-resistant gram-negative infections. Colistimethate (CSM) is a prodrug of colistin, which is used as a last resort for multi-drug resistant infections. The prepared CS-CSM gel was found to be cyto and hemocompatible, and further in vitro antibacterial and biofilm studies against ATTC strains and clinically isolated gram-negative strains such as *E. coli*, *K. pneumoniae*, and *P. aeruginosa* were conducted to validate its antibacterial and biofilm disruption efficacy. The prepared gel showed enhanced antibacterial activity against the ATTC strains, being

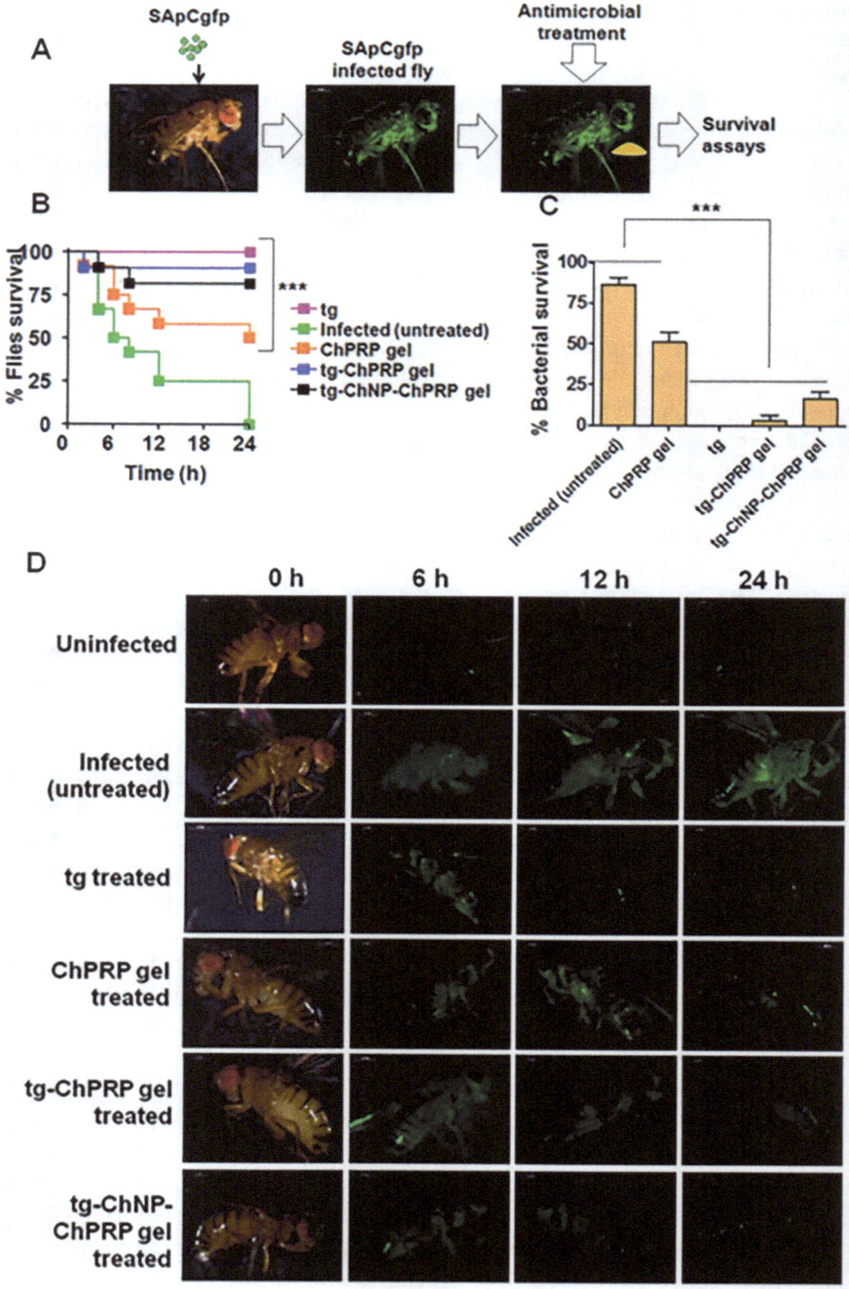

Fig. 9 In vivo antibacterial study in *D. melanogaster*. (A) *S. aureus* infected flies supplied with CS-PRP, tg-CS-PRP, and tg-CSNPs-CS-PRP gels (B) Graph depicting the survivability of the flies with respect to time (C) Quantification of bacterial load within the flies after administering CS-PRP, tg-CS-PRP, and tg-CSNPs-CS-PRP gels (*, $p < 0.05$; **, $p < 0.01$; ***, $p \leq 0.001$) and (D) Antibacterial activity of the distinct hydrogel systems captured by fluorescence microscopic imaging. Reproduced with permission from ACS [12]

most effective against *K. pneumoniae* (1.8 ± 0.08 cm) and was also found to be effective against their clinical strains and exhibited good biofilm disruption. CSM by virtue of its lipophilic and hydrophilic moieties acts by disrupting the outer membrane of the bacteria by targeting the cationic lipids in the bacterial membrane, resulting in membrane disruption, leakage of contents, and ultimately bacterial cell death [13].

The development of self-healing hydrogels generates tissues similar to those found in the body, due to their ability to recover by themselves from damage, thereby restoring their structural and functional characteristics. Mahalakshmi Pandian et al. documented a research study regarding the development and utility of a self-healing hydrogel, comprising of in situ silver nanoparticles (AgNPs) that are incorporated into *N, O*-carboxymethyl chitosan (*N, O*-CMC) hydrogel system, and the addition of ethylenediaminetetraacetic acid-ferric ion (EDTA: Fe^{3+}) complex in the hydrogel, via carboxyl-metal-ligand co-ordination bond. *N, O*-CMC is a modified derivative of CS, which imparts higher hydrophilicity and better physico--chemical properties than CS. The developed gel system was found to be injectable, adhesive, hemocompatible, biocompatible, conductive, elastic, and exhibited enhanced swelling ability. The self-healing property of the prepared gel system was tested by arranging two hydrogels side by side with a small gap between them, to monitor its self-healing capacity (Fig. 10). The final gel system exhibited improved self-healing behaviour and had a reduced healing time of 1 min. The elastic nature of the healed hydrogel was demonstrated when the hydrogel when physically dragged apart (Fig. 10). The closure of the gap created in the gel is due to the

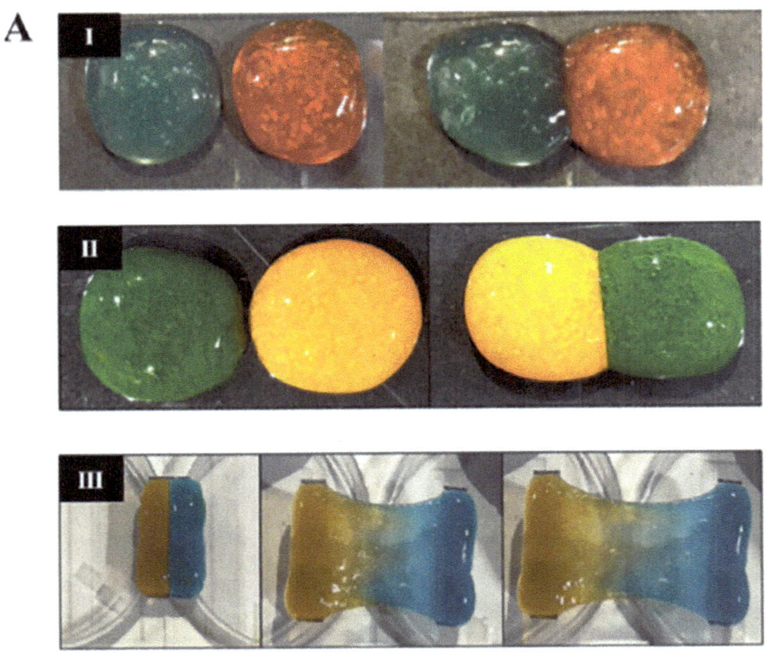

Fig. 10 Image displaying the self-healing capacity of (I) *N, O*-CMC, (II &III) *N, O*-CMC/AgNPs hydrogel. Reproduced with permission from Elsevier [14]

transport of free mass between the adjacent hydrogel, which occurs due to the interaction between the free Fe^{3+} of EDTA-Fe^{3+}complex (metal) and -COOH (ligand) of *N*, *O*-CMC. The presence of AgNPs in the hydrogel allowed for good antibacterial activity and biofilm disruption property against the ATCC and clinical isolates of *S. aureus*, *E. coli*, MRSA, *P. aeruginosa*, and *K. pneumoniae* [14].

The wide use of antibiotics has led to the development of antibiotic resistance, making most of the developed antibiotics futile against resistant infections. New research has been conducted on the efficacy of antiseptics against drug-resistant bacteria as an alternative to antibiotics, to overcome the antibiotic-associated resistance in infectious wounds. Recent research has focused on developing a CS-Poly(hexamethylene) Biguanide (PHMB) hydrogel (CS-PHMB) for the management of infectious wounds. PHMB is an antiseptic, having a wide spectrum activity against bacteria, and operates by impairing the microbial membrane by the interaction of positively charged amine groups on PHMB with negatively charged lipid membrane. It is also known to specifically bind DNA and inhibit bacterial cell division. The stable, biocompatible, and hemocompatible CS-PHMB gel showed great moisture-retaining capacity, which was validated by the water vapor permeability test. The release study of PHMB from the CS-PHMB gel showed a burst release, and excellent antibacterial efficacy against *E. coli*, *P. aeruginosa*, *K. pneumoniae*, *S. aureus*, and MRSA and clinical strains of *E. coli* and *S. aureus*. Hence, this CS-PHMB gel serves as an alternative to antibiotic-loaded gels and is effective against both gram-positive and gram-negative infections [15]. These studies demonstrate the utility of CS hydrogels loaded with antibiotics in nanoparticle formulation and antiseptics, as an excellent matrix for the treatment of infected wounds and for promoting wound healing and self-healing.

4 Application of CS Hydrogels in Tissue Engineering

4.1 Adipose Tissue

Adipose tissue, also known as fat tissue, is a type of soft tissue that stores fat inside the body. It plays a pivotal role in sourcing a large amount of stem cells, mainly adipose-derived stem cells, mesenchymal stem cells, etc. Patients usually undergo fat replacement surgeries for aesthetics or due to trauma or injury, and in some cases require adipose tissue grafts. The traditional surgical techniques involved in the reconstruction of adipose tissue inherit certain disadvantages like multiple surgical sessions, donor site complications, and a depletion of 40–55% of fat graft volume. CS hydrogel systems can be used as an alternative, to harbor stem cells and biomolecules within and avoid surgical therapy, due to its injectable nature [4, 16].

Dhanya Jaikumar et al. studied adipose tissue engineering by making use of an injectable alginate-*O*-carboxymethyl chitosan/nano-fibrin (AOF) composite hydrogel. The paper discussed the comparison between Alginate/*O*-carboxymethyl chitosan/nano (AOF) and Alginate/poly (vinyl alcohol) /nano fibrin (APF), and checked which hydrogel system served a higher performance in promoting adipogenesis. The results obtained demonstrated that the AOF composite hydrogel showed a better compressive strength, elastic moduli, swelling ratio, and degradation profile than APF and resembled the native adipose tissue. AOF scaffold was subjected to in vitro cell proliferation, viability, and attachment studies in human adipose-derived stem cells (ADSCs), and displayed good cell attachment and proliferation. The fibrin particles played a huge role in promoting cell adhesion and proliferation on the scaffold. AOF scaffold successfully allowed the differentiation of ADSCs into adipocytes, which was verified by the Oil Red O staining (Fig. 11). This AOF gel system can be an excellent choice for adipose tissue engineering [16]. Hence, composite CS hydrogel systems prove to be useful in promoting adipose tissue regeneration.

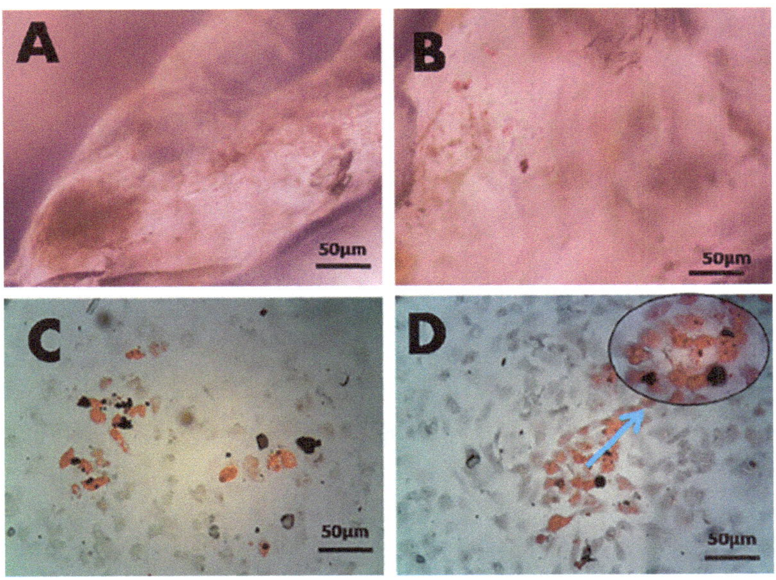

Fig. 11 Oil Red O staining of AOF hydrogel. (A) and (B) depicts the control media on the7th and 14th day, (C) and (D) indicates the adipogenic-media on the 7th and 14th day (Arrows indicate enlarged regions displaying fat lobule formation). Reproduced with permission from Elsevier [16]

4.2 Bone Tissue

Bone tissue engineering refers to an extensive field, which focuses on using biomaterials such as polymers, metals, ceramics, composite materials, and biomolecules to regenerate or restore damaged bone tissue. CS hydrogels have been considerably used for various bone regeneration applications due to their biocompatibility, biodegradability, and ability to mimic the natural ECM of bone, allowing osteoblast proliferation. The hydrogel form of CS can be useful for bone defects as the gel can be injected locally to the defect and can mould according to the defect shape [4, 17].

Research has highlighted the use of a CS hydrogel-hydroxyapatite (CS-HAp) composite membrane synthesised via a wet chemical synthesis process, for bone tissue engineering applications. HAp is a bioceramric and is the most abundantly found ceramic in the human bone, and it was being adsorbed on the hydrogel membrane by soaking the membranes in two different solutions, $CaCl_2$ (pH 7.4), and Na_2HPO_4 in an alternate manner for varying time points (Fig. 12). SEM and AFM studies proved the higher HAp deposition from nano to micro level (from 356 nm to 2.17 μm) with the higher number of immersion cycles. The CS-HAp composite hydrogel membrane was found to be biocompatible in MG-63 osteosarcoma cells, confirmed using the MTT assay [17].

Alveolar bone is an integral part of the periodontium and is being comprehensively researched in clinical dentistry due to the bone loss that occurs due to periodontitis, age, congenital conditions, etc. A study proposed the fabrication of a bilayered scaffold, comprising of poly(caprolactone)(PCL) electrospun membrane and a CS hydrogel-$CaSO_4$ scaffold that can simultaneously regenerate both the

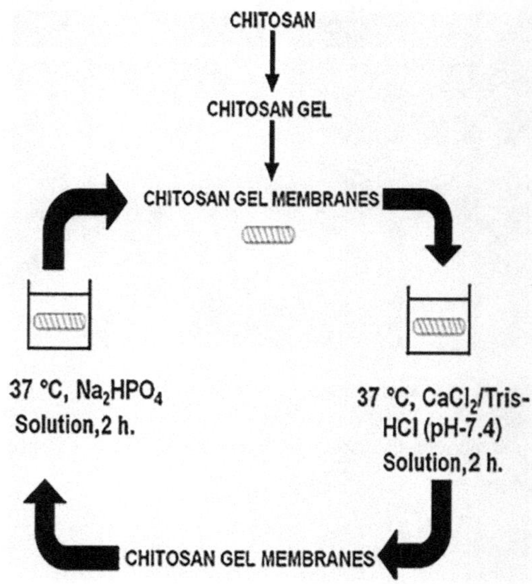

Fig. 12 Represents the soaking protocol of CS gel membrane to obtain HAp deposition. Reproduced with permission from Elsevier [17]

periodontal ligament and the alveolar bone, respectively. The bilayered construct demonstrated better protein adsorption and, porous structure when compared to the control CS hydrogel scaffold. It was found to be biocompatible with human dental follicle stem cells (hDFCs), and it exhibited impressive alkaline phosphatase activity on the 7th day compared to the controls. This indicates the osteogenic potential of the bilayered scaffold, indicating its impact in the field of dentistry [18]. These studies highlight the applicability of CS hydrogels in bone tissue engineering.

4.3 Cartilage Tissue

Cartilage tissue engineering refers to the development of biomaterials that can mimic the natural cartilage composition and enable cartilogenesis, in addition to promoting the differentiation of MSCs into chondrocytes to form new cartilage tissue. CS hydrogels, being injectable and in gel form prove to be suitable candidates for soft tissue regeneration, as the gel form resembles the natural cartilage tissue. CS, being similar to glycosaminoglycans found in cartilage, helps to mimic the cartilage composition [4, 19]. Literature reveals the development of nanostrontium ranelate incorporated injectable chitosan/alginate/fibrin nano-composite hydrogel for cartilage regeneration. Strontium ranelate is conventionally used for treating osteoporosis and is known to stimulate cartilage regeneration, and was loaded and encapsulated within CS nanoparticles with an efficiency of $40 \pm 10\%$ and $36 \pm 2\%$ respectively. The drug release% was checked for 10 days and followed a pattern of sustained release, after an initial burst release of 10%. The hydrogel was found to be biocompatible in umbilical cord-derived MSCs, and showed comparable cell proliferation to the controls and exhibited good cell attachment. The prepared gel system showed a significantly higher level of glycosaminoglycans (GAGs) than the control systems, which was verified by GAG quantification analysis. The final system also displayed higher sulphated GAG expression and collagen deposition, indicating that the gel system mimics the natural cartilage tissue. The prepared gel system also showed the highest chondrogenic potential, indicating this gel system to be a promising approach for cartilage regeneration [19].

Research regarding a cartilage-like tissue, nucleus pulposus tissue engineering, has been significantly studied due to issues such as intravertebral disk degeneration. A group of researchers presented a study on developing a CS–poly(hydroxybutyrate--co-valerate) composite hydrogel loaded with chondroitin sulfate nanoparticles, targeted for nucleus pulposus tissue engineering. The prepared gel system showed great water retaining capability, and the rheological measurements verified the properties of the gel similar to the native nucleus pulposus tissue. The hydrogel system also conferred considerable mechanical properties, able to withstand varying stress conditions. It was found to be cytocompatible and promoted good cell adhesion in adipose-derived rat mesenchymal stem cells (ADMSCs). The prepared gel system exhibited elevated levels of chondrogenic differentiation in ADMSCs,

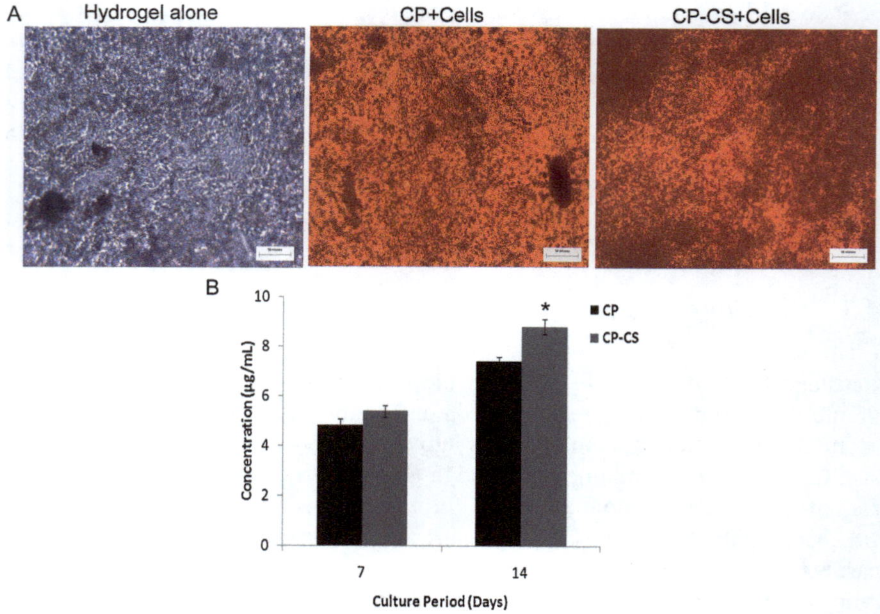

Fig. 13 Represents the chondrogenic differentiation of rat ADMSCs; (A) Safranin O staining depicting proteoglycan accumulation and (B) Sulfated GAG assay. Reproduced with permission from Elsevier [20]

when compared to the controls, which was confirmed via Safranin O staining and sulphated GAG deposition (Fig. 13), suggesting its potential in nucleus pulposus regeneration [20]. Therefore, the discussed studies point out the utility of CS hydrogels in cartilage regeneration.

4.4 Tendon and Ligament Tissue

Tendon and ligament tissue are both parts of the connective tissue system in the body, but they play diverse functions. Tendons are generally involved in connecting the muscle to the bone, whereas ligaments function as a connecting link between two bones. The ligament and tendon injuries are usually associated with athletes and the elderly, and improper healing of such injuries leads to compromised strength and other complications in the patients. Hence, various strategies have been put forward to manage and regenerate tendon and ligament tissues [4, 21].

Deepthi et al. conducted a study which focused on the development and application of a CS-hyaluronic acid composite hydrogel-coated poly(caprolactone) multiscale bilayer scaffold for promoting ligament regeneration. The composite hydrogel was coated on an electrospun PCL scaffold (CS-HA-PCL), which constituted both microfibers and random nanofibers, mimicking the structure of ligament tissue. According to mechanical studies conducted, the hydrogel-coated scaffold showed lower tensile strength and but showed protein adsorption on the coated scaffolds. The hydrogel-coated scaffold was also found to be biocompatible in rabbit ligament fibroblast cells, and showed enhanced cell attachment than the controls. It was seen that cell infiltration in the hydrogel-coated scaffold was significantly higher and promoted the elongation of the fibroblast cells, along the orientation of force in the arranged fibres, as seen in native ligament tissue. This CS-HA-PCL scaffold holds promise in the field of ligament tissue engineering [21].

There are some developments in the field of tendon tissue engineering, which include a study targeted on developing a tendon mimicking scaffold, with the help of electrospun aligned poly (l-lactic acid) (PLLA) nanofibers and depositing a CS-collagen (CS-CO) hydrogel layer within the PLLA fibers, that can replicate the natural tendon ECM and structure. To prevent the issue of peritendinous adhesion, the fabricated scaffold was rolled out and coated with alginate gel as an additional layer. The coated scaffolds exhibited lesser protein adsorption than the uncoated scaffolds, which can help to prevent the tendon tissue scar formation (peritendinous adhesion), as more protein adsorption can trigger the same. The hydrogel-coated scaffolds permitted good biocompatibility and cell proliferation in cultured tenocytes. It also exhibited good cell infiltration in the CS-CO hydrogel layers due to the porous structure of the CS-CO hydrogel layers, which was confirmed by confocal microscopy. This tendon mirroring construct imparts suitable properties that are required for tendon tissue regeneration [22].

More advancements in the field of tendon tissue regeneration took place, which led to better constructs that could mimic the tendon. One such study conducted by a group of researchers highlighted the CS hydrogel scaffold, strengthened by the introduction of aligned microfibrous bundle of twisted poly (L lactic acid) to mirror the structural characteristics of tendon (Fig. 14). The hydrogel-coated PLLA scaffold showed lower degradation rate than the uncoated scaffold, indicating the availability of the tendon scaffold to correlate with the degradation rate of natural tendon tissue. BCA assay was performed to calculate the protein adsorption, and found that the hydrogel-coated scaffold exhibited lower protein adsorption, which also correlates with the lower tenocyte attachment to the outer layer, which helps in hindering the peritendinous adhesion. But the CS hydrogel layer was found to allow the infiltration of tenocytes and support good proliferation of these cells, demonstrating the efficiency of this construct in tendon tissue engineering [23]. These papers portrayed the advantages of using CS hydrogels along with other polymers as a composite material to facilitate ligament and tendon tissue regeneration.

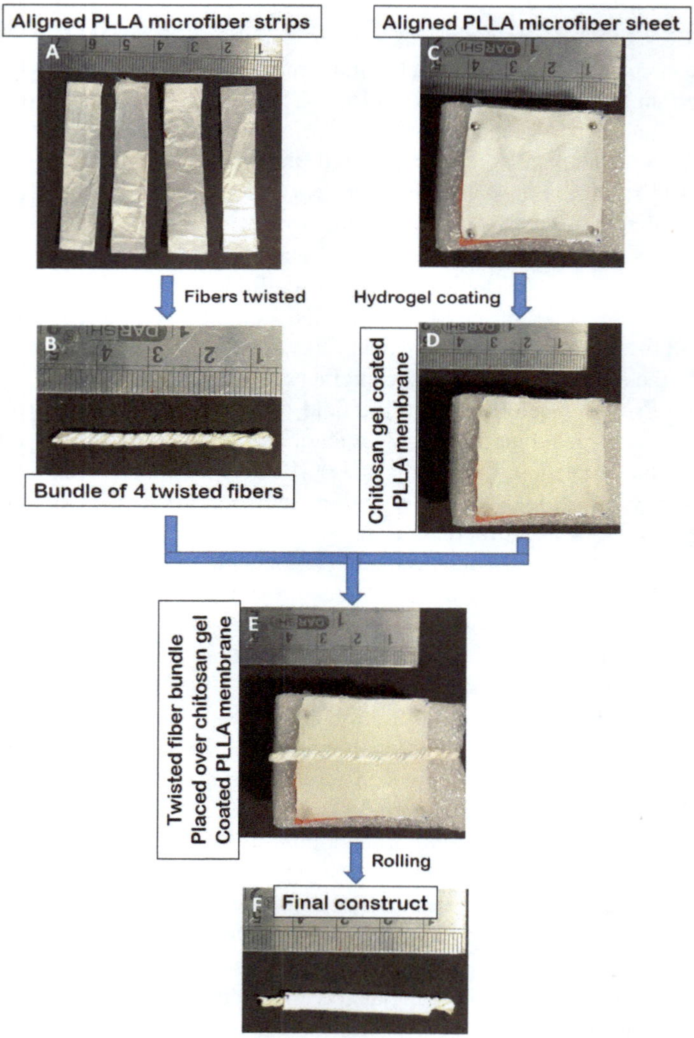

Fig. 14 Fabrication protocol of twisted PLLA composite scaffold with CS gel-coated PLLA membrane as outer covering. Reproduced with permission from Elsevier [23]

5 Conclusion

CS hydrogels have been extensively employed in the sector of tissue engineering and regenerative medicine due to their inherent characteristics such as biocompatibility, biodegradability, non-toxicity, antibacterial properties, etc. The injectable property, ease of gelation, high moisture-retaining capacity, ability to mimic the ECM of native tissues, mouldability to the defect site, in situ gelling property, etc., allow it to prosper over other physical forms of CS. The cationic nature and the presence of amine and hydroxyl groups play a crucial role in the bioactivity of CS. The ability to modify these functional groups opens a passage for versatility and allows for improving and altering the functionality of native CS to overcome the associated drawbacks. New research and developments in the field of CS hydrogels shed light on the field of stimuli-responsive hydrogels that can sense the signals from the surrounding environment and act based on the stimuli. Such 'smart hydrogels' can revolutionise the usage of CS hydrogels in tissue engineering and regenerative medicine, allowing the hydrogels to act like the native tissue based on biological cues. Many more advancements and breakthroughs are yet to come in the field of CS hydrogels to harness the true potential of CS hydrogels for tissue engineering and regenerative medicine area.

6 Summary

In this book chapter, we have briefly discussed about the applications of CS hydrogel in the fields of tissue engineering and regenerative medicine. The structure and properties of CS, the benefits of using CS hydrogels and their fabrication techniques have been mentioned. The employment of CS hydrogels as biomaterials in the sector of haemostasis, drug delivery, and wound healing has been comprehensively explored, and supporting literature has been provided. The adoption of CS hydrogels in the domain of tissue engineering has been extensively reviewed for different tissues such as adipose tissue, bone, cartilage, tendon and ligament tissues. This book chapter has outlined various research findings on using CS hydrogels for distinct fields of tissue engineering and regenerative medicine.

References

1. Dragan ES, Dinu MV (2020) Advances in porous chitosan-based composite hydrogels: synthesis and applications. React Funct Polym 146:104372
2. Ravishankar K, Dhamodharan R (2020) Advances in chitosan-based hydrogels: evolution from covalently crosslinked systems to ionotropically crosslinked superabsorbents. React Funct Polym 149:104517
3. Fu J, Yang F, Guo Z (2018) The chitosan hydrogels: from structure to function. New J Chem 42:17162–17180
4. Padmanabhan A, Nair LS (2016) Chitosan hydrogels for regenerative engineering. In: Regenerative engineering of musculoskeletal tissues and interfaces, pp 3–40
5. Ashok N, Pradeep A, Jayakumar R (2021) Synthesis-structure relationship of chitosan based hydrogels. Adv Polym Sci 287:105–129
6. Sundaram MN, Amirthalingam S, Mony U, Varma PK, Jayakumar R (2019) Injectable chitosan-nano bioglass composite hemostatic hydrogel for effective bleeding control. Int J Biol Macromol 129:936–943

7. Muthiah Pillai NS, Eswar K, Amirthalingam S, Mony U, Kerala Varma P, Jayakumar R (2019) Injectable nano whitlockite incorporated chitosan hydrogel for effective hemostasis. ACS Appl Bio Mater 2:865–873
8. Sundaram MN, Mony U, Varma PK, Jayakumar R (2021) Vasoconstrictor and coagulation activator entrapped chitosan based composite hydrogel for rapid bleeding control. Carbohydr Polym 258:117693
9. Joju JM, Sunil S, Visakhan M, Jayakumar R (2024) Applications of chitosan hydrogel in drug delivery. In: Chitosan in drug delivery, pp 213–233
10. Mohandas A, Rangasamy J (2021) Nanocurcumin and arginine entrapped injectable chitosan hydrogel for restoration of hypoxia induced endothelial dysfunction. Int J Biol Macromol 166:471–482
11. Vignesh S, Sivashanmugam A, Annapoorna M, Janarthanan R, Subramania I, Shantikumar VN et al (2018) Injectable deferoxamine nanoparticles loaded chitosan-hyaluronic acid coacervate hydrogel for therapeutic angiogenesis. Colloids Surf B Biointerfaces 161:129–138
12. Nimal TR, Baranwal G, Bavya MC, Biswas R, Jayakumar R (2016) Anti-staphylococcal activity of injectable nano tigecycline/chitosan-PRP composite hydrogel using Drosophila melanogaster model for infectious wounds. ACS Appl Mater Interfaces 8:22074–22083
13. Pradeep A, Ashok N, Priya V, Pillai AV, Menon RR, Kumar VA et al (2022) Colistimethate sodium-chitosan hydrogel for treating gram-negative bacterial wound infections. Int J Biol Macromol 214:610–616
14. Pandian M, Selvaprithviraj V, Pradeep A, Rangasamy J (2021) In-situ silver nanoparticles incorporated N,O-carboxymethyl chitosan based adhesive, self-healing, conductive, antibacterial and anti-biofilm hydrogel. Int J Biol Macromol 188:501–511
15. Antony IR, Pradeep A, Pillai AV, Menon RR, Kumar VA, Jayakumar R (2023) Antiseptic chitosan-poly(hexamethylene) biguanide hydrogel for the treatment of infectious wounds. J Function Biomater 14:528
16. Jaikumar D, Sajesh KM, Soumya S, Nimal TR, Chennazhi KP, Nair SV et al (2015) Injectable alginate-O-carboxymethyl chitosan/nano fibrin composite hydrogels for adipose tissue engineering. Int J Biol Macromol 74:318–326
17. Madhumathi K, Shalumon KT, Rani VVD, Tamura H, Furuike T, Selvamurugan N et al (2009) Wet chemical synthesis of chitosan hydrogel-hydroxyapatite composite membranes for tissue engineering applications. Int J Biol Macromol 45:12–15
18. Sundaram MN, Sowmya S, Deepthi S, Bumgardener JD, Jayakumar R (2016) Bilayered construct for simultaneous regeneration of alveolar bone and periodontal ligament. J Biomed Mater Res B Appl Biomater 104:761–770
19. Deepthi S, Abdul Gafoor AA, Sivashanmugam A, Nair SV, Jayakumar R (2016) Nanostrontium ranelate incorporated injectable hydrogel enhanced matrix production supporting chondrogenesis: in vitro. J Mater Chem B 4:4092–4103
20. Nair MB, Baranwal G, Vijayan P, Keyan KS, Jayakumar R (2015) Composite hydrogel of chitosan-poly(hydroxybutyrate-co-valerate) with chondroitin sulfate nanoparticles for nucleus pulposus tissue engineering. Coll Surf B Biointerf 136:84–92
21. Deepthi S, Nivedhitha Sundaram M, Deepti Kadavan J, Chennazhi KP, Jayakumar R (2015) Chitosan-hyaluronic acid hydrogel coated poly(caprolactone) multiscale bilayer scaffold for ligament regeneration. Chem Eng J 260:478–485
22. Deepthi S, Nivedhitha Sundaram M, Deepti Kadavan J, Jayakumar R (2016) Layered chitosan--collagen hydrogel/aligned PLLA nanofiber construct for flexor tendon regeneration. Carbohydr Polym 153:492–500
23. Nivedhitha Sundaram M, Deepthi S, Mony U, Shalumon KT, Chen JP, Jayakumar R (2019) Chitosan hydrogel scaffold reinforced with twisted poly(L lactic acid) aligned microfibrous bundle to mimic tendon extracellular matrix. Int J Biol Macromol 122:37–44

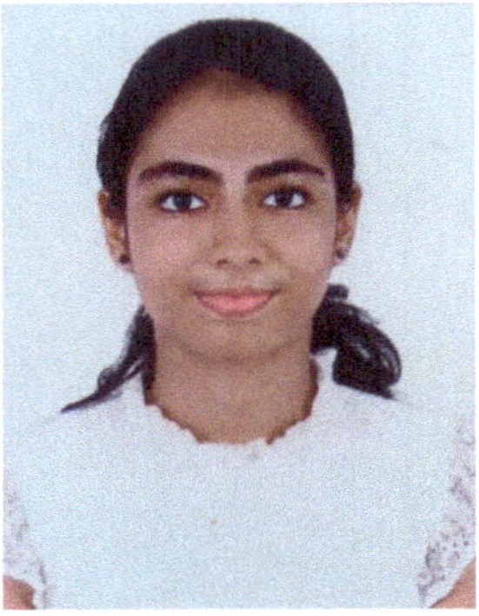

Sona Sunil

Jayakumar Rangasamy

Chitosan in Periodontal Healthcare

Wanpen Tachaboonyakiat

Abstract Chitosan, a natural biopolymer derived from chitin, has emerged as a versatile biomaterial with significant applications in periodontal healthcare. This chapter provides a comprehensive overview of unique properties of chitosan--including biocompatibility, biodegradability, mucoadhesiveness, and inherent antimicrobial activity—that make it particularly suitable for oral and periodontal applications. The chapter begins with a detailed examination of periodontal anatomy, including the structural components of the periodontium, followed by an exploration of periodontal pathology encompassing gingivitis and periodontitis, as well as periodontal therapy approaches. The antimicrobial properties of chitosan are discussed in relation to oral healthcare, specifically addressing its efficacy against dental caries and periodontal pathogens. Both non-surgical and surgical treatment using chitosan in the forms of mouthwash, local drug delivery systems, periodontal dressings, and guided tissue regeneration (GTR) membranes are described in this chapter.

Keywords Chitosan · Periodontal · Healthcare

1 Introduction

1.1 Chitosan and Its Properties [1]

Chitosan, a naturally derived biopolymer obtained through the deacetylation of chitin, has emerged as a promising biomaterial in periodontal healthcare applications. Chitin, the second most abundant polysaccharide in nature after cellulose, is primarily sourced from crustacean shells, fungal cell walls, and insect exoskeletons. When

W. Tachaboonyakiat (✉)
Department of Materials Science, Faculty of Science, Chulalongkorn University, Bangkok, Thailand
e-mail: wanpen.ta@chula.ac.th

S.-L. Wang, M. Anpo (eds.), *Chitin and Chitosan*, Springer Series on Polymer and Composite Materials, https://doi.org/10.1007/978-981-95-7457-5_10

chitin undergoes deacetylation-typically to a degree exceeding 50%, it transforms into chitosan, a cationic polysaccharide composed of randomly distributed β-(1,4)-linked D-glucosamine and *N*-acetyl-D-glucosamine units. Chitosan's molecular structure contains primary amino group at C-2 position, primary hydroxyl group at C-6 position and secondary hydroxyl group at C-3 position that provide the key sides for chemical modification into diverse derivatives suitable for biomedical applications. The chemical structure of chitn/chitosan and its derivatives mentioned in this chapter are shown in Fig. 1. At pH below pKa (pKa ~ 6.5), amino groups become protonated ($-NH_3^+$), conferring the polycationic character, exhibiting antimicrobial and mucoadhesive properties.

Chitosan possesses several unique physicochemical and biological properties that make it particularly valuable for periodontal applications as follows:

Biocompatibility and biodegradability: Chitosan exhibits excellent biocompatibility with oral tissues and degrades naturally through enzymatic hydrolysis by lysozyme and other enzymes present in the oral cavity. Its degradation products are non-toxic and can be readily absorbed or eliminated by the body, making it safe for long-term periodontal applications.

Antimicrobial activity: The polycationic nature of chitosan, arising from the protonated amino groups at acidic pH, enables it to interact electrostatically with negatively charged bacterial cell membranes. This interaction disrupts membrane integrity and inhibits the growth of various periodontal pathogens, including *Porphyromonas gingivalis* and *Aggregatibacter actinomycetemcomitans*.

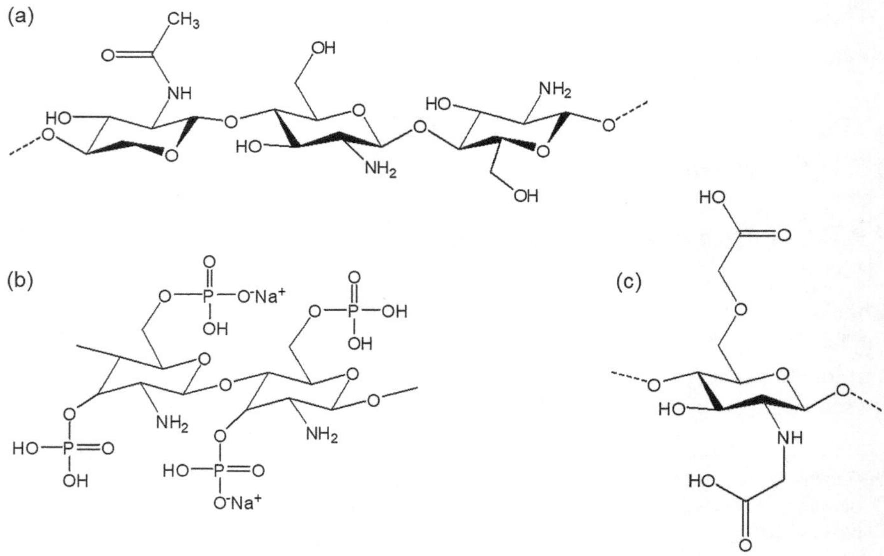

Fig. 1 Chemical structure of (**a**) chitin/chitosan and (**b**) phosphorylated chitosan, (**c**) *N,O*--carboxylmethyl chitosan

Mucoadhesive properties: Chitosan demonstrates strong adhesion to mucosal surfaces through ionic interactions between its positively charged amino groups and negatively charged residues on mucin. This mucoadhesive property allows for prolonged retention at periodontal sites, extending the duration of therapeutic action.

Hemostatic properties: Chitosan promotes blood clotting by activating platelets and providing a scaffold for clot formation. This property is particularly beneficial in periodontal surgery and during scaling and root planning procedures, where bleeding control is essential.

Wound healing enhancement: Chitosan accelerates tissue regeneration by stimulating fibroblast proliferation, promoting collagen deposition, and modulating inflammatory responses. It also enhances angiogenesis and supports the migration of cells involved in periodontal tissue repair.

1.2 Purpose in Periodontal Healthcare

The integration of chitosan into periodontal healthcare serves multiple therapeutic objectives. Periodontal diseases, including gingivitis and periodontitis, affect the supporting structures of teeth and are primarily caused by bacterial biofilm accumulation and subsequent inflammatory host responses. Traditional periodontal treatments often face challenges such as inadequate drug delivery to periodontal pockets, rapid clearance of therapeutic agents, and difficulty in maintaining sustained antimicrobial concentrations at disease sites.

Chitosan addresses these limitations through various applications: as a component of local drug delivery systems for sustained release of antimicrobials and anti-inflammatory agents into periodontal pockets; as a scaffold material for guided tissue regeneration membranes that prevent epithelial downgrowth while supporting periodontal ligament and bone regeneration; as an ingredient in oral care products including toothpastes, mouthwashes, and gels for daily plaque control.

Furthermore, chitosan's ability to chelate metal ions and scavenge free radicals contributes to reducing oxidative stress in inflamed periodontal tissues [2], while its capacity to modulate immune responses helps control excessive inflammation that leads to tissue destruction. The combination of antimicrobial, anti-inflammatory, regenerative, and drug delivery properties positions chitosan as a multifunctional biomaterial capable of addressing the complex pathophysiology of periodontal diseases while promoting tissue healing and restoration of periodontal health.

2 Periodontal Healthcare

2.1 Periodontium [3]

Periodontal health is fundamental to maintaining functional dentition and overall oral wellbeing. The periodontium comprises the tissues that surround and support the teeth, including the gingiva, periodontal ligament (PDL), cementum, and alveolar bone (Fig. 2).

The gingiva consists of stratified squamous epithelium supported by underlying connective tissue. Three distinct epithelial zones exist: the oral epithelium (keratinized), sulcular epithelium (non-keratinized), and junctional epithelium. The junctional epithelium is particularly important as it forms the biological attachment between the gingival tissue and the tooth surface through hemidesmosomes and basal lamina.

Healthy gingival tissue exhibits specific characteristics: coral pink coloration, firm consistency, stippled surface texture (resembling orange peel), and knife-edge margins that fit snugly around teeth. The gingival sulcus, a shallow space between the teeth and gingiva, measures 1–3 mm in health.

The lamina propria (connective tissue layer) contains primarily type I collagen fibers arranged in functional groups that provide structural support. These include dentogingival, dentoperiosteal, circular, and transseptal fiber groups. The connective tissue also contains blood vessels, nerves, and cellular components including fibroblasts, which synthesize and maintain the extracellular matrix.

The PDL comprises specialized connective tissue fibers, predominantly type I collagen (Sharpey's fibers) that anchor the tooth root to the alveolar bone.

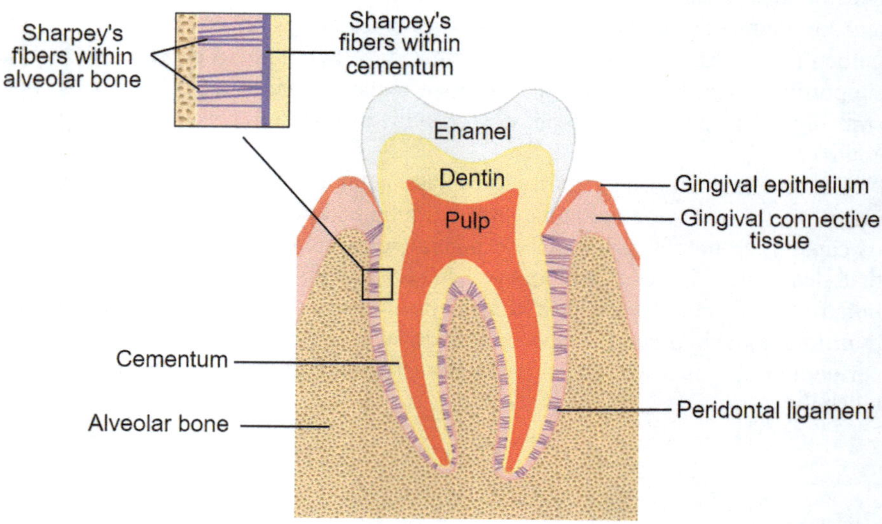

Fig. 2 Periodontium components

Sharpey's fibers are the anchoring portions of the principal collagen fiber bundles of the periodontal ligament that penetrate and embed into the mineralized tissues on either side of the periodontal space. One end of Sharpey's fibers is embedded in the cementum, the calcified tissue covering the tooth root. The opposite end of Sharpey's fibers is embedded in the alveolar bone. Between these two anchoring points—cementum and alveolar bone—lies the unmineralized middle portion of the periodontal ligament fibers. The mineralized Sharpey's fiber portions provide firm anchorage at both ends, while the unmineralized middle section allows for slight tooth movement and acts as a shock absorber during mastication. The cellular components include fibroblasts, osteoblasts, cementoblasts, and undifferentiated mesenchymal cells that enable tissue remodeling and repairing.

Cementum is a calcified tissue covering the tooth root, providing attachment for periodontal ligament fibers. Unlike bone, cementum is avascular and does not undergo continuous remodeling under normal conditions. Alveolar bone forms the tooth socket and depends on the presence of teeth for its development and maintenance. Bone resorption is a sign of periodontitis.

2.2 *Gingivitis vs Periodontitis [4]*

Initial bacterial accumulation at the gingival margin elicits inflammatory changes that manifest clinically as red, swollen, bleeding gingival tissues—the signs of gingivitis. This early stage, while nearly ubiquitous in populations with inadequate oral hygiene, remains fully reversible through plaque removal and resolution of inflammation. However, sustained bacterial presence, particularly by pathogenic species, combined with host susceptibility factors, can trigger progression to periodontitis. This advanced stage involves extension of inflammation into deeper tissues, destruction of connective tissue attachment, and irreversible loss of alveolar bone support.

Gingivitis is a reversible inflammation of gingival tissues without attachment loss or bone destruction. Periodontitis is an irreversible destruction of periodontal supporting structures characterized by clinical attachment loss, radiographic bone loss, and pocket formation.

Understanding this pathogenic sequence provides the scientific foundation for preventive and therapeutic strategies. Early detection and treatment of gingivitis, combined with modification of risk factors and maintenance of effective oral hygiene, can prevent the transition to destructive periodontal disease. When periodontitis develops, timely treatment can stop progression and preserve remaining tooth support, thereby maintaining oral function and contributing to overall health. The recognition that periodontal disease exists along a continuum from reversible inflammation to irreversible destruction emphasizes the critical importance of early treatment and sustained preventive care throughout the lifespan.

2.3 *Periodontal Therapy [5]*

Periodontal disease represents a chronic inflammatory condition affecting the supporting structures of teeth, driven primarily by bacterial biofilm accumulation and subsequent host immune response. Therapeutic processes have evolved significantly, emphasizing a sequential approach that prioritizes non-surgical treatments before progressing to surgical treatments when indicated.

2.3.1 Non-surgical Treatments

Non-surgical therapy constitutes the foundation of periodontal treatment and remains effective for mild to moderate disease. The cornerstone procedure is scaling and root planing (SRP), which involves mechanical removal of subgingival plaque, calculus, and contaminated cementum. This treatment disrupts bacterial biofilms and creates a biologically compatible root surface conducive to periodontal healing. Modern non-surgical approaches increasingly incorporate adjunctive therapies. Locally delivered antimicrobials, such as chlorhexidine or doxycycline gels, provide sustained drug release into periodontal pockets, enhancing bacterial suppression. The antimicrobial properties of chitosan relevant to oral healthcare in terms of mouth wash as well as local drug delivery were described in Sect. 3.

2.3.2 Surgical Treatments

When non-surgical therapy fails to achieve adequate pocket reduction or anatomical defects prevent proper maintenance, surgical therapy becomes necessary. Surgical approaches offer direct access for debridement, tissue modification, and regenerative procedures. Regenerative surgery represents an alternative way toward restoring lost periodontal attachment apparatus. Guided tissue regeneration (GTR) employs barrier membranes to exclude epithelial cells while promoting repopulation by periodontal ligament cells, cementoblasts, and osteoblasts. Chitosan relevant to mucogingival surgery and regenerative surgery in the form of periodontal dressing and GTR membrane were described in Sect. 4.

3 Antimicrobial Properties Relevant to Oral Healthcare

Chitosan exhibits potent antimicrobial activity against a wide range of microorganisms, including bacteria, fungi, and certain viruses [6]. The antimicrobial potency of chitosan is influenced by several factors, including its molecular weight, degree of deacetylation, pH of the environment, and concentration. The antibacterial mechanism of chitosan is a complex phenomenon that varies between Gram-positive and

Gram-negative bacteria, primarily due to distinct structural differences in their cell wall architecture and surface properties.

The principal mechanisms govern chitosan's antibacterial activity, dependent on molecular weight. High molecular weight chitosan functions by forming an impermeable polymeric layer on the bacterial cell surface. This external coating acts as a physical barrier that blocks the entry of essential nutrients into the cell, effectively starving the bacteria and inhibiting growth. While low molecular weight chitosan operates through a penetrative mechanism, wherein the smaller chitosan molecules cross the cell membrane via electrostatic interactions with negatively charged cell surface components. Once inside, these molecules interfere with intracellular metabolic processes, disrupting normal cellular functions and leading to bacterial death.

The relationship between molecular weight and antimicrobial efficacy differs markedly depending on the bacterial target. For Gram-positive bacteria, increasing molecular weight enhances antibacterial effectiveness. The larger chitosan polymers are particularly adept at forming encapsulating films around Gram-positive bacterial cells, creating a barrier that prevents nutrient transport and ultimately inhibits bacterial survival. Conversely, for Gram-negative bacteria, reducing the molecular weight of chitosan improves antimicrobial activity. This enhanced effectiveness can be attributed to the unique surface characteristics of Gram-negative organisms, which possess cell walls with greater hydrophilicity and more abundant negatively charged surface groups, particularly lipopolysaccharides in the outer membrane. The positively charged chitosan molecules experience stronger electrostatic attraction to these negatively charged Gram-negative surfaces, facilitating more efficient adhesion and interaction. The smaller size of low molecular weight chitosan enables better penetration through the complex outer membrane structure characteristic of Gram-negative bacteria, allowing the polymer to reach and disrupt critical intracellular targets more effectively [6].

The objective of utilizing chitosan in oral healthcare is to hold its antimicrobial properties through various formulations, encompassing both oral hygiene products for routine use and local drug delivery systems for therapeutic applications in periodontal treatment.

3.1 Oral Hygiene Products

Periodontitis demonstrates a well-established correlation with bacterial plaque accumulation. The antimicrobial properties of chitosan offer significant potential for preventing infectious complications in the periodontal environment. It has been used in various forms as an antimicrobial agent in oral care. Chitosan has been incorporated into diverse formulations as an antimicrobial agent for oral healthcare applications. When formulated as gels and hydrogels, chitosan serves as an effective antimicrobial component in various oral hygiene products including toothpastes, oral rinses, and chewing gums, demonstrating the ability to reduce microbial populations within the oral environment. Research has shown that chitosan possesses substantial antibacterial activity against common oral bacteria, notably *Streptococcus mutans* (*S. mutans*) and *Streptococcus sobrinus*, while also effectively preventing the development of bacterial biofilms. Chitosan also exhibits antiplaque activity against several oral pathogens including *Porphyromonas gingivalis* and *Aggregatibacter actinomycetemcomitans* [7].

Dental caries and periodontal diseases represent the most prevalent infectious conditions affecting the oral cavity in humans, both of which are closely associated with dental plaque accumulation.

A chitosan mouthwash was found to be effective in reducing both plaque formation and levels of salivary *Streptococcus* and *Lactobacillus* in an in vivo study. The mouthwash showed a strong inhibitory effect on these bacteria and can be considered a natural alternative for oral hygiene, controlling plaque biofilms without negatively affecting other oral microorganisms [8]. Different molecular weights of chitosan mouthwash formulations have been investigated against *S. mutans*. Chitosan mouthwash with lower molecular weight (30 kDa) demonstrated more effective reduction of established *S. mutans* biofilms compared to its higher molecular weight chitosan (890 kDa). The enhanced performance of low molecular weight chitosan can be attributed to its smaller molecular size, which likely provides improved mobility and facilitates better interaction with bacterial cell walls relative to larger molecular structures. Furthermore, when compared to chlorhexidine mouthwash, chitosan formulations exhibited superior antibiofilm properties [9]. In addition, the effectiveness of a novel mouthwash formulation containing chlorhexidine (0.1%) combined with bioadhesive chitosan (0.5%) in reducing dental plaque bacteria, comparing its efficacy to that of chlorhexidine or chitosan used individually has been investigated. This innovative mouthwash formulation combining chlorhexidine and chitosan demonstrates superior efficacy in controlling *Streptococcus sanguinis* oral biofilm formation compared to the use of either antimicrobial agent alone [10].

Phosphorylated chitosan (PCTS), a water-soluble derivative, is synthesized through phosphorylation of chitosan [11]. It has demonstrated significant potential in combating dental caries through its remineralization. This modified biopolymer exhibits therapeutic effects on both enamel and dentin, the primary mineralized tissues of teeth. Research has extensively documented the bioremineralization capacity of phosphorylated chitosan on dental hard tissues. When formulated as nanoparticles combined with amorphous calcium phosphate, phosphorylated chitosan promotes enamel remineralization with a structural pattern closely resembling that of healthy, intact enamel. Notably, this remineralization pattern differs distinctly from the crystalline structure produced by conventional fluoride treatments, suggesting an alternative biomimetic mechanism of action [12]. In dentin tissue, phosphorylated chitosan demonstrates the ability to modify demineralized collagen matrices, creating a favorable environment for the precipitation and deposition of calcium and phosphate ions. This process facilitates the restoration of mineral content in compromised dentin structures [13]. The dual action of phosphorylated chitosan on both enamel and dentin is as a promising biomaterial for comprehensive dental caries management, offering a biocompatible approach to enhance remineralization and potentially reduce tooth decay progression through mechanisms that complement or provide alternatives to traditional fluoride-based therapies.

3.2 Local Drug Delivery

Periodontal disease management has traditionally relied on mechanical debridement procedures such as scaling, root planing, and soft tissue curettage. However, these conventional approaches sometimes yield limited success in achieving complete bacterial elimination and disease resolution. This challenge has prompted researchers to explore alternative therapeutic strategies, including systemic antibiotic administration for more comprehensive bacterial control.

Despite their potential benefits, systemically administered drugs present several limitations, including suboptimal bioavailability at the site of infection, systemic side effects, and the development of antibiotic resistance [14]. These concerns have driven the development of local drug delivery systems as a more targeted approach to periodontal therapy. Local delivery offers distinct advantages by positioning therapeutic agents directly within the periodontal pocket, where gingival crevicular fluid serves as a natural reservoir, allowing for sustained drug release and enhanced local bioavailability while minimizing systemic exposure.

Chitosan has emerged as a promising biomaterial for local drug delivery applications due to its biocompatibility, biodegradability, mucoadhesive properties, and inherent antimicrobial activity. Historically utilized as a carrier system for various pharmaceutical agents, its unique characteristics make it particularly well-suited for periodontal applications. Recent studies have demonstrated that a 1% w/w resorbable chitosan membrane can function effectively as a local drug delivery system when used as adjunctive periodontal therapy in the non-surgical management of chronic periodontitis [15].

Recent advances in drug delivery have focused on developing smart systems that respond to specific microenvironmental triggers, enabling targeted therapeutic release with reduced adverse effects. The periodontal pocket presents an ideal site for such targeted delivery strategies due to its unique biochemical environment. During periodontitis progression, various enzymes are secreted into the gingival crevicular fluid and saliva by host cells and pathogenic bacteria residing in periodontal pockets. Alkaline phosphatase (ALP) has emerged as a particularly significant enzyme in this context, serving both as a reliable diagnostic biomarker for disease progression and as a prognostic indicator for periodontitis outcomes. Notably, ALP concentrations increase substantially in crevicular fluid during active periodontal disease and possess the catalytic ability to accelerate the degradation of polyphosphoester compounds. Concerning with this enzyme-rich microenvironment, an innovative enzyme-responsive chitosan membrane incorporating polyphosphoester and minocycline hydrochloride has been developed. This bioresponsive delivery system is specifically designed for targeted antibiotic release within infected periodontal pockets while simultaneously supporting alveolar bone regeneration. The formulation containing 0.6% (w/v) minocycline hydrochloride achieves sustained drug release and demonstrates potent antibacterial efficacy while preserving excellent biocompatibility. This enzyme-mediated approach represents a promising strategy for localized antibiotic delivery in periodontal therapy, offering site-specific treatment that responds dynamically to the pathological microenvironment of diseased periodontal tissues [16].

Effective local drug management in periodontitis treatment faces multiple obstacles, including the limitations of systemic drug administration, the constant flow dynamics of gingival crevicular fluid, and the complex morphology of periodontal pockets. The anatomical challenges posed by deep and irregularly shaped periodontal pockets have generated significant interest in injectable thermo-sensitive chitosan-based hydrogels as innovative drug delivery platforms. This smart hydrogel system is formulated by combining chitosan with β-glycerophosphate (β-GP), creating a solution that undergoes sol-gel transition through electrostatic interactions when exposed to physiological body temperature. This unique temperature-responsive behavior enables chitosan/β-GP hydrogels to remain liquid at room temperature for easy injection, then solidify in situ upon contact with body temperature within the periodontal pocket. Once gelled, the system provides controlled release of incorporated therapeutic agents directly at the disease site. The injectable nature of chitosan/β-GP hydrogels makes them particularly suitable for accessing and conforming to the irregular geometry of periodontal pockets, ensuring comprehensive coverage and sustained drug delivery [17].

4 Periodontal Applications

Due to its chelating properties and versatility in fabrication into various forms including sponges, scaffolds, and membranes, chitosan has been extensively applied in periodontal dressing and guided tissue regeneration (GTR) as described below.

4.1 Periodontal Dressing

Periodontal dressing serves as a protective covering applied to gingival and periodontal tissues following periodontal surgery. The primary objective of periodontal surgery is to eliminate periodontal pockets and restore gingival architecture that enables effective patient-controlled plaque management. Applied to the necks of teeth and adjacent tissues, periodontal dressings protect the surgical wound during the critical healing phase. While these dressings possess no inherent curative properties, they facilitate healing by providing physical protection rather than delivering healing factors. Specifically, periodontal dressing minimizes the risk of postoperative infection and hemorrhage, control bleeding, and prevents surface trauma during mastication, thereby creating an optimal environment for tissue repair.

Optimal periodontal dressing materials must meet several essential requirements: the materials should be non-toxic and non-irritating to tissues while minimizing patient discomfort. Periodontal dressings are classified into three categories: hard packs, soft packs, and light-curable gels. Soft packs are the most used products in clinical practice. These materials are prepared by mixing equal lengths of paste from an accelerator tube and a base tube until a uniform color is achieved, indicating homogeneous mixing.

Traditional periodontal dressings consist of three main components: organic carboxylic acids, metal oxide, and a polymeric resin containing carboxylic groups. Solidification or setting of the dressing occurs through a reaction between the carboxylic acids and zinc oxide (ZnO), which proceed rapidly. The solidification time can be shortened by adjusting the accelerator amount (ZnO). However, to prolong the setting time and achieve a clinically manageable working period, polymeric resins containing carboxylic groups are incorporated as reaction rate-modifying agents. These polymers function effectively due to their macromolecular structure and the availability of functional groups that form complexes with ZnO, thereby slowing the setting reaction.

Despite their functional effectiveness, conventional polymeric resins exhibit moderate to severe cytotoxicity toward human gingival fibroblasts (HGFs), rendering them unsuitable components for periodontal dressings that remain in direct contact with healing tissues. This cytotoxicity concern has prompted research efforts to identify safer alternative materials that can replace these toxic polymers while maintaining the necessary rate-modifying properties.

Phosphorylated chitosan (PCTS) was investigated as a promising alternative to conventional toxic polymeric resins in periodontal dressing formulations [18]. This research focuses on the complex formation between PCTS and ZnO, which is central to the setting mechanism. ZnO exhibits hydroxyl species at its surface, formed through the dissociative chemisorption of water molecules. PCTS forms metal complexes with ZnO through multiple interaction mechanisms: the nitrogen atoms of amino groups and the oxygen atoms in hydroxyl or phosphate groups donate lone pairs of electrons to interact with the protonated hydroxyl species on the ZnO surface. This complexation effectively modulates the setting reaction, providing the necessary working time for clinical application.

Besides, direct contact testing with primary HGFs demonstrated that PCTS exhibits excellent biocompatibility with HGFs. Significantly, PCTS not only proved non-toxic itself but also reduced the cytotoxicity of ZnO through complexation in a dose-dependent manner. This dual benefit—maintaining functional effectiveness while enhancing biological safety—positions PCTS as an ideal candidate to replace the toxic polymeric resins currently employed as reaction rate-modifying agents in periodontal dressings.

4.2 Guided Tissue Regeneration (GTR) Membrane

In GTR procedures, chitosan functions as a barrier membrane to prevent epithelial downgrowth and allow selective repopulation of the periodontal defect by cells from the periodontal ligament and alveolar bone. The barrier membrane is placed at the interface between the soft tissue and the periodontal defect, serving to impede the infiltration of soft tissue while creating an undisturbed space for periodontal regeneration. The successful designs of barrier membranes in GTR depend on multiple factors from the inherent material properties and biological characteristics. An ideal GTR membrane must satisfy several critical requirements that can be categorized into biological compatibility, mechanical functionality, barrier properties, and degradation behavior [19, 20].

1. Biological Compatibility and Host Response: Biocompatibility represents a fundamental prerequisite for any GTR membrane, as the material must integrate harmoniously with the host tissue without triggering adverse immune responses or chronic inflammation that could compromise the regenerative process. The membrane should elicit minimal foreign body reaction while promoting favorable tissue-material interactions. Complementing this requirement, the membrane must possess inherent antibacterial properties to prevent microbial colonization and subsequent infection at the surgical site, which remains one of the primary causes of GTR failure. The ability to resist bacterial adhesion and proliferation is particularly crucial given the oral cavity's polymicrobial environment and the proximity of periodontal defects to the sulcular region.
2. Mechanical Properties and Structural Integrity: The mechanical characteristics of GTR membranes directly influence their clinical performance and handling properties. The membrane must exhibit sufficient mechanical strength and stiffness to withstand tissue forces and maintain the space necessary for bone ingrowth, thereby preventing collapse into the defect site that would impede regeneration. This space-making function is essential for allowing selective repopulation by osteogenic cells. Additionally, adequate tear strength is imperative to facilitate surgical manipulation and enable secure fixation through suturing without material rupture during implantation. The membrane must also demonstrate tissue integration capability, achieving stable attachment to surrounding tissues to ensure wound stabilization and prevent micro-movements that could disrupt the healing cascade. This mechanical stability must be sustained throughout the critical initial healing phase, typically 4–6 weeks for periodontal tissue regeneration, during which the membrane guides tissue growth and maintains its barrier function before undergoing controlled degradation.
3. Barrier Function and Selective Permeability: A defining characteristic of GTR membranes is their ability to function as selective barriers that guide tissue regeneration through controlled cellular access. The membrane must exhibit cellular occlusivity, effectively excluding the rapid migration of epithelial cells and gingival fibroblasts into the defect site while permitting slower-growing periodontal ligament cells and osteoprogenitor cells to repopulate the wound area.

This selective exclusion is critical for preventing soft tissue infiltration that would otherwise interfere with periodontal regeneration. Paradoxically, while maintaining this barrier effect, the membrane must simultaneously possess a porous microarchitecture that permits bidirectional diffusion of nutrients, oxygen, and metabolic waste products. This porosity also facilitates cellular adaptation and supports neovascularization, which is essential for sustaining the metabolic demands of regenerating tissues and ensuring long-term graft survival.

4. Bioactivity and Osteogenic Properties: Beyond serving as a passive physical barrier, modern GTR membranes are increasingly expected to exhibit bioactive properties that actively promote tissue regeneration. Osteoconductivity, the ability to support osteoblast attachment, migration, and proliferation, represents a desirable characteristic that can accelerate bone formation. Bioactive membranes may also stimulate cellular differentiation and matrix synthesis, thereby enhancing the quality and quantity of regenerated periodontal tissues. These bioactive properties should complement rather than compromise the membrane's primary barrier function.
5. Degradation Kinetics and Temporal Functionality: The degradation profile of a GTR membrane must be carefully calibrated to match the kinetics of new tissue formation. An ideal membrane maintains its structural integrity and barrier function throughout the initial critical healing period of approximately 4–6 weeks, during which epithelial exclusion and space maintenance are important. Following this period, the membrane should undergo controlled biodegradation at a rate that parallels tissue maturation, gradually transferring mechanical load to the newly formed tissue while being resorbed through physiological processes. This eliminates the need for secondary surgical processes to remove the membrane, thereby reducing patient morbidity and healthcare costs. The degradation products must be biocompatible and readily metabolized or eliminated by the host without inducing inflammatory responses or leaving residual foreign material that could interfere with long-term tissue remodeling.

Understanding these fundamental requirements guides the development of improved membrane systems through material modification, fabrication, and functionalization strategies [21]. Chitosan-based membranes, with their inherent biocompatibility, antibacterial properties, and tunable degradation characteristics, represent promising candidates that can be engineered to meet many of these essential design criteria. Figure 3 illustrates regenerative therapy using guided tissue regeneration membrane.

Chitosan membranes demonstrate adequate mechanical strength for space maintenance, controlled biodegradation rates that align with tissue regeneration timelines, and the ability to support cellular attachment and proliferation. Nonwoven chitosan membranes have been investigated in created one-wall intrabony defects in beagle dogs, demonstrating superior formation of new cementum and bone compared with polylactic acid bioabsorable membranes, with notably dense arrangements of cementoblasts and osteoblasts along newly formed bone surfaces [22]. In addition, the dual component designed composite of chitosan membranes have been

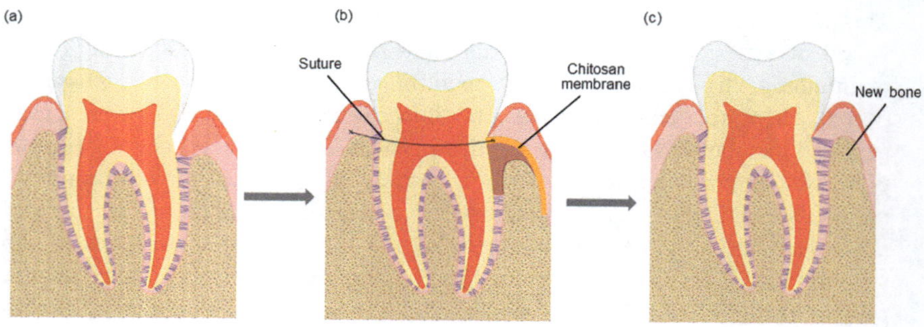

Fig. 3 Illustration of guided tissue regeneration: (**a**) vertical bone loss, (**b**) chitosan-based membrane placement, and (**c**) new bone growth

fabricated by coating collagen sponges with chitosan solution, chitosan serves as the functional barrier, providing critical tissue exclusion and regenerative guidance, while the collagen sponge functions merely as a structural scaffold [23]. Both investigations confirm that chitosan membranes effectively promote organized periodontal tissue regeneration in animal models, supporting their potential clinical application in treating intrabony periodontal defects.

Carboxymethyl chitosan (CMC) represents a strategic modification of chitosan designed to overcome its poor solubility. The carboxymethylation targets specific sites on the chitosan molecule, introducing carboxymethyl groups at either the C-6 hydroxyl position or the C-2 amino (NH_2) group. This selective modification yields three distinct derivatives of *N*-carboxymethyl chitosan (*N*-CMC), *O*-carboxymethyl chitosan (*O*-CMC), or the dual-substituted *N,O*-carboxymethyl chitosan (*N,O*-CMC). A particularly promising application involves chitosan-carboxymethyl chitosan blending membranes for guided periodontal tissue regeneration. These composite membranes demonstrate excellent biocompatibility, showing no toxic effects on surrounding tissues. While an initial inflammatory response occurs following membrane insertion, characterized by macrophage infiltration, this reaction is temporary. Then, the inflammatory processes diminishes progressively as the membrane naturally degraded, indicating a favorable biological response that supported the healing process [24].

5 Conclusion

Chitosan has demonstrated remarkable versatility and clinical potential as a multifunctional biomaterial in periodontal healthcare, offering innovative solutions across the spectrum of periodontal disease prevention, management, and regenerative therapy.

Throughout this chapter, the diverse applications of chitosan in periodontal healthcare have been systematically explored, ranging from preventive methods to advanced regenerative approaches. As an antimicrobial agent, chitosan has shown superior efficacy in mouthwash formulations, effectively disrupting and eliminating mature biofilms of periodontal pathogens, demonstrating enhanced antibiofilm activity compared to conventional agents such as chlorhexidine. The development of chitosan-based local drug delivery systems, including resorbable membranes, enzyme-responsive platforms, and injectable thermo-sensitive hydrogels, represents a significant advancement in targeted periodontal therapy. These systems enable sustained release of therapeutic agents directly within periodontal pockets while minimizing systemic exposure and side effects.

In surgical periodontal therapy, chitosan has proven invaluable both as a protective periodontal dressing that promotes wound healing and controls postoperative bleeding, and as a scaffold material for GTR membranes. Furthermore, chitosan derivatives such as phosphorylated chitosan have extended the application spectrum to include dental hard tissue remineralization, offering promising strategies for managing dental caries through biomimetic approaches that complement or provide alternatives to conventional fluoride treatments. Additionally, the development of smart, stimulus-responsive chitosan systems that can dynamically respond to the pathological microenvironment of diseased periodontal tissues represents an exciting frontier for precision periodontal therapy.

References

1. Thangavelu A, Stelin KS, Vannala V, Mahabob N, Hayyan FMB, Sundaram R (2021) An overview of chitosan and its role in periodontics. J Pharm Bioall Sci 13(S1):S15–S18
2. Wu Y, Chen C, Lu S, Song W, Cheng X, Zhang LW, Wang Y, Wang Y (2025) Tailored modulation of reactive oxygen species and anti-inflammatory responses in periodontitis using in situ diselenium-crosslinked chitosan hydrogel. Nano Today 61:102588
3. Knight NN (2023) Overview of the periodontium. In: Niekrash CE, Ferneini EM, Goupil MT (eds) Dental science for the medical professional. Springer, pp 21–30
4. Niekrash CE (2023) Periodontal infections: Gingivitis and periodontitis. In: Niekrash CE, Ferneini EM, Goupil MT (eds) Dental Science for the Medical Professional. Springer, pp 89–93
5. Kwon T, Lamster IB, Levin L (2021) Current concepts in the management of periodontitis. Int Dent J 71(6):462–476
6. Tachaboonyakiat W (2017) Antimicrobial applications of chitosan. In: Jennings JA, Bumgardner JD (eds) Chitosan based biomaterials Volume 2. Woodhead Publishing, pp 245–274
7. Pichayakorn W, Boonme P (2013) Evaluation of cross-linked chitosan microparticles containing metronidazole for periodontitis treatment. Mater Sci Eng C Mater Biol Appl 33:1197–1202
8. Indu SRA, Sargod SS, Bhat SS, Manikkoth S, Ramakrishnan N (2019) Effectiveness of chitosan mouthwash on plaque formation and levels of salivary *Streptococcus* and *Lactobacillus* count: an *in vivo* study. Int J Bas Clin Pharm 8:2416–2422
9. Yawilat P, Niamsiri N, Surarit R, Phonghanyudh A (2021) Effect of different molecular weight of chitosan mouthwash formulations against *Streptococcus mutans*. Mahidol Dent J 41(3):205–212
10. Archana V, Prabhuji MLV, Karthikeyan BV, Arul KS (2013) Control of *Streptococcus sanguis* oral biofilm by novel chlorhexidine-chitosan mouthwash: an *in vitro* study. J Exp Integr Med 3:165–169

11. Tachaboonyakiat W, Netswasdi N, Srakaew V, Opaprakasit M (2010) Elimination of inter- and intramolecular crosslinks of phosphorylated chitosan by sodium salt formation. Polym J 42:148–156
12. Zhang X, Li Y, Sun X, Kishen A, Deng X, Yang X, Wang H, Cong C, Wang Y, Wu M (2014) Biomimetic remineralization of demineralized enamel with nano-complexes of phosphorylated chitosan and amorphous calcium phosphate. J Mater Sci Mater Med 25(12):2619–2628
13. Xu Z, Neoh KG, Lin CC, Kishen A (2011) Biomimetic deposition of calcium phosphate minerals on the surface of partially demineralized dentine modified with phosphorylated chitosan. J Biomed Mater Res B Appl Biomater 98(1):150–159
14. Sah AK, Dewangan M, Suresh PK (2019) Potential of chitosan-based carrier for periodontal drug delivery. Colloids Surf B: Biointerfaces 178:185–198
15. Babrawala IS, Mlv P, Bv K, Khanna D (2016) A novel approach using natural 1% (w/w) chitosan as a local drug delivery system in the management of non-surgical periodontal treatment: a pilot study. J Int Acad Periodontol 18(4):129–133
16. Li N, Jiang L, Jin H, Wu Y, Liu Y, Huang W, Wei L, Zhou Q, Chen F, Gao Y, Zhu B, Zhang X (2019) An enzyme-responsive membrane for antibiotic drug release and local periodontal treatment. Colloids Surf B: Biointerfaces 183:110454
17. Ma S, Lu X, Yu X, Du Y, Xu S, Li M, Peng C, Liu Z, Deng J (2022) An injectable multifunctional thermo-sensitive chitosan-based hydrogel for periodontitis therapy. Biomater Adv 142:213158
18. Srakaew V, Ruangsri P, Suthin K, Thunyakitpisal P, Tachaboonyakiat W (2012) Sodium-phosphorylated chitosan/zinc oxide complexes and evaluation of their cytocompatibility: an approach for periodontal dressing. J Biomater Appl 27(4):403–412
19. Bee S, Abdul Hamid ZA (2025) Chitosan-based dental barrier membrane for periodontal guided tissue regeneration and guided bone regeneration: a review. Int J Biol Macromol 295:139504
20. Xu C, Lei C, Meng L, Wang C, Song Y (2012) Chitosan as a barrier membrane material in periodontal tissue regeneration. J Biomed Mater Res Part B 100(5):1435–1443
21. Tachaboonyakiat W (2021) Physical and chemical modification of chitin/chitosan for functional wound dressings. In: Jayakumar R, Prabaharan M (eds) Chitosan for biomaterials III. Advances in polymer science, vol 287. Springer, pp 257–300
22. Yeo YJ, Jeon DW, Kim CS, Choi SH, Cho KS, Lee YK, Kim CK (2005) Effects of chitosan nonwoven membrane on periodontal healing of surgically created one-wall intrabony defects in beagle dogs. J Biomed Mater Res B Appl Biomater 72:86–93
23. Park JS, Choi SH, Moon IS, Cho KS, Chai JK, Kim CK (2003) Eight-week histological analysis on the effect of chitosan on surgically created one-wall intrabony defects in beagle dogs. J Clin Periodontol 30:443–453
24. Zhuang ZX, Lin ZY, Cao JF (2003) Biocompatibility of chitosan-carboxymethylchitosan as membrane for periodontal guided tissue regeneration. Shanghai J Stomatol 12:362–365

Wanpen Tachaboonyakiat

Nanomaterials Based on Chitosan as Potential Drug Delivery Carriers

Anh Dzung Nguyen, Van Bon Nguyen, Vu Hoa Tran, Chien Thang Doan, and San-Lang Wang

Abstract Chitosan, a natural polysaccharide obtained from chitin—a major byproduct of seafood processing—has emerged as a highly versatile biomaterial for biomedical applications. Its intrinsic properties, including biocompatibility, biodegradability, non-toxicity, and antibacterial activity, make it particularly attractive for drug delivery systems. These advantages are further enhanced when chitosan is formulated into nanoparticles or integrated into nanocomposites. Chitosan nanoparticles can be synthesized using several approaches, such as ionic gelation, solvent evaporation, coacervation, spray drying, and other advanced techniques. Likewise, chitosan nanocomposites can be developed by blending chitosan with natural or synthetic polymers to improve mechanical strength, stability, and functional performance. Each method presents specific benefits, limitations, and potential biomedical applications. This chapter reviews the fundamental principles of these synthesis techniques, highlights their relevance in drug delivery and broader biomedical fields, and discusses the existing challenges along with future directions for the development of chitosan-based nanoparticles and nanocomposites.

Keywords Chitosan · Nanoparticles · Nanocomposites · Drug delivery · Biomedical applications

---

A. D. Nguyen (✉) · V. B. Nguyen · V. H. Tran
Institute of Biotechnology and Environment, Tay Nguyen University, Buon Ma Thuot, Vietnam
e-mail: nadzung@ttn.edu.vn; nvbon@ttn.edu.vn

C. T. Doan
Department of Natural Science and Technology, Tay Nguyen University, Buon Ma Thuot, Vietnam
e-mail: dcthang@ttn.edu.vn

S.-L. Wang
Department of Chemistry, Tamkang University, New Taipei, Taiwan
e-mail: sabulo@mail.tku.edu.tw

S.-L. Wang, M. Anpo (eds.), *Chitin and Chitosan*, Springer Series on Polymer and Composite Materials, https://doi.org/10.1007/978-981-95-7457-5_11

1 Introduction

Chitosan, a biopolymer of glucosamine and N-acetylglucosamine, is produced from seafood by-products, for example shrimp and crab shells, and from the cell walls of fungi. Chitosan, a natural polysaccharide derived from chitin—a major byproduct of seafood processing—offers multiple advantages for drug delivery applications. Structurally, both chitin and chitosan share similarities with cellulose, providing a versatile framework for functional modifications and the development of advanced delivery systems" [1]. Chemical structure of chitin, chitosan và cellulose as Fig. 1.

Chitosan and its derivatives are natural, non-toxic, biocompatible, and biodegradable polymers, and natural polycationic agents. Owing to their biocompatibility, biodegradability, and non-toxicity, chitin, chitosan, and their derivatives have been widely applied in agriculture, pharmaceutics, medicine, industry, and environmental protection. The biological activity of chitin and chitosan is critically influenced by parameters such as molecular weight, degree of deacetylation, and net charge [2]. These parameters, in turn, are substantially affected by the source of raw materials and the specific production or extraction processes employed. For instance, variations in molecular weight and deacetylation degree have been shown to modulate encapsulation efficiency and release kinetics in microparticle systems and to impact mechanical, thermal, and solubility properties in chitosan nanocomposites. Consequently, ensuring consistent quality and stability of chitin and chitosan

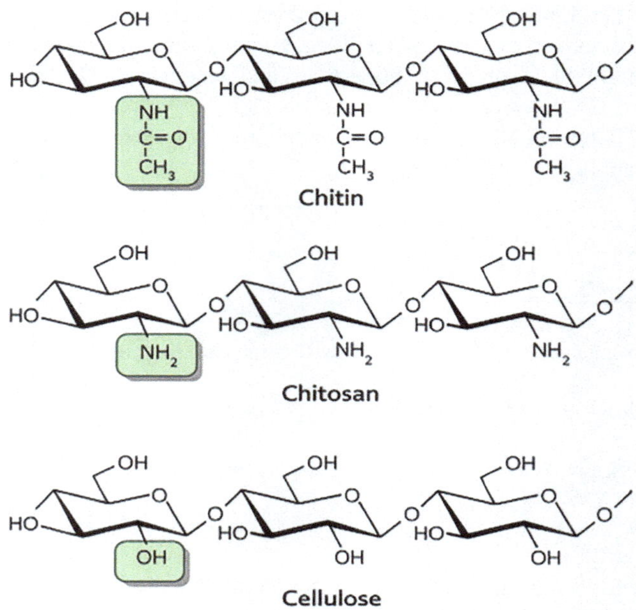

Fig. 1 Chemical structure of chitin, chitosan và cellulose

remains a considerable challenge, which has in turn limited their widespread application across diverse scientific and industrial domains.

Chitosan-based nanomaterials such as chitosan nanoparticles or chitosan/other polymer nanocomposites have attracted considerable attention by forming stable matrix, cross linking co-polymers. They are as promising drug delivery carriers due to their unique properties, including biocompatibility, biodegradability, non-toxicity, and the ease with which they can be chemically modified.

2 Chitosan Nanoparticles and Their Application

2.1 Methods for Preparation of Chitosan Nanoparticles

Chitosan nanoparticles (CSNPs) can be synthesized using various techniques, including ionic gelation, solvent evaporation, coacervation, and spray drying. Among these approaches, the particle size can be precisely controlled, which plays a critical role in modulating drug release kinetics. A comparative analysis of the advantages and disadvantages of each method is summarized in Table 1. In general, the primary methods employed for the preparation of CSNPs are described as follows.

2.1.1 Ionic Gelation Method

Ionic (ionotropic) gelation of chitosan uses electrostatic crosslinking between the positively charged amine groups of chitosan (protonated at acidic pH) and a multivalent anionic crosslinker (most commonly sodium tripolyphosphate, TPP). When a TPP aqueous solution is added to a chitosan solution under stirring, ionic bridges form and spherical nanoparticles spontaneously appear. This is a mild, solvent-free, room-temperature method ideal for encapsulating labile biomolecules. This is one of the most widely used methods for preparing chitosan nanoparticles. It involves the interaction of chitosan with anionic agents (e.g., glutaraldehyde, tripolyphosphate, fatty acids) under specific conditions, leading to the formation of nanoparticles. The size and shape of the nanoparticles can be controlled by adjusting parameters like the molecular weight, concentration of chitosan and the crosslinking agent, pH and stirring speed, etc. Normally, the size of chitosan nanoparticles synthesized by this method ranges from 50 to 200 nm depended on molecular weight, concentration of chitosan, TPP, and stirring speed. Higher molecular weight and concentration of chitosan is bigger size and lower zeta potential [3, 4] (Fig. 2).

Although this method is simple and straightforward to implement, the stability of chitosan nanoparticles (CNS) remains limited such as weak ionic crosslinking, batch variability, burst drug release, short shelf-life, and difficulty purification.

Table 1 Comparison of common methods for preparing chitosan nanoparticles

Method	Principle	Advantages	Disadvantages	References
Ionic gelation	Electrostatic interaction between chitosan (positive) and polyanions (e.g., TPP)	Simple, mild, eco-friendly Biocompatible Cost-effective and scalable High efficiency for charged drugs Good mucoadhesion	Limited stability Weak ionic crosslinking Batch variability Burst drug release Short shelf-life	[3, 4]
Solvent evaporation	Chitosan dissolved in organic solvent → emulsification → solvent removal	High drug loading for hydrophobic drugs Stable nanoparticles Good control over size/release Versatile for many drug types	Requires toxic solvents/surfactants Harsh processing (may denature biomolecules) More costly Difficult to scale up	[5]
Coacervation	Phase separation induced by adding a coacervating agent → nanoparticle formation	Simple process Can encapsulate both hydrophilic and hydrophobic drugs High drug loading capacity	Often requires organic solvents Limited particle size control May produce polydisperse nanoparticles Less stable	[6]
Spray drying	Atomization of chitosan solution into hot air → solvent evaporation → solid nanoparticles	Scalable, industrially applicable Produces dry, stable nanoparticles Suitable for heat-stable drugs Long shelf-life	High temperature may degrade sensitive biomolecules Expensive equipment Lower encapsulation efficiency for hydrophilic drugs	[7–9]

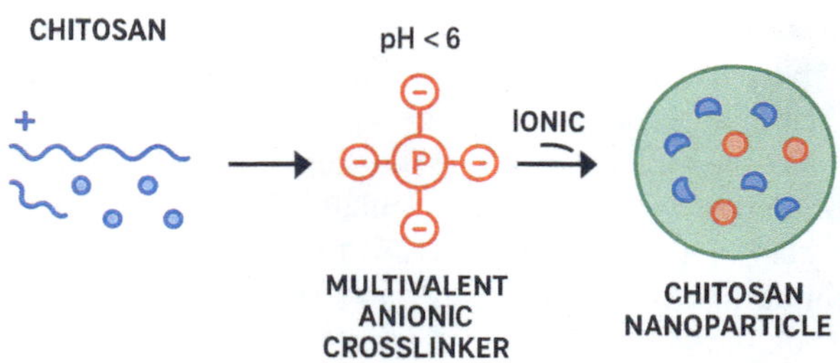

Fig. 2 Ionic gelation method for preparation of chitosan nanoparticles

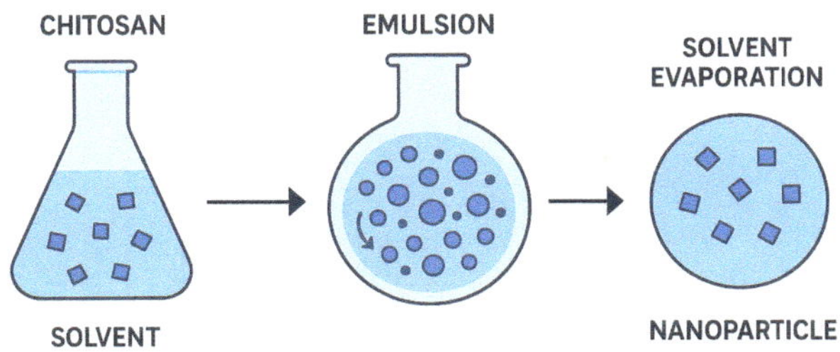

Fig. 3 Solvent evaporation method for chitosan nanoparticles

2.1.2 Solvent Evaporation Method for Chitosan Nanoparticles

In this method, chitosan is dissolved in a suitable organic or aqueous solvent along with the drug, typically a hydrophobic compound. The mixture is then emulsified, and upon solvent evaporation, chitosan precipitates to form nanoparticles encapsulating the drug molecules as Fig. 3.

Advantages of this method are simple and cost-effective process requiring no specialized equipment, highly suitable for encapsulating hydrophobic drugs, enabling efficient delivery of poorly water-soluble compounds, and particle size can be controlled by adjusting emulsifier concentration, stirring speed, or solvent properties [5].

Conversely, solvent evaporation method gets some limitations as potential toxicity from residual solvents, batch-to-batch variability affecting reproducibility, and low encapsulation efficiency for hydrophilic drugs.

Chitosan nanoparticles are synthesized by this way can be applied as some following fields:

Drug delivery: Particularly effective for cancer therapy, antibiotics, and anti-inflammatory drugs.

Gene delivery: Especially when functionalized with targeting ligands.

Vaccine adjuvants: Enhances antigen stability and immune responses.

2.1.3 Coacervation Method for Chitosan Nanoparticles

The coacervation method involves phase separation of a polymer-rich phase from a polymer-poor phase to form nanoparticles. In this process, chitosan is first dissolved in an appropriate aqueous medium. A coacervating agent—often a polyanion such as sodium sulfate or tripolyphosphate is then added under controlled conditions (e.g., pH, temperature, stirring rate). The interaction between the cationic amino groups of chitosan and the anionic groups of the coacervating agent or anionic polymer induces electrostatic interactions that lead to phase separation.

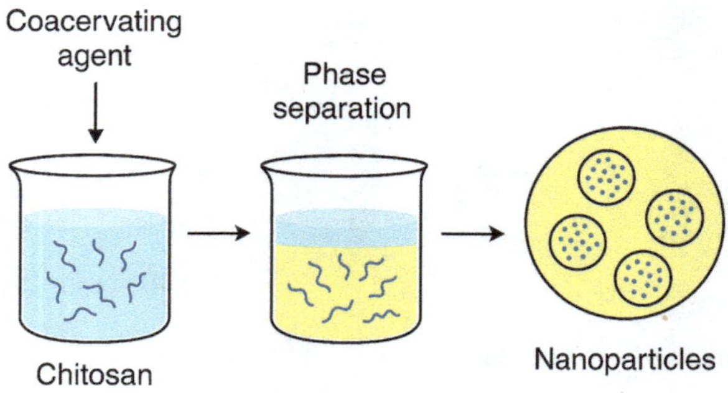

Fig. 4 Coacervation method for chitosan nanoparticles

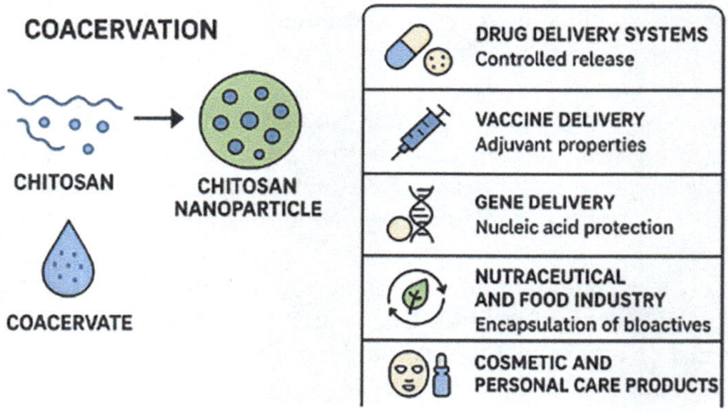

Fig. 5 Application of chitosan nanoparticles prepared by coacervation method

As the polymer-rich phase precipitates out, nanoparticles spontaneously form. These particles can be further crosslinked using agents like glutaraldehyde or tripolyphosphate to improve stability (Fig. 4) [6].

This is a simple and cost-effective process and mild reaction conditions suitable for sensitive bioactive compounds. Although, this method may get some disadvantages are possible particle aggregation and limited control over size distribution compared to other methods.

Application of Chitosan Nanoparticles Prepared by Coacervation Method
The coacervation method for preparing chitosan nanoparticles has gained significant attention because it is simple, mild, and suitable for encapsulating sensitive bioactive compounds. Its main applications include: Drug Delivery Systems, Vaccine Delivery, Gene Delivery, Nutraceutical and Food Industry, Cosmetic and Personal Care Products, Sustained Release of Active Ingredients, Enhanced Skin Penetration (Fig. 5). The advantages of the coacervation method are a simple process, high drug-loading capacity, and controlled release.

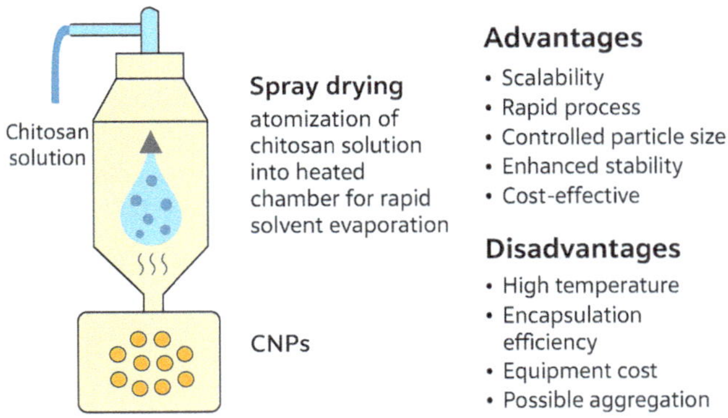

Fig. 6 Spray drying, its advantages and limitations

2.1.4 Spray Drying

Spray drying has been widely used for the preparation of micro- and nanoparticles for use by the food and pharmaceutical industries, because, in comparison with other techniques, it is relatively low-cost technology, rapid, and easy to scale up. The process is flexible, adaptable to commonly used processing equipment, and produces particles with good properties [7–9]. Microencapsulation of such antibiotics as ampicillin, amoxicillin, and vancomycin, by chitosan, by use of a spray drying method, to enhance the efficiency of their antibacterial activity has been widely used in industrial applications.

Chitosan nanoparticles (CNPs) produced using a spray dryer have several advantages and disadvantages (Fig. 6) and Table 1.

Spray drying is a promising way to scale up and apply widely for food, biomedicine and pharmacy because of:

– Controlled Particle Size: Spray drying allows for the production of uniform and fine nanoparticles, improving consistency in applications.
– Scalability: It is a continuous and scalable process suitable for industrial production.
– Fast Drying Process: The method is rapid, reducing processing time compared to other nanoparticle production techniques.
– Good Encapsulation Efficiency: It can effectively encapsulate bioactive compounds, improving their stability and bioavailability.
– Improved Stability: Spray drying helps in maintaining the stability of chitosan nanoparticles, preventing agglomeration.
– Cost-Effective: Compared to other nanoparticle synthesis techniques (e.g., ionic gelation, solvent evaporation), spray drying is relatively economical.
– Solvent-Free or Minimal Solvent Use: Reduces the need for toxic solvents, making it environmentally friendly and safe for pharmaceuticals and food applications.

Although, use of spray drying to prepare chitosan nanoparticles and micro encapsulation of medicine or bioactive compounds has some Disadvantages such as:

- High Energy Consumption: The process requires significant energy input, especially for drying and atomization.
- Particle Agglomeration: Small nanoparticles may aggregate during drying, affecting their functional properties.
- Limited Solubility Control: The solubility and dispersion of spray-dried chitosan nanoparticles can be challenging, especially in aqueous environments.
- Possible Thermal Degradation: High drying temperatures may degrade heat-sensitive bioactive compounds such as enzymes, vitamins, other bioactive compounds encapsulated within the nanoparticles.
- Loss of Functional Properties: The high shear forces in the spray drying process may affect the molecular structure of chitosan, impacting its bioactivity.
- Need for Post-Processing Steps: Additional steps like surface modification or dispersion aids may be required to improve nanoparticle performance.

The spray drying method for chitosan nanoparticles (CNPs) is suitable for applications where rapid production, controlled particle size, and enhanced stability are critical. Here are the main areas where spray-dried CNPs are commonly used:

Drug Delivery Systems

- Oral drug delivery: Produces dry, stable nanoparticles for easy incorporation into tablets or capsules.
- Pulmonary drug delivery: Fine particles with controlled size for inhalation therapies.
- Controlled release: Encapsulation of drugs for sustained and targeted release.

Vaccine Delivery

- Mucosal vaccines: Spray-dried CNPs can be administered via nasal or oral routes, enhancing antigen stability and immune response.
- Thermostable vaccines: Dry powder form improves shelf-life, reducing the need for cold-chain storage.

Nutraceutical and Food Industry

- Encapsulation of bioactive compounds such as vitamins, probiotics, and antioxidants are protected against heat, moisture, and oxidation.
- Functional foods: Controlled release of nutrients in the gastrointestinal tract.

Cosmetic and Personal Care Products

- Controlled release of active ingredients: Used in creams, lotions, and face masks for sustained delivery.
- Stability enhancement: Improves the shelf-life of sensitive cosmetic ingredients.

Agricultural Applications

- Controlled release fertilizers or pesticides: Minimizes environmental impact and enhances efficiency.
- Encapsulation of plant growth regulators: Protects bioactive compounds from degradation.

2.2 Application of Chitosan Nanoparticles for Biomedicine

The particle size of CSNPs can be carefully controlled, which is critical for their applications, especially in drug delivery. Smaller nanoparticles have better penetration through cell membranes, while larger nanoparticles may have sustained drug release characteristics. The size also impacts the biodistribution, targeting ability, and stability of nanoparticles.

The control of particle size in Chitosan-based nanoparticles (CSNPs) is indeed crucial for optimizing their performance in various applications, particularly in drug delivery. Here's a breakdown of how particle size influences their behavior and effectiveness:

Penetration Through Cell Membranes: Smaller nanoparticles (typically in the range of 1–200 nm) are generally more efficient in penetrating cell membranes due to their increased surface area-to-volume ratio. This enhances cellular uptake, which is particularly important for delivering drugs into specific cells or tissues.

Sustained Drug Release: Larger nanoparticles often have a slower release rate, which is beneficial for controlled and sustained drug delivery. This size range can provide a reservoir for drugs that are gradually released over time, reducing the frequency of administration and improving therapeutic outcomes.

Biodistribution: The size of the nanoparticles affects their circulation time in the bloodstream and their ability to reach target tissues. Smaller nanoparticles tend to have longer circulation times due to reduced recognition and clearance by the immune system, while larger nanoparticles may accumulate in specific tissues (e.g., tumors) more efficiently.

Targeting Ability: Nanoparticles can be functionalized with targeting ligands (like antibodies or peptides) to bind to specific receptors on target cells. The particle size can influence how well these functionalized nanoparticles can target particular cells or tissues. Smaller particles can diffuse more easily through tissues, while larger ones might be more effective in targeted drug delivery where accumulation at a specific site is desired.

Stability: The physical stability of nanoparticles can be influenced by their size. Smaller nanoparticles may be prone to aggregation or degradation due to their larger surface area, whereas larger particles may exhibit more robust structural integrity, making them more stable in certain environments.

In summary, controlling the particle size of CSNPs is essential for tailoring their properties to meet the specific demands of drug delivery systems, ensuring effective and targeted therapeutic delivery while optimizing stability and release kinetics.

Advantages of CSNPs for Biomedicine Applications are:

Positive Charge and Cellular Uptake: The inherent positive charge of chitosan nanoparticles allows them to interact more easily with the negatively charged components of the cell membrane. This charge interaction enhances cellular uptake, making CSNPs suitable for drug delivery applications. The positive charge may also improve the internalization of nanoparticles by endocytosis, facilitating better delivery to the target site [7].

Biodegradability and Biocompatibility: Chitosan is derived from chitin, a natural polymer, and it is biodegradable and biocompatible. This makes CSNPs attractive for biomedical applications, including drug delivery and tissue engineering.

Controlled Drug Release: Due to their ability to encapsulate various drugs, chitosan nanoparticles provide a controlled and sustained release, minimizing side effects and improving therapeutic outcomes. The release rate can be modulated by adjusting the properties of the nanoparticles, such as size, surface charge, and crosslinking.

Targeted Delivery: Chitosan nanoparticles can be further modified (e.g., with ligands or antibodies) to target specific cells or tissues. This feature makes CSNPs promising candidates for targeted drug delivery systems, including cancer therapy.

Drug Loading: Chitosan nanoparticles can load a variety of drugs, including hydrophobic and hydrophilic ones, making them versatile carriers for a range of therapeutic agents.

Chitosan nanoparticles are highly versatile drug delivery systems due to their ability to load both hydrophobic and hydrophilic drugs. This versatility makes them suitable for a wide array of therapeutic applications. Chitosan's structure can be modified to encapsulate hydrophobic drugs, typically by forming micelles or nanoemulsions. The nanoparticle surface can interact with the drug molecules through hydrophobic interactions, allowing these drugs to be solubilized and delivered effectively. Chitosan can also form stable aqueous solutions or gels, allowing the encapsulation of hydrophilic drugs. The polysaccharide chains of chitosan provide a suitable environment for the absorption and delivery of water-soluble drugs. The ability to load different types of drugs allows for enhanced stability and bioavailability, ensuring that the therapeutic agents are delivered at the desired location in the body, often improving the drug's effectiveness and reducing side effects. By modifying the surface properties of chitosan nanoparticles (e.g., adding targeting ligands), these particles can be engineered for targeted drug delivery, enhancing the therapeutic outcomes and minimizing systemic toxicity [4, 7, 8].

Thus, the broad drug-loading capability of chitosan nanoparticles opens up opportunities for treating a wide range of diseases, including cancer, infections, and chronic conditions.

2.3 Safety and Biocompatibility of Chitosan Nanoparticles

Biodegradability: Chitosan is broken down by enzymes in the body, reducing the risk of accumulation and long-term toxicity. This is a key benefit for use in drug delivery systems.

Non-toxic: Chitosan is considered non-toxic at appropriate concentrations and does not induce significant immune responses, making it suitable for human use.

Low Immunogenicity: Since chitosan is a natural biopolymer, it has a low potential for triggering immune responses, which is important for the development of long-term, safe drug delivery systems [10].

3 Chitosan Nanocomposites and Biomedical Application

Chitosan nanoparticles, based on the linear polysaccharide chitosan, have a larger contact surface area and positive charge than regular chitosan. Recently, studies have demonstrated that chitosan nanoparticles are very potent drug carriers and delivery nanomaterials. They can protect drugs from being broken down by enzymes in the intestinal tract. Due to their small size, they can penetrate deeply into the body, delivering drugs to the correct target tissues and cells to improve treatment effectiveness. Many studies have used chitosan to carry antibiotics, such as amoxicillin, ciprofloxacin, chlortetracycline, and gentamycin, enhancing efficiency and fighting drug resistance in pathogenic bacteria. However, chitosan nanoparticles are limited by their high cost and instability in acidic pH conditions, which restricts their application in the nanoparticle pharmaceutical field. Due to these limitations, chitosan was incorporated with other polymers such as alginate, cellulose, and starch [11, 12].

Meanwhile, natural branched polysaccharides, is a safe, low-cost, and biocompatible polymers such as starch, cellulose, dextran, and alginat that are widely applied in the food and pharmaceutical fields. Therefore, they may be blended with chitosan or synthetic polymer such as PVA (poly vinyl acetate), PMA (poly methylmethacrylate), PEG (poly ethylen glycol) to improve their physical properties for pharmaceutical applications [13–15].

3.1 Methods for Prepare Chitosan Nanocomposites

3.1.1 Solution Blending/Solution Casting (Most Common for Films & Membranes)

Principle: dissolve both polymers in a common solvent (or compatible solvents), mix, cast and dry to form a blended film or membrane.

When to use: films, coatings, membranes, and bulk composites. Works well when both polymers are water-soluble (or one soluble in water and the other in a miscible solvent).

Typical conditions and steps as follows (Fig. 7).

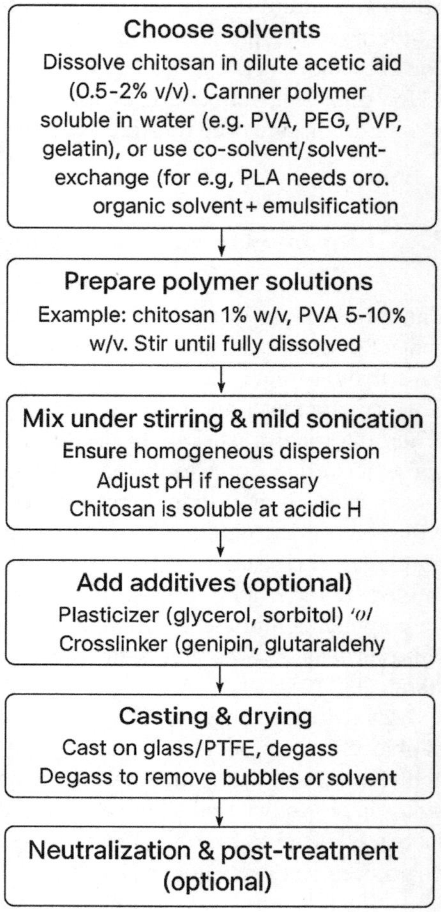

Fig. 7 Typical conditions and steps to prepare blending films

3.1.2 Coacervation /Polyelectrolyte Complexation (PEC)

Principle: chitosan (cationic) forms complexes with oppositely charged polymers (anionic) like alginate, pectin, hyaluronic acid, carrageenan, or polyacrylic acid. Complexation yields nanoparticles, beads, or membranes without organic solvents or toxic crosslinkers.

When to use: hydrogel beads, microparticles, nanoparticles, membranes with ionic bonding.

Typical steps: (1) Make chitosan solution in acetic acid (0.1–0.5% w/v). Prepare anionic polymer solution (e.g., sodium alginate 0.5% w/v) in water. (2) Add one solution dropwise into the other under stirring to form coacervates. Order and rate control size/morphology. (3) Optionally further crosslink (e.g., $CaCl_2$ for alginate) or freeze-dry to make porous scaffolds.

3.1.3 Interpenetrating Polymer Network (IPN)/Semi-IPN

Principle: one polymer network is formed in the presence of another polymer (which may or may not be crosslinked), producing mechanically interlocked networks. Can be sequential (first network then second) or simultaneous.

When to use: hydrogels, mechanical reinforcement, stimuli-responsive systems.

Typical steps: (1) Prepare solution of polymer A (e.g., chitosan) and polymer B (e.g., polyacrylamide). (2) Initiate polymerization/crosslinking for network A or B using appropriate initiator (e.g., APS/TEMED for acrylamide) while the other polymer is present. (3) Cure, wash to remove unreacted monomers, and dry or store hydrated.

3.1.4 Graft Copolymerization/Chemical Modification

Principle: chemically graft polymer chains onto chitosan backbone (via —NH_2 or —OH). This improves compatibility with hydrophobic polymers and tailors solubility. Common grafts: poly(ε-caprolactone) (PCL), polyethylene glycol (PEG), acrylics.

When to use: need for covalent bonding for stability, compatibilization, and tailored properties.

Typical steps: (1) Activate chitosan (e.g., with carbodiimide chemistry EDC/NHS for carboxyl-containing monomers) or use ring-opening polymerization from chitosan amino groups for PCL. (2) Control grafting density and purification (dialysis, precipitation).

3.1.5 Melt Blending/Extrusion (for Thermoplastics)

Principle: mix chitosan (often dried powder or granules of chitosan derivative) into molten thermoplastic (PLA, PCL, polypropylene) using twin-screw extruder. Good for producing films, fibers, pellets [15] (Fig. 8).

When to use: industrial-scale production, thermoplastic composites.

3.1.6 In Situ Polymerization/In Situ Nanoparticle Formation within Polymer Matrix

Principle: polymerization of partner monomer inside chitosan network or reduction/synthesis of nanoparticles in presence of both polymers to embed fillers and create strong interactions.

When to use: conductive composites (chitosan + polyaniline), metal NP–polymer composites, or when fillers must be generated inside the matrix.

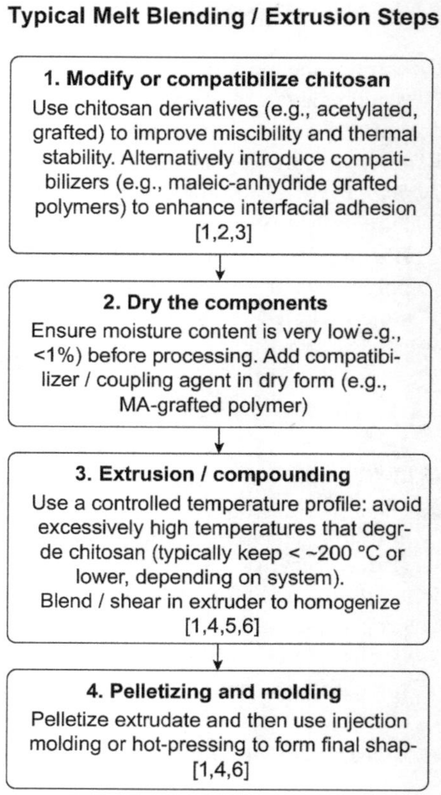

Fig. 8 Typical conditions and steps to prepare Melt Blending

Typical steps: (1) Disperse chitosan in solution; add monomer (aniline) or metal salt. (2) Initiate polymerization or reduction (oxidant for aniline, reducer for metal) under controlled conditions. (3) Purify, cast, or form desired shape.

3.1.7 Electrospinning for Nanofibers

Principle: create nanofibers by electrospinning a polymer solution or blend. Chitosan alone is hard to electrospin; blending with spinnable polymers (PVA, PEO, PCL with chitosan derivative) is common.

When to use: tissue engineering scaffolds, wound dressings, filtration membranes.

Typical steps: (1) Prepare chitosan solution (often low concentration) and blend with PVA or PEO (PVA 8–12% w/v). Ratio often chitosan:PVA 10–50 wt% chitosan. (2) Adjust viscosity and conductivity; degas. (3) Electrospin with appropriate voltage (10–25 kV), tip–collector distance (10–20 cm), and flow rate (0.2–1.0 mL/h). (4) Crosslink fibers (heat, genipin, or chemical crosslinkers) to make insoluble.

3.1.8 Emulsion/Solvent Evaporation (for Core–Shell Nanoparticles)

Principle: prepare nanoparticles with one polymer forming core and another forming shell (e.g., PLGA core, chitosan shell) using emulsion and solvent evaporation or solvent diffusion.
When to use: drug delivery where controlled release is needed (chitosan shell can provide mucoadhesion).
Typical steps: (1) Dissolve hydrophobic polymer (PLGA) and drug in organic solvent (e.g., dichloromethane). (2) Emulsify into aqueous chitosan solution (with surfactant) under sonication/homogenization to form O/W emulsion. (3) Evaporate solvent under stirring; PLGA precipitates as nanoparticles with chitosan adsorbing/complexing to form shell. (4) Collect by centrifugation, wash, and optionally crosslink.

3.2 Drug Encapsulation and Loading

The loading of therapeutic molecules into the chitosan nanocomposite is a critical step that dictates both efficacy and release behavior. Drugs may be incorporated through physical entrapment, electrostatic interactions, or covalent conjugation. The method of encapsulation influences drug stability, release kinetics, and bioavailability.

- *Hydrophilic drugs* are generally entrapped within the polymeric network, taking advantage of the porous nature of chitosan.
- *Hydrophobic drugs* can be incorporated by modifying chitosan with amphiphilic polymers or embedding them within lipid–chitosan hybrid nanocomposites.
- *Biological macromolecules* (e.g., siRNA, plasmid DNA, or proteins) are stabilized by the strong positive charge of chitosan, which enhances loading capacity and protects sensitive molecules from degradation.

The encapsulation efficiency is typically high due to the protonated amino groups of chitosan, which provide strong binding affinity. This encapsulation not only improves solubility and prolongs circulation time but also reduces drug leakage during systemic transport.

Surface Functionalization and Targeting The surface of chitosan nanocomposites can be engineered to optimize biological interactions, circulation half-life, and tissue specificity. Functionalization strategies include:

- PEGylation (polyethylene glycol modification): Enhances hydrophilicity, reduces opsonization by serum proteins, and prolongs systemic circulation.
- Ligand conjugation: Attaching ligands such as folic acid, transferrin, peptides, or antibodies allows for receptor-mediated targeting of tumor cells, inflamed tissues, or bacterial biofilms.
- Charge modulation: Adjusting surface charge enhances mucosal adhesion or promotes uptake by specific cell types.

Surface engineering not only confers stealth and targeting ability but also minimizes off-target effects, thus enhancing therapeutic efficacy while reducing systemic toxicity.

3.3 *Mechanism of Drug Delivery*

Once administered, chitosan nanocomposites (CNCs) follow a multistage drug delivery pathway:

Circulation and stability: Functionalized CNCs remain stable in physiological fluids, preventing aggregation and avoiding premature clearance by the mononuclear phagocyte system (MPS).

Targeting and adhesion: Surface modifications enable active targeting through ligand–receptor interactions (e.g., folate, transferrin, or antibody-mediated targeting) or passive targeting via the enhanced permeability and retention (EPR) effect and mucoadhesive properties.

Cellular internalization: CNCs enter cells predominantly by endocytosis, although direct membrane penetration can also occur. Their nanoscale size and cationic surface charge (protonated amino groups of chitosan) enhance interactions with negatively charged cell membranes, facilitating uptake.

Drug release: The release of encapsulated drugs is controlled by several mechanisms, including:

- Diffusion through the polymer matrix.
- Polymer swelling and erosion.
- Enzymatic degradation of chitosan by lysozymes.
- Stimuli-responsiveness, such as pH-triggered release in acidic tumor microenvironments or redox-triggered release in the cytosol.

This multistage mechanism enables both temporal and spatial control of drug delivery, thereby maximizing therapeutic efficacy while minimizing systemic side effects.

3.4 *Biomedical Applications of Chitosan Nanocomposites*

The versatility of chitosan nanocomposites has led to their application across a broad range of therapeutic areas:

3.4.1 Cancer Therapy

Targeted delivery of chemotherapeutics reduces systemic toxicity and overcomes multidrug resistance. One of the most promising applications of chitosan nanocomposites (CNCs) is in cancer therapy, where they address critical challenges such as systemic toxicity, multidrug resistance (MDR), and poor bioavailability of chemotherapeutics [16, 17].

Targeted Delivery and Tumor Accumulation: CNCs can be surface-functionalized with ligands such as folic acid, transferrin, hyaluronic acid, or monoclonal antibodies that specifically bind to receptors overexpressed on tumor cells (e.g., folate receptor, CD44, HER2). This ligand-mediated targeting enhances selective accumulation at the tumor site, reducing off-target effects in healthy tissues. Additionally, CNCs can exploit the enhanced permeability and retention (EPR) effect, passively accumulating in tumor tissues due to leaky vasculature and poor lymphatic drainage (Fig. 9).

CNCs can be surface-functionalized with ligands that specifically bind to receptors overexpressed on tumor cells. This ligand-mediated targeting enhances selective accumulation at tumor site, reducing off-target effects in healthy tissues.

Overcoming Multidrug Resistance (MDR): Many cancer cells overexpress P-glycoprotein (P-gp) efflux pumps that actively expel chemotherapeutic drugs, reducing intracellular drug concentrations. Chitosan nanocomposites can bypass this by facilitating endocytosis-based uptake, which circumvents efflux mechanisms. CNCs can also co-deliver P-gp inhibitors, siRNA, or gene silencing agents to suppress MDR pathways, restoring chemosensitivity.

Controlled and Stimuli-Responsive Release: CNCs provide pH-responsive release, as chitosan dissolves or swells in the acidic tumor microenvironment or within endosomes/lysosomes. Stimuli-responsive modifications (e.g., redox-sensitive, enzyme-degradable, temperature-sensitive) further allow on-demand drug release specifically at tumor sites. This minimizes systemic exposure while ensuring high intracellular drug concentration in cancer cells.

CNCs can be engineered as multidrug carriers, allowing simultaneous delivery of chemotherapeutics with other agents (e.g., photosensitizers for photodynamic therapy, siRNA for gene silencing, or immunomodulators for combination therapy). Such synergistic approaches improve therapeutic efficacy and reduce the likelihood of recurrence. Studies have demonstrated that CNC-based formulations of doxorubicin, paclitaxel, cisplatin, and curcumin significantly enhance tumor inhibition compared to free drugs, with markedly reduced systemic toxicity. The versatility of chitosan also allows combination with metallic nanoparticles (e.g., gold, iron

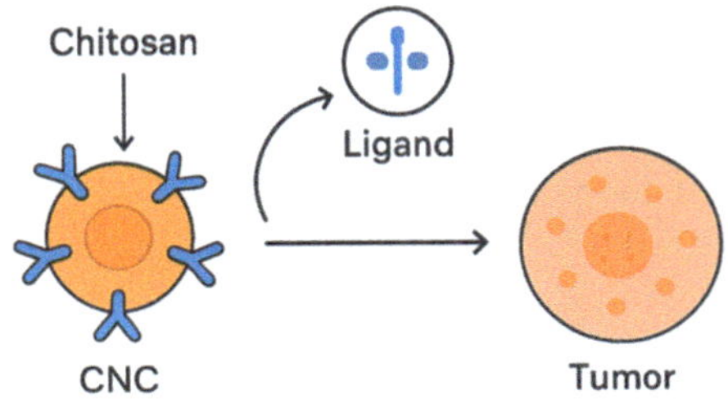

Fig. 9 Targeted delivery and tumor accumulation

oxide), enabling theranostic applications where therapy and imaging are integrated. Chitosan nanocomposites enable precise, efficient, and safe chemotherapy delivery by combining passive and active targeting, overcoming drug resistance, and offering controlled release in tumor-specific environments. This multifunctionality positions CNCs as powerful candidates for next-generation cancer nanomedicine.

3.4.2 Antimicrobial and Antifungal Treatments

Sustained release of antibiotics, combined with chitosan's inherent antimicrobial activity, enhances efficacy against resistant pathogens.

Chitosan nanocomposites are particularly effective in antimicrobial and antifungal therapy, owing to their ability to combine sustained drug release with the inherent antimicrobial properties of chitosan. This dual functionality is highly advantageous in combating resistant pathogens and biofilm-associated infections.

Intrinsic Antimicrobial Properties of Chitosan: Chitosan itself exhibits broad-spectrum antimicrobial activity against Gram-positive and Gram-negative bacteria, as well as fungi. Its cationic amino groups interact electrostatically with negatively charged microbial cell membranes, leading to increased permeability, leakage of intracellular components, and eventual cell lysis. Chitosan can also bind to microbial DNA, interfering with replication and protein synthesis.

Sustained and Controlled Release of Antibiotics/Antifungals: CNCs act as carriers for antibiotics (e.g., ampicillin, ciprofloxacin, and vancomycin) or antifungal agents (e.g., amphotericin B, fluconazole). The nanocomposite structure provides protection against premature degradation and enables controlled release kinetics, ensuring prolonged drug exposure at the infection site. Localized release reduces the required dosage, minimizing systemic side effects and slowing the development of resistance [11, 18].

Overcoming Antimicrobial Resistance (AMR): CNCs can bypass efflux pumps and disrupt bacterial biofilms, which are major contributors to antimicrobial resistance. Chitosan's mucoadhesive nature ensures prolonged retention at infection sites, improving therapeutic concentration and efficacy. Co-delivery of antibiotics with antimicrobial peptides, essential oils, or metallic nanoparticles (Ag, ZnO, CuO) enhances synergistic antimicrobial action.

Antifungal Applications: CNCs are highly effective against *Candida* spp., *Aspergillus* spp., and *Cryptococcus* spp., where conventional antifungal therapies often fail due to resistance or toxicity. Encapsulation of antifungal drugs within CNCs reduces their toxicity (e.g., amphotericin B nephrotoxicity) while maintaining or enhancing antifungal efficacy [19, 20].

Chitosan nanocomposites act as dual-action antimicrobial platforms, providing both intrinsic microbial inhibition and sustained antibiotic/antifungal release. This synergy makes them promising candidates for addressing antimicrobial resistance, persistent infections, and biofilm-related diseases.

3.4.3 Gene Therapy

Chitosan nanocomposites (CNCs) are emerging as effective non-viral vectors for gene delivery, offering a safer alternative to viral systems for precision medicine. Their strong electrostatic interaction with negatively charged nucleic acids (DNA, siRNA, mRNA, CRISPR–Cas components) enables efficient condensation, protection, and intracellular delivery [21] (Fig. 10).

Electrostatic Complexation and Protection: The positively charged amino groups of chitosan readily bind to the negatively charged phosphate backbone of nucleic acids, forming stable polyelectrolyte complexes (polyplexes). This protects genetic material from enzymatic degradation (e.g., nucleases) during systemic circulation. The nanoscale complexes also enhance solubility and facilitate transport across physiological barriers.

Cellular Uptake and Intracellular Delivery: CNCs enter cells via endocytosis, bypassing extracellular degradation. The proton-sponge effect of chitosan enables endosomal escape, ensuring that nucleic acids are released into the cytoplasm or nucleus for therapeutic action. This mechanism is particularly valuable for siRNA-mediated gene silencing and CRISPR-mediated genome editing.

Controlled Release and Targeting: Chemical modifications (e.g., PEGylation, ligand conjugation with folic acid, transferrin, or peptides) improve cell-specific delivery and reduce off-target effects. pH and redox-sensitive CNCs ensure stimuli-responsive release of nucleic acids within the intracellular environment.

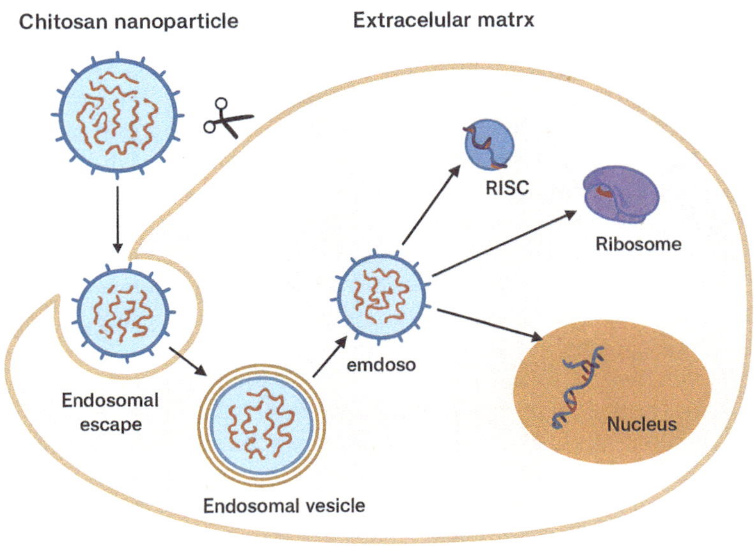

Fig. 10 Non-viral vectors for gene delivery of chitosan nanocomposites

Applications in Precision Medicine:

- siRNA delivery: CNCs silence overexpressed oncogenes, inflammatory mediators, or viral genes.
- DNA plasmid delivery: Promotes sustained expression of therapeutic proteins or growth factors.
- CRISPR/Cas systems: Chitosan-based nanocomposites protect and deliver CRISPR–Cas9 ribonucleoprotein complexes for targeted genome editing.

Advantages Over Viral Vectors: CNCs avoid the risks of viral vectors such as immunogenicity, mutagenesis, and limited loading capacity. They provide biocompatibility, repeat-dosing potential, and ease of large-scale synthesis, which are crucial for clinical translation.

In generation, chitosan nanocomposites offer a versatile and safe platform for gene therapy applications, enabling efficient encapsulation, protection, and intracellular delivery of DNA, siRNA, and CRISPR components. Their tunable properties and low toxicity make them a promising alternative to viral vectors, paving the way for precision and personalized medicine.

3.4.4 Tissue Engineering and Regenerative Medicine

Chitosan nanocomposites (CNCs) are widely applied in tissue engineering and regenerative medicine due to their biocompatibility, biodegradability, mucoadhesive properties, and structural similarity to glycosaminoglycans (GAGs) found in the extracellular matrix (ECM). These features make CNCs ideal as both drug/growth factor carriers and scaffold components for tissue repair and regeneration.

Scaffold and Matrix Support: CNCs can be incorporated into hydrogels, nanofibers, or 3D porous scaffolds to mimic the native ECM microenvironment. Such scaffolds provide mechanical support, cell adhesion sites, and guidance for tissue remodeling. Integration with inorganic nanomaterials (e.g., hydroxyapatite, bioactive glass, silica) further improves mechanical strength and osteoconductivity.

Controlled Release of Bioactive Molecules: CNCs act as carriers for growth factors, cytokines, and signaling peptides such as vascular endothelial growth factor (VEGF), bone morphogenetic proteins (BMPs), and epidermal growth factor (EGF). Sustained and localized release of these molecules enhances cell proliferation, angiogenesis, and tissue regeneration.

Wound Healing and Skin Regeneration: CNCs exhibit intrinsic antimicrobial and hemostatic activity, preventing infections and accelerating wound closure. Chitosan-based dressings maintain a moist wound environment, while nanocomposite incorporation ensures controlled delivery of antibiotics or regenerative agents.

Bone and Cartilage Regeneration: CNCs combined with calcium phosphate, hydroxyapatite, or silica nanoparticles provide osteoinductive scaffolds that promote osteoblast adhesion, proliferation, and differentiation. These composites enhance bone mineralization and have shown success in preclinical bone defect repair. In cartilage engineering, CNC hydrogels support chondrocyte growth and extracellular matrix deposition.

Nerve and Cardiovascular Regeneration: CNC scaffolds with aligned nanofibers can guide neurite outgrowth, supporting peripheral nerve regeneration. Functionalized CNCs also deliver angiogenic factors to improve vascularization in ischemic tissues, which is crucial for cardiac repair.

Advantages for Clinical Translation: CNC-based scaffolds are biodegradable and non-immunogenic, allowing them to degrade into harmless by-products after tissue regeneration. Their tunable porosity, mechanical strength, and bioactive surface modifications enable customization for specific tissue applications. Chitosan nanocomposites serve dual roles in tissue engineering: as structural scaffolds mimicking the extracellular matrix and as carriers for bioactive molecules that regulate cellular behavior. Their applications span wound healing, bone and cartilage repair, nerve regeneration, and cardiovascular tissue engineering, underscoring their promise in regenerative medicine.

3.4.5 Vaccination and Immunotherapy

Chitosan nanocomposites (CNCs) are emerging as promising adjuvants and delivery vehicles for vaccines and immunotherapies due to their ability to enhance antigen stability, improve immune recognition, and induce both humoral and cellular immune responses [22–24].

Antigen Encapsulation and Protection: CNCs can encapsulate or adsorb protein antigens, peptides, DNA vaccines, or mRNA within their matrix. This protects sensitive antigens from degradation (e.g., enzymatic digestion in mucosal tissues or systemic circulation), ensuring that intact immunogenic molecules reach target immune cells.

Mucosal Vaccine Delivery: The mucoadhesive properties of chitosan allow CNCs to adhere to mucosal surfaces (nasal, oral, pulmonary, or intestinal), prolonging antigen residence time. This promotes mucosal immunity, which is critical for protection against pathogens entering through respiratory or gastrointestinal routes.

Adjuvant Properties: Chitosan itself acts as an immunostimulatory agent, activating antigen-presenting cells (APCs) such as dendritic cells and macrophages. CNCs enhance uptake by APCs through endocytosis and facilitate antigen presentation via MHC class I and II pathways, stimulating both Th1 (cell-mediated) and Th2 (humoral) responses.

Chitosan possesses inherent immunostimulatory activity by activating antigent-presenting cells (APC), including dendritic cells and macrophages. CNCs enhance cellular uptake through endocytosis and promote antigen presentation via both MHC class I and II pathways, thereby stimulatinf Th1 (cell-mediated) and Th2 (humoral) immune responses (Fig. 11).

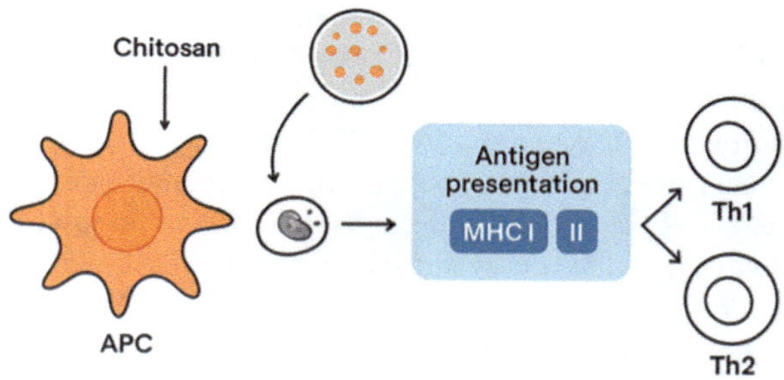

Fig. 11 Adjuvant properties of chitosan nanocomposites (CNCs)

Targeted and Controlled Antigen Release: CNCs provide sustained release of antigens, ensuring prolonged immune stimulation with fewer booster doses. Functionalization with ligands can direct CNCs to lymphoid tissues or specific immune cell subsets, improving vaccine efficacy.

Advantages Over Conventional Systems: Unlike aluminum salts or oil-in-water emulsions, CNCs are biodegradable, non-toxic, and versatile. They enable needle-free vaccination routes (oral, nasal, pulmonary), increasing patient compliance and reducing costs.

Chitosan nanocomposites function as dual-purpose vaccine platforms: protecting and delivering antigens while simultaneously acting as immunostimulatory adjuvants. Their ability to enhance both mucosal and systemic immunity, along with their tunability for cancer or infectious disease applications, positions CNCs as powerful candidates for next-generation vaccines and immunotherapies.

4 Challenges and Future Directions

Regulatory Issues: Although chitosan is widely recognized as biocompatible and safe, regulatory approval for chitosan-based drug delivery systems still requires comprehensive preclinical and clinical studies to establish long-term safety and efficacy.

Scalability: Large-scale manufacturing of chitosan-based nanomaterials with consistent quality remains a key challenge. Factors include:

Raw material variability: Chitosan is derived from natural sources such as crustacean shells, and variations in source, processing, and molecular weight distribution can result in batch-to-batch inconsistencies.

Reproducible synthesis: Nanoparticle size, morphology, and surface characteristics are highly sensitive to synthesis parameters (pH, temperature, mixing speed, concentration). Ensuring reproducibility at an industrial scale is complex.

Solubility and stability: Chitosan shows limited solubility at neutral and basic pH, which impacts formulation stability and reduces its application range.
Process scalability: Techniques commonly used at the laboratory scale (e.g., ionic gelation, microemulsion) are often difficult to scale up without compromising uniformity in particle size and charge.
Regulatory and quality control: Compliance with pharmaceutical, biomedical, or food industry standards requires strict monitoring of purity, sterility, and biocompatibility.
Cost-effectiveness: Large-scale production must balance efficiency and cost, particularly when using high-purity chitosan and specialized equipment.

Strategies to Address Challenges: Potential solutions include optimization of raw material sourcing, refinement of synthesis protocols, incorporation of advanced process monitoring, and the adoption of scalable techniques such as microfluidics or spray drying.

Chitosan-based nanomaterials hold immense promise in drug delivery due to their versatility, biocompatibility, and biodegradability. They are well-suited for controlled and targeted delivery of diverse therapeutic agents. However, further research is needed to overcome scalability issues, improve reproducibility, and achieve regulatory approval to enable successful translation into clinical applications.

References

1. Baharlouei P, Rahman A (2022) Chitin and chitosan: prospective biomedical applications in drug delivery, cancer treatment, and wound healing. Mar Drugs 20:460. https://doi.org/10.3390/md20070460
2. Yanat M, Betemariam G, Workneh TS (2021) Preparation methods and applications of chitosan nanoparticles: a review. Int J Biol Macromol 178:350–369. https://doi.org/10.1016/j.ijbiomac.2021.12.041
3. Gan Q, Wang T, Cochrane C, McCarron P (2005) Modulation of surface charge, particle size and morphological properties of chitosan-TPP nanoparticles intended for gene delivery. Colloids Surf B Biointerfaces 44(2–3):65–73. https://doi.org/10.1016/j.colsurfb.2005.06.001
4. Nguyen AD, Nguyen TNH, Dang THV, Nguyen TLP, Nguyen TNQ, Dinh MH, Le VH (2011) Chitosan nanoparticle as a novel delivery system for a/H1N1 influenza vaccine: safe property and immunogenicity in mice. Int J Biotechnol Bioeng 5(12):915–922
5. Nagpal K, Singh S, Mishra D (2010) Chitosan nanoparticles: a promising system in novel drug delivery. Chem Pharm Bull 58:1423–1430. https://doi.org/10.1248/cpb.58.1423
6. Klein RS, de Almeida DA, de Oliveira AC, Bonafé EG, Monteiro JP, Sabino RM, Martins AF (2025) Iota-carrageenan/chitosan nanoparticles via Coacervation: achieving stability for tiny particles. Nanomaterials 15:161. https://doi.org/10.3390/nano15030161
7. Nguyen TV, Nguyen TTH, Wang S-L, Vo TPK, Nguyen AD (2017) Preparation of chitosan nanoparticles by TPP ionic gelation combined with spray drying, and the antibacterial activity of chitosan nanoparticles and a chitosan nanoparticle–amoxicillin complex. Res Chem Intermed 43:3527–3537. https://doi.org/10.1007/s11164-016-2428-8
8. La TKN, Wang S-L, Dinh H, Phung ML, Nguyen TV, Nguyen AD (2014) Preparation of chitosan nanoparticles by spray drying and their antibacterial activity. Res Chem Intermed 40(6):3887–3900. https://doi.org/10.1007/s11164-013-1440-2

9. Wei Y, Wang P, Xu J (2020) Investigations of the influences of processing conditions on size, shape and surface morphology of spray-dried chitosan-TPP particles. Sci Rep 10:1155. https://doi.org/10.1038/s41598-020-58184-3
10. Frigaard J, Jensen JL, Galtung HK, Hiorth M (2022) The potential of chitosan in nanomedicine: an overview of the cytotoxicity of chitosan-based nanoparticles. Front Pharmacol 13:880377. https://doi.org/10.3389/fphar.2022.880377
11. Nguyen VN, Wang S-L, Nguyen TH, Nguyen VB, Doan MD, Nguyen AD (2024) Preparation and characterization of chitosan/starch nanocomposites loaded with ampicillin to enhance antibacterial activity against *Escherichia coli*. Polymers 16:2647. https://doi.org/10.3390/polym16182647
12. Bashir SM, Rather GA, Patrício A, Haq Z, Sheikh AA, Shah MZuH, Singh H, Khan AA, Imtiyaz S, Ahmad SB et al (2022) Chitosan nanoparticles: a versatile platform for biomedical applications. Materials 15:6521. https://doi.org/10.3390/ma15196521
13. Gong X, Gao Y, Shu J, Zhang C, Zhao K (2022) Chitosan-based nanomaterial as immune adjuvant and delivery carrier for vaccines. Vaccine 10:1906. https://doi.org/10.3390/vaccines10111906
14. Milenkova S, Ambrus R, Mukhtar M, Pilicheva B, Marudova M (2024) Spray-dried chitosan hydrogel particles as a potential delivery system for benzydamine hydrochloride. Gels 10:189. https://doi.org/10.3390/gels10030189
15. Quiroz-Castillo JM, Rodríguez-Félix DE, Grijalva-Monteverde H, Del Castillo-Castro T, Plascencia-Jatomea M, Rodríguez-Félix F, Herrera-Franco PJ (2014) Preparation of extruded polyethylene/chitosan blends compatibilized with polyethylene-graft-maleic anhydride. Carbohydr Polym 101:1094–1100
16. Gonciarz W, Balcerczak E, Brzeziński M, Jeleń A, Pietrzyk-Brzezińska AJ, Narayanan VHB, Chmiela M (2025) Chitosan-based formulations for therapeutic applications: a recent overview. J Biomed Sci 32:62. https://doi.org/10.1186/s12929-025-00815-2
17. Malamatari M, Charisi A, Malamataris S, Kachrimanis K, Nikolakakis I (2020) Spray drying for the preparation of nanoparticle-based formulations: review and practical considerations. PRO 8:788. https://doi.org/10.3390/pr8070788
18. Amaral R, Concha T, Vítor J, Almeida AJ, Calado C, Gonçalves LM (2025) Chitosan nanoparticles for enhanced immune response and delivery of multi-epitope *helicobacter pylori* vaccines in a BALB/c mouse model. Pharmaceutics 17:132. https://doi.org/10.3390/pharmaceutics17010132
19. Antoniou V, Mourelatou EA, Galatou E, Avgoustakis K, Hatziantoniou S (2024) Gene therapy with chitosan nanoparticles: modern formulation strategies for enhancing cancer cell transfection. Pharmaceutics 16:868. https://doi.org/10.3390/pharmaceutics16070868
20. Naghib SM, Ahmadi B, Kangarshahi BM, Mozafari MR (2024) Chitosan-based smart stimuli-responsive nanoparticles for gene delivery and gene therapy: recent progresses on cancer therapy. Int J Biol Macromol 278(3):134542
21. Wang J, Zhao Z, Wang Q, Shi J, Wong DWC, Cheung JCW (2024) Advancements in nanoparticle-based adjuvants for enhanced tuberculosis vaccination: a review. Vaccine 12:1335. https://doi.org/10.3390/vaccines12121335
22. Bugybayeva D, Dumkliang E, Patil V, Yadagiri G, Suresh R, Singh M, Schrock J, Dolatyabi S, Shekoni OC, Yassine HM (2024) Evaluation of Efficacy of Surface Coated versus Encapsulated Influenza Antigens in Mannose–Chitosan Nanoparticle-Based Intranasal Vaccine in Swine. Vaccine 12:647. https://doi.org/10.3390/vaccines12060647
23. Zhao T, Li Q, Zhang Y (2023) Vaccine adjuvants: mechanisms and platforms. Signal Transduct Target Ther 8:237. https://doi.org/10.1038/s41392-023-01303-8
24. Jiang Y, Wang X, Liu C (2025) Polysaccharide nanoparticles as potential immune adjuvants. Mater Sci Eng C Mater Biol Appl 149:115472. https://doi.org/10.1016/j.msec.2024.115472

Anh Dzung Nguyen

Van Bon Nguyen

Vu Hoa Tran

Chien Thang Doan

San-Lang Wang

Thiolated Chitosan: A Versatile Mucoadhesive Polymer for Enhanced Mucosal Drug Delivery

Manasa N, Swornahuti Panda, Jerifa Sonowal, Sudarshan Naidu Chilamakuri, and Subramanian Natesan

Abstract Thiolated chitosan is a versatile and extremely promising mucoadhesive polymer for sophisticated mucosal drug delivery applications. This chapter provides an in-depth overview of its synthesis, functionalization, and mechanism of enhanced mucoadhesion through the formation of covalent disulfide bonds with mucosal glycoproteins. This enhances drug retention, bioavailability, and permeation through various routes, including oral, nasal, buccal, ocular, and vaginal delivery. Comparative analyses with other mucoadhesive polymers are used to support the description of key applications in the context of hydrogels, films, injectable systems, and nanoparticles for targeted and sustained release. Emphasis is placed on safety considerations, thiol oxidation stability issues, and prospects for intelligent, stimuli-responsive drug delivery systems, highlighting the clinical utility and translational potential of thiolated chitosan in next-generation polymer-based medicines.

Keywords Thiolated chitosan · Mucoadhesion · Disulfide bond · Drug delivery · S-protection · Oxidation stability

1 Introduction

Due to limitations associated with conventional dosage forms, such as low bioavailability, rapid clearance, and the breakdown of therapeutic agents before they reach their site of action, the development of sophisticated drug delivery systems

Manasa N · S. Panda · J. Sonowal · S. N. Chilamakuri · S. Natesan (✉)
Department of Pharmaceutics, Advanced Formulation Laboratory, National Institute of Pharmaceutical Education and Research, Kolkata, West Bengal, India
e-mail: snatesan@niperkolkata.ac.in

S.-L. Wang, M. Anpo (eds.), *Chitin and Chitosan*, Springer Series on Polymer and Composite Materials, https://doi.org/10.1007/978-981-95-7457-5_12

remains a primary focus in pharmaceutical research [1]. For macromolecules like peptides, proteins, and nucleic acids which are extremely vulnerable to enzymatic degradation and have limited permeability across mucosal barriers these difficulties are especially noticeable. Using mucoadhesive polymers, which can adhere to mucosal surfaces and prolong the time that drug carriers remain in contact with the absorption site, is a promising tactic. The mucosa, which covers a broad range of drug administration routes, provides numerous opportunities for local and systemic delivery, including oral, nasal, ocular, buccal, vaginal, and rectal pathways. Such adhesion increases the chance of drug absorption, decreases clearance by mucociliary action or fluid turnover, and lengthens residence time at the absorption site [2]. Chitosan has been identified as one of the most promising natural polymer for mucoadhesion. The deacetylation of chitin yields chitosan, a linear cationic polymer composed of glucosamine and N-acetylglucosamine units. It is a flexible excipient in mucosal drug delivery due to its biodegradability, biocompatibility, low toxicity, and natural mucoadhesive qualities [3].

Native chitosan has serious disadvantages in spite of these benefits. Protonated amino groups and negatively charged sialic acid or sulfate groups in mucins mainly rely on hydrogen bonding and ionic interactions to maintain mucoadhesion. Despite their advantages, these non-covalent interactions have limited adhesion strength due to their comparative weakness and reversibility. Moreover, chitosan's pH-dependent solubility limits its use in physiological settings where neutral or alkaline pH predominates, as it is only soluble in acidic environments [4]. To overcome these drawbacks, various chemical modifications have been investigated. Thiolation has proven to be one of the most successful tactics among them. This method uses covalent conjugation to add thiol groups (-SH) to the chitosan backbone. Covalent disulfide bonds can be formed between the resulting thiomers or thiolated chitosan and mucus glycoproteins with cysteine-rich subdomains. Compared to the reversible forces of unmodified chitosan, this produces a significantly stronger and more stable interaction, resulting in longer retention at mucosal sites. Furthermore, the thiol groups have the capacity to undergo oxidative cross-linking, imparting cohesive gel-forming ability and enhancing the mechanical strength of dosage forms. All of these characteristics together make thiolated chitosan a next-generation mucoadhesive polymer that can overcome many of the drawbacks of both unmodified chitosan and traditional delivery methods [5].

Thiolated chitosan has additional functional benefits besides adhesion. It can prevent molecules like peptides from degrading too quickly by inhibiting enzymes like proteases. Additionally, by preserving tight junction opening, it raises epithelial permeability. When combined, these characteristics result in notable enhancements in the bioavailability of drugs. The functionalization, synthesis, characterisation, and mechanisms of thiolated chitosan, as well as its exceptional mucoadhesive qualities, are thoroughly covered in this chapter. Its use in oral, nasal, buccal, ocular, vaginal, injectable, and nanoparticulate drug delivery systems is thoroughly explained. Additionally, case studies demonstrating improvements in bioavailability and pharmacokinetics are provided to highlight its clinical significance.

2 Functionalization of Chitosan

The versatile natural biopolymer known as chitosan (CS), which is produced by deacetylating chitin, is distinguished by the presence of reactive amino and hydroxyl groups along its backbone. These functional groups enable a range of chemical modifications that enhance chitosan's solubility, mechanical strength, mucoadhesion, and biocompatibility, thereby customising it for specific drug delivery applications. Covalent modifications that target amino and hydroxyl groups and non-covalent modifications that involve electrostatic interactions and hydrogen bonding with small molecules, proteins, or metal ions are the two main categories of functionalization [6].

2.1 *Covalent Functionalization Strategies*

2.1.1 Amine Functionalization

- Amide Bond Formation: The most frequent covalent modification is the creation of amide bonds between the primary amines in chitosan and molecules that contain carboxyl. Typically, carbodiimide coupling agents, such as EDC, are used to achieve this. In mild aqueous or organic conditions, HCl is frequently mixed with NHS or HOBt to create stabilised active esters, which facilitate the effective grafting of bioactive molecules, polymers, or peptides. Examples include attaching targeting peptides for cancer treatment, conjugating to PEG derivatives for increased solubility, and functionalizing with hydrophobic groups to create micelles for drug encapsulation.
- Anhydride-mediated Conjugation: Chitosan amines and cyclic anhydrides, such as succinic and maleic anhydrides, engage in ring-opening reactions that release carboxyl groups that improve solubility and offer locations for additional modification. For instance, N-succinyl chitosan exhibits enhanced hydrophilicity and biocompatibility, making it easier to use for purposes such as intestinal absorption enhancement and gene delivery.
- Phthaloylation: By protecting amino groups, phthaloylation makes chitosan more soluble in organic solvents and permits the selective modification of hydroxyl groups prior to deprotection. This approach facilitates multifunctionalization and has been used to create nucleic acid carriers and antifouling membranes.
- Urea and Alkylation Reactions: Amines react with carbonyldiimidazole or isocyanates to form urea bonds, which are then used to graft cationic moieties like polyethyleneimine (PEI) to improve gene delivery. By adding tiny alkyl or functional groups, such as quaternized ammonium moieties (trimethyl chitosan), alkylation via nucleophilic substitution enhances the substance's water solubility and mucoadhesive properties.

- Ring Opening of Epoxides and Aziridines: Amines' nucleophilic attack on epoxides or aziridines creates stable bonds, resulting in derivatives that can self--assemble into nanostructures appropriate for gene and medication delivery applications.
- Imines and Michael Additions: Reversible imine bonds are created when chitosan amines react with aldehydes; these bonds are helpful for cross-linked networks and self-healing materials. Aza-Michael additions with α, β-unsaturated carbonyls increase the diversity of functional groups by facilitating covalent conjugation in mild conditions.

2.1.2 Hydroxyl Functionalization

After protecting amino groups, hydroxyl groups mostly at the C6 position can be modified selectively. Reagents like carbonyldiimidazole for carbamates and anhydrides for esters are frequently used in the formation of carbamate and ester bonds. Drug carriers can have customised release profiles thanks to these derivatisations, which change solubility and degradability. Additional methods for functionalizing hydroxyl groups with bio-responsive moieties include transacetalization and nucleophilic substitution.

2.1.3 Oxidative Cleavage

By cleaving sugar rings along the backbone, periodate oxidation produces reactive aldehyde groups and provides conjugation sites through reductive amination with amines or peptides. This produces derivatives of functionalized chitosan that can be used in targeted gene and drug delivery systems.

2.2 Non-Covalent Functionalization

Stable ionic interactions with negatively charged molecules such as proteins, polymers, and metal ions are made possible by chitosan's positive charge at physiological pH. Without requiring significant chemical modification, these non-covalent assemblies improve stability, mechanical strength, and sustained release. Examples include metal ion cross-linking for antimicrobial activity, electrostatic complexation with alginate to form hydrogels, and encapsulation of nanoparticles to stop them from aggregating and becoming toxic.

3 Thiolated Chitosan

Thiolated chitosan represents a significant advancement in the development of multifunctional polymers for drug delivery, owing to its superior stability, enhanced mucoadhesiveness, and ability to improve permeation compared to native chitosan. The ability of thiolated chitosan to form robust covalent disulfide bonds with the cysteine-rich domains of mucus glycoproteins, thereby extending the residence time and promoting effective drug absorption at mucosal surfaces, is the primary justification for its utility in drug delivery systems.

Using sulfhydryl-bearing reagents like cysteine, thioglycolic acid, or specialised thiol ligands, thiol groups are attached to the primary amino positions of chitosan's glucosamine units. This chemical modification is usually accomplished by forming amide or amidine bonds. Achieving the ideal level of modification and preserving the stability of thiol groups are crucial factors that are often controlled by closely monitoring synthesis parameters, such as pH, reaction temperature, and oxygen exclusion. To ensure successful functionalization and the intended performance in drug delivery systems, thorough characterisation is carried out after synthesis. This includes spectrophotometric quantification of thiol groups, assessment of mucoadhesion, and evaluation of physicochemical characteristics.

The primary reactive site for the attachment of a thiol group, which is essential for the improved mucoadhesive qualities of thiolated chitosan, is the principal amino groups located in the second position of the glucosamine subunits of chitosan. Commonly, thiolation is accomplished by coupling cysteine or thioglycolic acid to these amino groups through carbodiimide-mediated reactions, which create amide bonds. 2-Iminothiolane is used in a different one-step process to create amidine bonds. Synthesis is typically conducted in inert atmospheres or at acidic pH values below 5 to minimise the formation of unwanted disulfide bonds and prevent the oxidation of sensitive thiol groups. Because alkyl thiol groups have higher pKa values (8–10) and are more susceptible to oxidation, this pH control is particularly crucial. As opposed to alkyl thiol or unmodified chitosan derivatives, aryl thiol groups, which have lower pKa values (5–7), show higher reactivity and stabilise protein folding, improving the polymer's mucoadhesive and functional properties. If disulfides are present, Ellman's reagent can be used to precisely quantify the amount of thiol, both free and oxidised, providing a trustworthy indicator of successful thiolation. All things considered, these synthesis protocol considerations guarantee the production of strong thiolated chitosan polymers with exceptional bioadhesive capabilities for drug delivery applications [7]. Three majorly used thiolated chitosan derivatives enhanced mucoadhesion in drug delivery are shown in Fig. 1.

The primary amino groups located at the 2-position of the chitosan backbone act as nucleophiles for conjugation. Thioglycolic acid (HS-CH_2-COOH) serves as the source of reactive thiol groups. The carboxyl group of thioglycolic acid is first activated by the carbodiimide coupling agent EDC.HCl (1-ethyl-3-(3--dimethylaminopropyl) carbodiimide hydrochloride) in the presence of

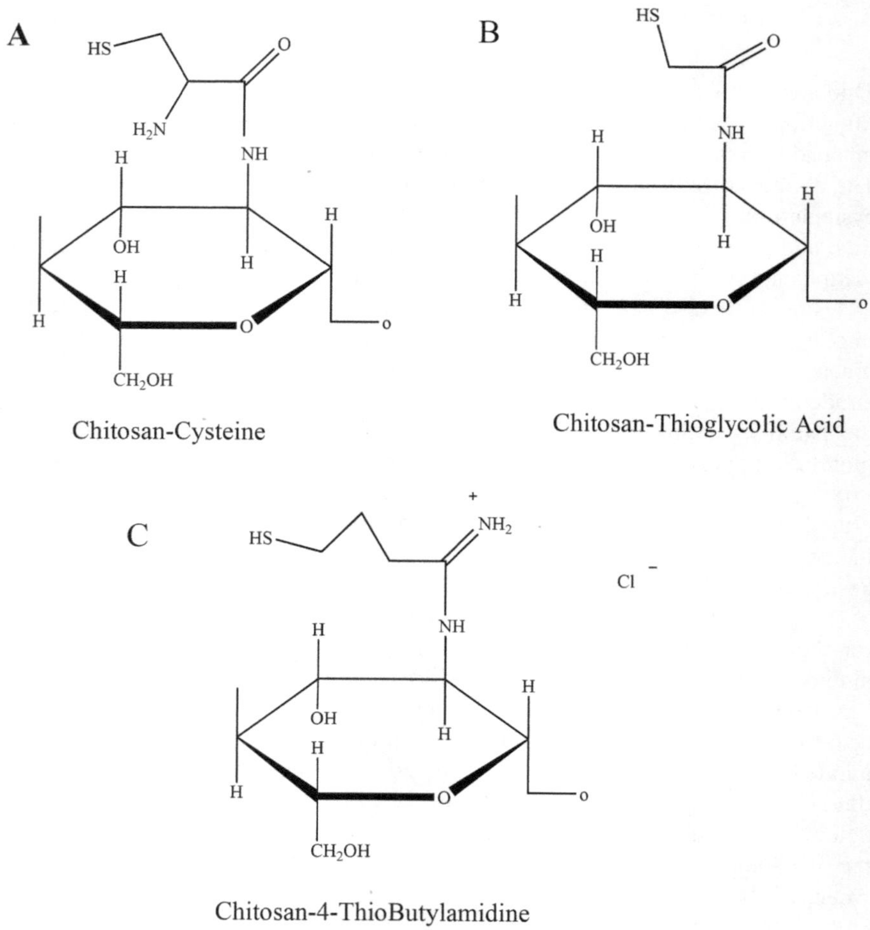

Fig. 1 Structures of three thiolated chitosan derivatives used for enhanced mucoadhesion and drug delivery: (**a**) chitosan–cysteine, where cysteine introduces thiol groups via amide linkage; (**b**) chitosan–thioglycolic acid, in which thioglycolic acid is conjugated to chitosan for additional thiol functionality; and (**c**) chitosan–4-thio-butylamidine, featuring a thiol-containing amidine group for superior mucoadhesiveness and drug carrier potential

N-hydroxysuccinimide (NHS), forming an active ester intermediate in water (Fig. 2). This intermediate then reacts with the primary amines of chitosan under mildly acidic conditions (pH ≈ 5), facilitating efficient amide bond formation and the covalent attachment of the thiol moiety to the chitosan polymer. Because the slightly acidic pH lowers the prevalence of reactive thiolate ions and inhibits undesired disulfide formation, this process guarantees high coupling efficiency while minimising oxidation of thiol groups. Usually, dialysis against diluted acid and water is used to complete purification, and characterisation is carried out using

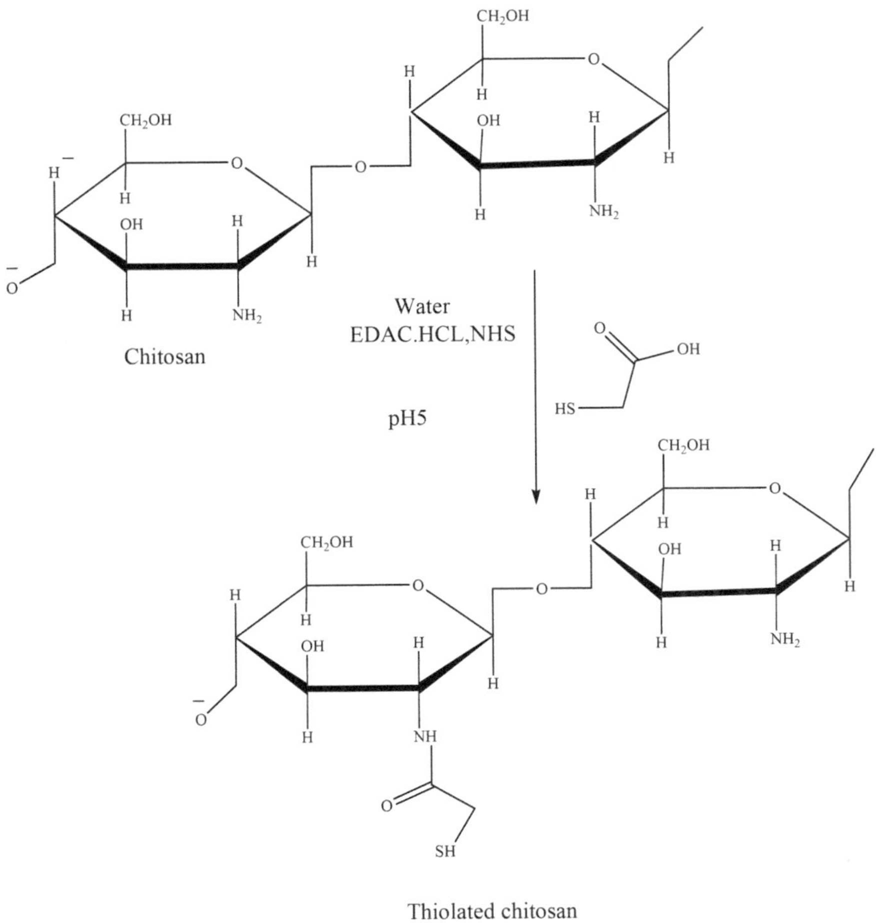

Fig. 2 Diagrammatic illustration of the process of creating thiolated chitosan by covalently conjugating thioglycolic acid to chitosan's amino groups using EDC. When used as coupling agents in an aqueous medium (pH 5), HCl and NHS introduce thiol (-SH) groups to improve mucoadhesive qualities

mucoadhesion tests, IR spectroscopy to confirm chemical structure, and Ellman's reagent to measure thiol content [8].

In another method, thiolated chitosan was synthesised by the interaction between chitosan and reactive NHS ester with a thiol group, which was prepared earlier. NHS ester functions as an activated carboxyl derivative, forming amide bonds with chitosan's primary amino groups very effectively. The NHS ester selectively couples to the amino sites on the chitosan backbone in mildly acidic conditions (pH 5 in water), resulting in thiolated chitosan and the release of NHS as a by-product (Fig. 3). Compared to in situ activation, this technique has better reproducibility,

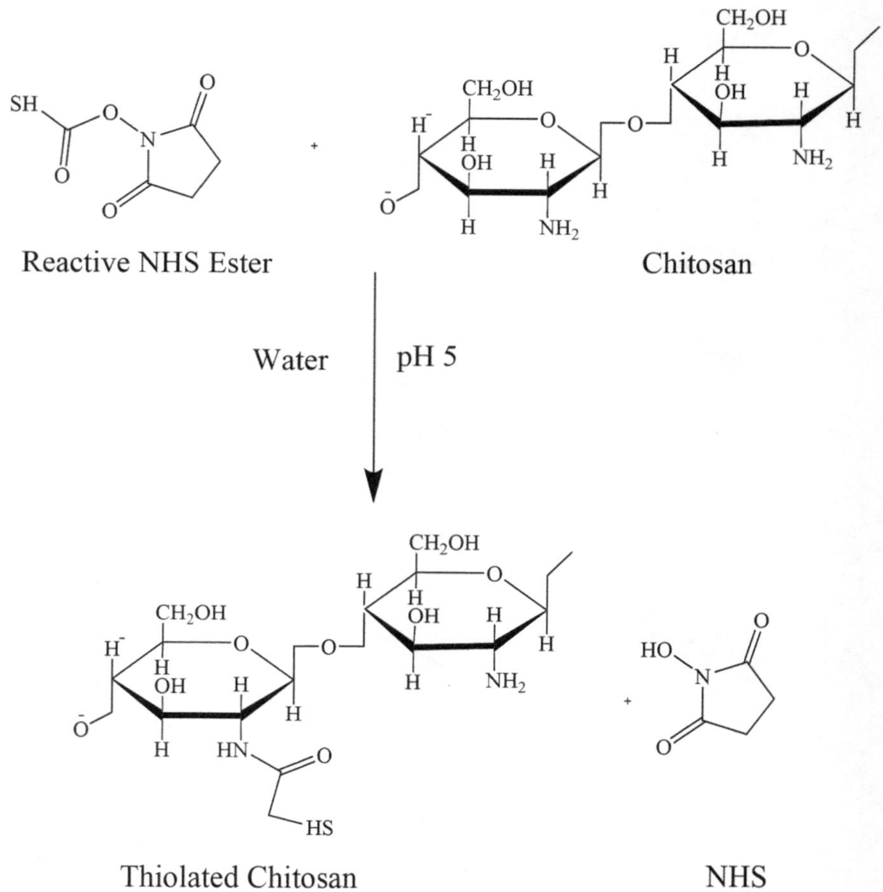

Fig. 3 Diagrammatic representation of the synthesis of thiolated chitosan with a reactive NHS ester serving as the thiol donor. NHS is released as a by-product of the reaction between the NHS ester and the amino groups in chitosan, which takes place in an aqueous medium at pH 5

higher conjugation yields, and a lower chance of thiol oxidation or adverse reactions. Pendant thiol groups in the resultant thiolated chitosan are accessible for mucoadhesive interactions and additional chemical alterations [8].

Thiolated chitosan was synthesised through the formation of thioethyl amidine conjugate using 2-iminothiolane (Traut's reagent). This one-step process involves an aqueous reaction between 2-iminothiolane hydrochloride and the primary amino groups of chitosan. 2-Iminothiolane is a bifunctional reagent: its imidoester moiety forms a stable amidine bond with the amine group of chitosan, and simultaneously introduces a thiol group at the same location (Fig. 4). To maximise

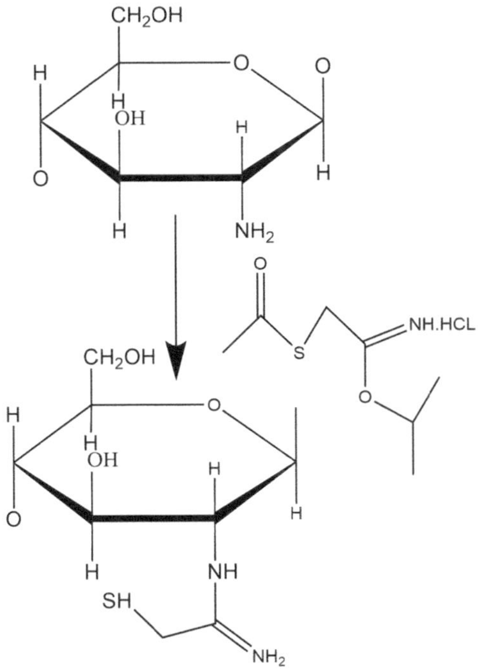

Fig. 4 Diagrammatic illustration of the functionalization of chitosan using 2-iminothiolane (Traut's reagent), which forms amide bonds to add thiol (-SH) groups to the chitosan backbone for thiolated chitosan synthesis

conjugation and minimise thiol oxidation, the procedure is typically carried out at a pH of 6–7, which is mildly acidic to neutral. The reaction effectively adds free thiol groups to the chitosan backbone while also streamlining the synthesis by eliminating the need for harsh conditions or separate activation steps. Thioethylamidine side chains, present in the resulting thiolated chitosan, provide reactive thiol groups for additional functionalization or to enhance the mucoadhesive properties of drug delivery systems. This method offers a simplified alternative to carbodiimide-mediated amide coupling, with fewer synthetic steps and mild reaction conditions, providing comparable or superior levels of thiol functionality [9].

Carbodiimide-mediated coupling is used to synthesise the chitosan-1--mercaptomethyl-cyclopropane conjugate. This method involves the reaction of chitosan and mercaptomethyl-cyclopropane acetic acid for 24 h at 37 °C with NHS and EDC (1-ethyl-3-(3-dimethylaminopropyl) carbodiimide). The thiol-containing acid's carboxyl group is activated by EDC, creating an intermediate that effectively couples with the chitosan's primary amine groups to form an amide bond and produce the thiolated chitosan derivative (Fig. 5). The product has pendant mercaptomethyl-cyclopropane side chains with free thiol (-SH) groups that can be oxidised to create disulfide cross-links inside or between chitosan chains [10].

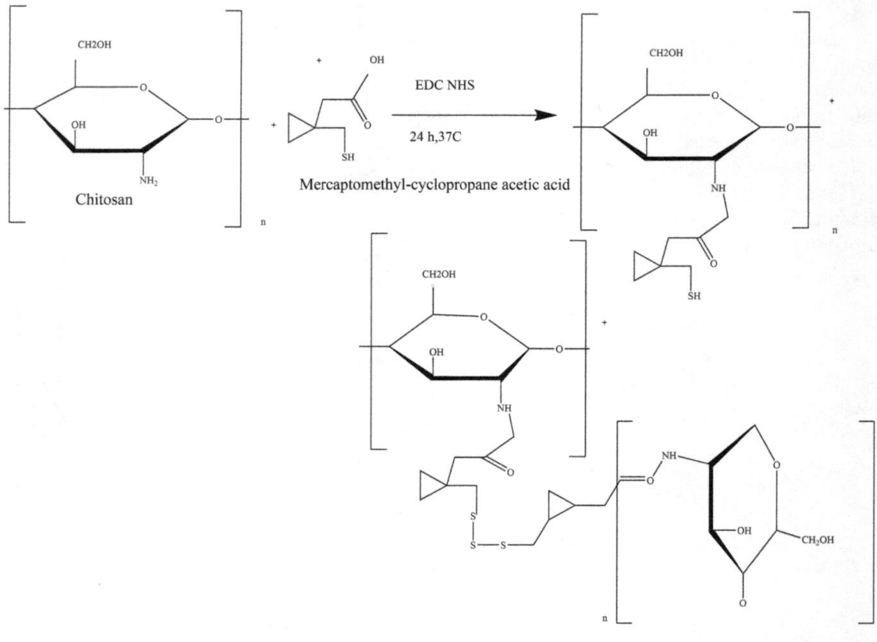

Fig. 5 Diagram of mercaptomethyl-cyclopropane acetic acid functionalization of chitosan. The carboxyl group of mercaptomethyl-cyclopropane acetic acid is coupled to the primary amines of chitosan in an aqueous medium (pH ~5) under EDC/NHS activation for 24 h at 37 °C, resulting in thiolated chitosan. Following mild oxidation of free thiols, modified chitosan chains form intermolecular disulfide cross-links, resulting in the formation of chitosan-1-mercaptomethyl-cyclopropane derivatives

4 Mechanism of Mucoadhesion

The term "mucoadhesion" describes a polymeric material's capacity to stick to mucosal surfaces, which helps to improve the localisation, absorption, and retention of the drug delivery system over time. Extended residence time can greatly boost both bioavailability and therapeutic efficacy, a property that has been widely utilised in formulations intended for oral, buccal, nasal, ocular, vaginal, and rectal routes. Due to the high permeability of mucosal membranes, mucoadhesive systems offer several significant advantages over traditional drug delivery methods. These include improved bioavailability by avoiding some physiological defence mechanisms, prolonged drug residence at the site of application, and increased systemic uptake. Mucoadhesive polymers are particularly useful in oral controlled release dosage forms because they temporarily adhere to the epithelial surfaces or the gastrointestinal mucosa. Secondary interactions, such as hydrogen bonds and Van der Waals forces, between the polymer and the mucosal membrane play a major role in mediating this adhesion. Thus, mucoadhesive polymers represent a distinct subclass of bioadhesive materials [11].

Mucoadhesive behaviour arises from several physicochemical mechanisms, which are explained by various theories, including wetting, semi-empirical, fracture, electronic, and adsorption. The various physicochemical mechanisms of mucoadhesive polymer are shown in Fig. 6.

1. **Wetting Theory**: This theory highlights a mucoadhesive polymer's capacity to permeate surface imperfections, where it subsequently solidifies to establish mucoadhesion. The contact angle measurement is used to assess the degree of surface affinity [12].
2. **Semi-Empirical Theories (Diffusion, Mechanical, Fracture):**
    1. **Diffusion theory:** According to diffusion theory, the concentration gradient and the amount of time needed for polymer chains to diffuse into the mucus glycoprotein network determine mucoadhesion. Diffusion happens in both directions, and the process begins with the formulation of an interpenetration layer. Strong adhesion is established when this layer has a thickness of about 0.2–0.5 μm. factor such as the concentration gradient, molecular weight of

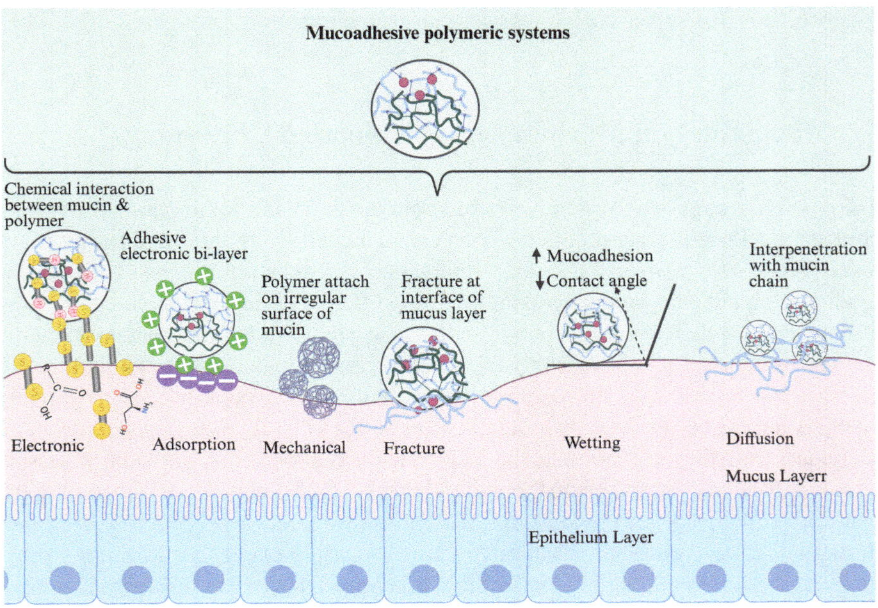

Fig. 6 An illustration of the main characteristics and mechanisms of mucoadhesive polymers in drug delivery systems. The figure illustrates the various phases of mucoadhesion, including diffusion/interpenetration with mucin chains, wetting, interfacial fracture, mechanical interlocking, adsorption, and electronic interactions. Every step enhances the capacity of polymer-based drug carriers to interact with and adhere to mucous and epithelial layers, thereby improving drug absorption and retention at mucosal surfaces

adhesive macromolecules, as well as the size, mobility, flexibility, and length of polymer chains, significantly influence the thickness of this layer.

2. **Mechanical theory:** According to the mechanical theory, adhesion happens when a mucoadhesive material penetrates and fills the irregularities of rough surfaces. These surface imperfections enhance the contact area, making this stage an essential step in the mucoadhesion process.
3. **Fracture theory:** According to fracture theory, the force that holds two surfaces together and the force needed to separate them depend on the presumption that separation happens as a result of c rack formation. The required detachment force is expressed as $\sigma = \frac{\sqrt{(E \cdot \varepsilon)}}{L}$, where E is young's modulus of elasticity, ε is the fracture energy, and L is the critical crack length.

3. **Electronic theory:** This postulates that variations in the electronic structures of contacting adhering surfaces cause electron transfer. It is suggested that this will cause an electrical double layer to form at the interface, followed by adhesion due to attractive forces.
4. **Adsorption theory**: It explains how adhesives adhere based on van der Waals forces and hydrogen bonds. It has been suggested that the sticky contact is primarily caused by these pressures. The chemisorption theory, a subset of this, postulates that strong covalent bonds cause an interaction across the surface [13].

5 Mucoadhesion Mechanisms of Thiolated Chitosan

Thiol (-SH) groups are covalently grafted onto the main amino functionalities of the chitosan backbone to create thiolated chitosan, a chemically modified form of chitosan. Due to its unique dual-mode interaction ability with mucosal tissues, this modification enhances the mucoadhesive properties of chitosan. Thiolated chitosan also forms covalent disulfide bonds with the cysteine-rich subdomains of mucin chains, in contrast to native chitosan, which primarily adheres to mucus through electrostatic interactions between its protonated amino groups and the negatively charged sialic acid residues on mucin glycoproteins [7]. Initial physical contact and secondary bonding are steps in the mucoadhesion process of thiolated chitosan, which is followed by the critical chemical cross-linking step. Under physiological conditions, the thiol groups on modified chitosan interact with mucosal cysteines through oxidative coupling reactions or thiol-disulfide exchange, forming strong, long-lasting covalent bonds. This covalent bonding significantly improves the adhesive strength, lengthens the residence time on mucosal linings, and provides resistance to fluid flow or mucus turnover-induced mechanical clearance, which is a major drawback of many traditional mucoadhesive systems [14].

Additionally, by temporarily widening the tight junctions between epithelial cells, thiolated chitosan regulates paracellular permeability. Reversible inhibition of protein tyrosine phosphatases mediates this effect, increasing the paracellular transport of macromolecular and hydrophilic medications that would otherwise have low mucosal permeability. An extra defence against early drug loss and enzymatic degradation is provided by the thiolated chitosan's inherent capacity to form in situ gels through the formation of intra- and intermolecular disulfide bonds, which complements the improved permeation [7].

The pH stability range of thiolated chitosan, which preserves functional thiol groups across a wide physiological pH range, contributes to its superior bioadhesive properties. However, in order to maintain mucoadhesive functionality during formulation and storage, thiol oxidation sensitivity necessitates strategies like S-protection or stabilisation with antioxidants. All of these characteristics make thiolated chitosan a versatile polymeric excipient for mucosal drug delivery applications, such as oral, nasal, buccal, vaginal, and ocular, that require extended retention, improved permeability, and controlled release [15].

6 Comparative Evaluation of Mucoadhesive Polymers

Rational formulation design in drug delivery requires an understanding of the unique physicochemical characteristics and functional benefits of different mucoadhesive polymers. In this section, thiolated chitosan is compared to other widely used polymers, including alginate, HPMC, Carbopol, and unmodified chitosan. Key attributes related to mucoadhesion, permeation enhancement, adhesion strength, biocompatibility, enzyme inhibition, pH sensitivity, gel-forming ability, sustained release capability, solubility, oxidative stability, cost, and typical applications are the focus of comparison [7, 3, 17] and these are summarised in Table 1, which helps choose the best one for a given therapeutic use.

7 Applications of Thiolated Chitosan in Drug Delivery

The chemically modified form of chitosan, thiolated chitosan, has also garnered significant interest in the fields of drug delivery and biomedical applications due to its enhanced mucoadhesive properties and biocompatibility. The chitosan itself is a natural polysaccharide derived from the deacetylation of chitin, which possesses biodegradability, low toxicity, and is also naturally mucoadhesive. But the weakness of electrostatic interactions and also pH dependence usually limits the mucoadhesive capacity of native chitosan, and this limits its adhesion to acidic environments. To avoid such shortcomings, thiolation of chitosan incorporates thiol (-SH) groups into the backbone of the polymer, enhancing its mucoadhesive capacity tremendously through the establishment of covalent disulfide bonds with the cysteine-rich regions of mucosal glycoproteins. This modification enables a stronger attachment to the mucosal surfaces compared to the non-covalent interaction of unmodified chitosan. The addition of thiol groups increases the residence time of the drug in mucosal sites, permitting sustained and targeted release of the drug, which is

Table 1 Key physicochemical attributes and performance traits of popular mucoadhesive polymers in drug delivery systems

Characteristic	Thiolated chitosan	Unmodified chitosan	Carbopol (Polyacrylic Acid)	HPMC (Hydroxypropyl Methylcellulose)	Alginate
Mucoadhesion mechanism	Covalent disulfide bonds with mucins; hydrogen bonding and ionic interaction	Ionic interactions and hydrogen bonding	Hydrogen bonding, ionic interactions	Hydrogen bonding, chain entanglement	Hydrogen bonding, gel formation
Strength of adhesion	Very high	Moderate	High	Moderate	Moderate
Permeation enhancement	Strong (opens tight junctions)	Weak	Low to none	Low to none	Low to none
Enzyme inhibition	Yes	Minimal	No	No	No
Biocompatibility	High	High	Moderate (can be irritant)	High	High
pH sensitivity	Functional over a wide pH range	Soluble only in acidic pH	Effective mostly in neutral pH	Stable across pH ranges	Stable across pH ranges
Gel-forming ability	Moderate (in situ gelling via disulfide bonds)	Moderate	High	High	High
Sustained release ability	Excellent	Moderate	High	Moderate	Moderate
Solubility	Soluble in acidic and neutral pH (better than unmodified)	Soluble in acidic media	Soluble in water	Soluble in water	Soluble in water
Stability (oxidation)	Sensitive to oxidation (can be stabilised via S-protections)	Stable	Stable	Stable	Stable
Cost	Moderate to High	Low	Low to moderate	Low	Low
Common applications	Buccal, nasal, ocular, vaginal, intestinal delivery	Buccal, nasal, ocular	Topical gels, oral, vaginal	Oral, ocular, topical	Oral, topical

beneficial in various transmucosal delivery routes, including oral, nasal, vaginal, ocular, and intravesical delivery [7, 16].

Thiolated chitosan conjugated with thiol-containing agents, which include cysteine, thioglycolic acid, N-acetyl cysteine and 3-mercaptopropionic acid. These thiomers have varied pharmaceutical applications, particularly in nanoparticle formulations, hydrogels, films, and tablets. Thiolated chitosan nanoparticles exhibit enhanced stability and adhesiveness to mucosal tissues due to increased interactions through disulfide bonding. Thiolated chitosan nanoparticles loaded with amoxicillin were designed to deliver the drug to Helicobacter pylori infection with greater efficacy in the acidic conditions in the stomach, causing the bacteria to grow, thus displaying the potential in the treatment of gastric ulcers. Thiolated chitosan nanoparticles have also been found useful in the field of colonic delivery [16]. Genipin cross-linked thiolated chitosan nanoparticles, surface-coated with anti-cancer compounds, were shown to exhibit better mucoadhesion and increased cytotoxicity against colorectal cancer cell lines. Such formulations facilitate longer retention at the colon site, which is essential in local therapy. On the same note, insulin-loaded thiolated chitosan nanoparticles in fast-disintegrating tablets and films exhibited greater insulin loading and mucoadhesiveness as well as long-term hypoglycaemic effects in diabetic animals, which implies they can be used for the oral administration of insulin [19]. Thiolated chitosan hydrogel systems have been tested as vaginal gels, exhibiting enhanced gelation behaviour and mucoadhesivity in comparison with non-modified ones. These thermoresponsive hydrogels enhance delayed drug delivery and exhibit excellent biocompatibility, making them a promising option for the treatment of vaginal infections and atrophic diseases [22]. Genistein, a natural product, has been shown to induce vaginal tissue repair in preclinical models when loaded into thiolated chitosan hydrogels, highlighting the regenerative capabilities of thiolated preparations [21].

Other delivery systems that have been used in buccal delivery include thiolated chitosan films and compacts to increase the availability of drugs. Drugs such as risedronate sodium and pamidronate disodium, used to treat osteoporosis, exhibited better mucoadhesion, prolonged drug release, and improved mechanical characteristics. Thiolated niosomal gels have been investigated as nasally administered agents, exhibiting prolonged retention and enhanced delivery of various drugs, such as duloxetine [25]. Thiolated chitosan-based membranes were found to be better in antimicrobial activity and mechanical strength than unmodified chitosan composites in periodontal therapy. These membrane formulations extend the drug release duration to several days and are biodegradable; therefore, they could be used as effective agents in the treatment of periodontal disease [19].

Furthermore, thiolated chitosan nanoparticles have enhanced vaccine delivery, as they demonstrate improved mucosal retention and an immune response following intranasal delivery. For example, thiolated chitosan nanoparticles loaded with an influenza vaccine enhanced mucin adsorption and provided greater protection against viral infections, indicating that they may be useful in mucosal vaccination programmes [21].

Other functional moieties that can be added to thiolated chitosan derivatives include acrylates, methacrylates, catechols, boronates, and maleimides to enhance the mucoadhesive and controlled release characteristics. There are several benefits associated with these derivatives, including improved cellular permeability, enhanced drug encapsulation and targeting capability. However, stability during physiological conditions and potential oxidation are also factors to be considered.

Taken altogether, the years of research during the past decade on thiolated chitosan underline an excellent mucoadhesive behaviour of chitosan in comparison to non-covalent chitosan within its covalent disulfide bonding with the thiol moieties of mucosal residues. This leads to a longer residence time, targeted therapy, enhanced efficacy of drugs, and a reduction in the dosing schedule. The fact that it is versatile in creating different pharmaceutical dosage forms, ranging from nanoparticles to gels and films (Table 2), in various mucosal routes underscores its clinical translation promise [16].

Table 2 Applications of thiolated chitosan in different routes of administration

Route of administration	Drug/Compound	Role of thiolated chitosan	Key results	Reference
Oral	Insulin (human, oral tablets)	Mucoadhesive, permeation enhancer, sustained release matrix (chitosan-6-MNA)	Mucoadhesion ≈ 80× vs unmodified chitosan; sustained release 5 h (20 kDa) and 8 h (400 kDa); absolute oral BA 0.73% (20 kDa) and 0.62% (400 kDa); AUC improvement 4.8× (20 kDa) and 21.0× (400 kDa).	[17]
Nasal	Insulin (intranasal nanoparticles)	Thiolated chitosan (chitosan-NAC) as mucoadhesive nanoparticle coating & permeation enhancer	Particle size 140–210 nm; zeta +19.5 to +31.7 mV; drug loading 13–42%; mucin-binding >1.8× vs unmodified; enhanced nasal absorption in rats (improved hypoglycemic response).	[18]
Buccal	Insulin nanoparticles in mucoadhesive buccal film (DMEC--Cys = thiolated derivative)	Provides strong mucoadhesion in film, increases residence time and permeation across buccal mucosa	Optimised film: Tensile strength 5.81 kg/mm^2; bioadhesive force 2.47 N. Ex vivo permeation through rabbit buccal mucosa: Chitosan 17.1%, DMEC 67.89%, DMEC-Cys (thiolated) 97.18%.	[19]
Ocular	Various ophthalmic drugs (model drug in study; nanoparticles, gels)	Thiolated chitosan (and chitosan-NAC) increases mucoadhesion, retention on ocular surface, and can enhance corneal permeation and protect sensitive drugs.	Reported improved mucin interaction and ocular retention vs non-thiolated systems; chitosan-NAC shown to improve ocular compatibility and mucoadhesion.	[20]

Table 2 (continued)

Route of administration	Drug/Compound	Role of thiolated chitosan	Key results	Reference
Vaginal	Tenofovir (vaginal microbicide nanoparticles)	Thiolated chitosan enhances mucoadhesion, retention in vaginal mucus and controlled release, improving local drug concentrations for prevention	Comparative study showed thiolated NPs had superior mucoadhesion and retention vs non--thiolated counterparts (improved distribution/retention relevant for microbicide efficacy).	[21]
Injectable	Doxorubicin (in situ-forming thiolated chitosan hydrogel)	Thiolated chitosan (CSSH) forms injectable, thermosensitive/pH-responsive gels that gel in situ and provide local sustained release and tumour-pH--triggered release	Gelation ≈ 8–12 min at 37 °C; pH-responsive release (higher release at pH 5.5 vs 7.4); first 24 h ~ 10% dox release at acidic pH; effective cancer cell inhibition in vitro.	[22]
Nanoparticulate	Docetaxel (core-shell thiolated chitosan-coated nanoparticles)	Thiolated chitosan shell improves mucoadhesion, protects hydrophobic payload, and enhances permeation/transport across mucosa	Mean diameter < 200 nm; entrapment efficiency ≈ 90%; sustained release up to 10 days; ex vivo showed significantly increased intestinal transport vs free drug.	[23]

8 Safety and Challenges of Thiolated Chitosan

Before being widely used in pharmaceuticals, thiolated chitosan must overcome several stability, reproducibility, and safety issues, despite its promising mucoadhesive and permeation-enhancing qualities.

8.1 Oxidation of Thiol Groups

The vulnerability of free thiol (-SH) groups to oxidation is a significant drawback of thiolated chitosan. This can lead to the early formation of disulfide bonds with atmospheric oxygen or within the polymer during handling and storage. The mucoadhesive function may be compromised as a result of this oxidation, as it decreases the availability of free thiols required for covalent bonding with mucin. In order to combat this, techniques like adding antioxidants to formulations or S-protecting thiols using reversible masking groups have been used to maintain thiol functionality under physiological and storage conditions [24].

8.2 Stability Under Storage and Physiological Conditions

The degree of thiolation, the synthesis process, and the presence of protective agents all impact the stability of thiolated chitosan formulations. Environmental factors that significantly impact polymer integrity include pH, temperature, and oxygen exposure. Compared to regular thiolated chitosan, S-protected chitosan and freeze-dried formulations show improved stability. However, it remains challenging to ensure consistent performance in biological environments and maintain reproducible quality during large-scale formulation [7].

8.3 Scale-Up and Reproducibility Issues

Scaling up the production of thiolated chitosan is made more challenging by the intricacy of the synthesis, which involves amidine formation or carbodiimide chemistry, as well as sensitivity to oxidation. Strong process control and consistent analytical characterisation are required because batch-to-batch variations in the degree of thiolation, molecular weight distribution, and cross-link density can impact mucoadhesive strength and permeation enhancement. Furthermore, meticulous validation is necessary for regulatory approval of the purification and elimination of residual reactants [16].

8.4 Biocompatibility and Toxicity Aspects

Numerous in vitro and in vivo investigations show that thiolated chitosan has low toxicity profiles and favourable biocompatibility, which are on par with those of unmodified chitosan. It is generally well tolerated by mucosal tissues and is non-immunogenic. However, the thiol content and potential oxidative by-products need to be taken into account, particularly when administering the medication repeatedly or over an extended period. To determine safety margins and facilitate clinical translation, cytotoxicity tests, histopathological examinations, and immunogenicity assessments remain crucial in formulation development.

9 Future Perspectives

Thiolated chitosan is evolving into a class of intelligent, multipurpose polymers with applications that extend well beyond mucoadhesion. To achieve targeted, controlled drug release and enhanced therapeutic efficacy, emerging thiolated polymers incorporate stimuli-responsive elements (pH, temperature, and light), redox-sensitive linkers, and enzyme-inhibiting functionalities. These developments are enabling the delivery of complex biologics and vaccines through mucosal routes, where targeted absorption and stability are crucial.

An expanding patent portfolio, which reflects advancements in chemical modifications, delivery methods, and therapeutic indications, supports the growing clinical translation of thiolated chitosan formulations. By integrating thiolated chitosan into precision medicine platforms, drug delivery can be customised to each patient's physiology, disease progression, and responsiveness, improving clinical outcomes, and reducing adverse effects.

10 Conclusion

Thiolated chitosan is an extremely promising and adaptable mucoadhesive polymer excipient. It performs better at drug delivery than traditional systems due to its unique ability to form covalent disulfide bonds with mucosal glycoproteins, temporarily alter epithelial tight junctions, and undergo stimulus-triggered gelation. New developments in production and stabilisation methods are overcoming long-standing obstacles and opening up new possibilities for drug delivery through the mouth, nose, buccal, eyes, and vagina. In the near future, thiolated chitosan is expected to have a significant impact on polymer-based drug delivery science, thanks to a strong clinical pipeline and planned integration with personalised medicine.

References

1. Laffleur F, Keckeis V (2020) Advances in drug delivery systems: work in progress still needed? Int J Pharm 590:119912
2. Khutoryanskiy VV (2011) Advances in mucoadhesion and mucoadhesive polymers. Macromol Biosci 11:748–764
3. Ways M, T. M., Lau, W. M. & Khutoryanskiy, V. V. (2018) Chitosan and its derivatives for application in Mucoadhesive drug delivery systems. Polymers (Basel) 10:267
4. Tekade M et al (2019) Chapter 13: Thiolated-chitosan: a novel mucoadhesive polymer for better-targeted drug delivery. In: Tekade RK (ed) Biomaterials and bionanotechnology. Academic Press, pp 459–493. https://doi.org/10.1016/B978-0-12-814427-5.00013-5

5. Bernkop-Schnürch A (2005) Thiomers: a new generation of mucoadhesive polymers. Adv Drug Deliv Rev 57:1569–1582
6. Nicolle L, Journot CMA, Gerber-Lemaire S (2021) Chitosan functionalization: covalent and non-covalent interactions and their characterization. Polymers (Basel) 13:4118
7. Federer C, Kurpiers M, Bernkop-Schnürch A (2021) Thiolated Chitosans: a multi-talented class of polymers for various applications. Biomacromolecules 22:24–56
8. Zhu X et al (2012) Synthesis of thiolated chitosan and preparation nanoparticles with sodium alginate for ocular drug delivery. Mol Vis 18:1973–1982
9. Prinz M (2006) Chitosan-thio-amidine conjugates and their cosmetic as well as pharmaceutic use
10. Kafedjiiski K, Hoffer M, Werle M, Bernkop-Schnürch A (2006) Improved synthesis and in vitro characterization of chitosan-thioethylamidine conjugate. Biomaterials 27:127–135
11. Khan AB, Mahamana R, Pal E (2015) Review on Mucoadhesive drug delivery system: novel approaches in modern era. RGUHS J Pharm Sci 4:128–141
12. Jawadi Z, Yang C, Haidar ZS, Santa Maria PL, Massa S (2022) Bio-inspired Muco-adhesive polymers for drug delivery applications. Polymers (Basel) 14:5459
13. Poddar RK, Rakha P, Singh SK, Mishra DN (2010) Bioadhesive polymers as a platform for drug delivery: possibilities and future trends. Research Journal of Pharmaceutical Dosage Forms and Technology 2:1–6
14. Bravo-Osuna I, Vauthier C, Farabollini A, Palmieri GF, Ponchel G (2007) Mucoadhesion mechanism of chitosan and thiolated chitosan-poly(isobutyl cyanoacrylate) core-shell nanoparticles. Biomaterials 28:2233–2243
15. Dünnhaupt S et al (2012) S-protected thiolated chitosan: synthesis and in vitro characterization. Carbohydr Polym 90:765–772
16. Abu Elella MH, Kolawole OM (2024) Recent advances in modified chitosan-based drug delivery systems for transmucosal applications: a comprehensive review. Int J Biol Macromol 277:134531
17. Millotti G et al (2014) In vivo evaluation of thiolated chitosan tablets for oral insulin delivery. J Pharm Sci 103:3165–3170
18. Wang X et al (2009) Chitosan-NAC nanoparticles as a vehicle for nasal absorption enhancement of insulin. J Biomed Mater Res B Appl Biomater 88:150–161
19. Mortazavian E, Dorkoosh FA, Rafiee-Tehrani M (2014) Design, characterisation and ex vivo evaluation of chitosan film integrating insulin nanoparticles composed of thiolated chitosan derivative for buccal delivery of insulin. Drug Dev Ind Pharm 40:691–698
20. Zhu X et al (2012) Synthesis of thiolated chitosan and preparation of nanoparticles with sodium alginate for ocular drug delivery. Mol Vis 18:1973–1982
21. Meng J, Zhang T, Agrahari V, Ezoulin MJ, Youan B-BC (2014) Comparative biophysical properties of tenofovir-loaded, thiolated and nonthiolated chitosan nanoparticles intended for HIV prevention. Nanomedicine (Lond) 9:1595–1612
22. Li N et al (2022) Temperature- and pH-responsive injectable chitosan hydrogels loaded with doxorubicin and curcumin as long-lasting release platforms for the treatment of solid tumors. Front Bioeng Biotechnol 10:1043939
23. Saremi S, Atyabi F, Akhlaghi SP, Ostad SN, Dinarvand R (2011) Thiolated chitosan nanoparticles for enhancing oral absorption of docetaxel: preparation, in vitro and ex vivo evaluation. Int J Nanomedicine 6:119–128
24. Sakloetsakun D, Bernkop-Schnürch A (2010) Thiolated chitosans. J Drug Delivery Sci Technol 20:63–69
25. Maryam kMF, Sohail G, Shahnaz F, ur Rehman Fakhar-ud-Din A, ur Rehman N, Ullah U, Amin GM, Khan KU, Shah (2019) Development and Evaluation of Optimized Thiolated Chitosan Proniosomal Gel Containing Duloxetine for Intranasal Delivery AAPS PharmSciTech 20(7). https://doi.org/10.1208/s12249-019-1484-y

Manasa N

Swornahuti Panda

Jerifa Sonowal

Sudarshan Naidu Chilamakuri

Subramanian Natesan

Application of Chitosan in Textile Science and Engineering

Ying Zhang, Mei-Chen Lin, Ching-Wen Lou, Ting-Ting Li, and Jia-Horng Lin

Abstract Chitin and its deacetylated derivatives, as a kind of abundant, renewable and biodegradable natural amino polysaccharide, have shown great application potential and research value in the field of textile science and engineering. This chapter systematically introduces the applications of chitin and chitosan in the textile field, covering the fundamental physicochemical properties of chitin and its advantages in textile applications, such as fiber morphology, mechanical characteristics, moisture absorption and retention properties, and spinnability. Besides, the preparation methods for chitin fibers, various technical approaches for preparation and application of chitin fibers in textile processing are also available, such as

Y. Zhang
School of Light Industry and Materials, Chengdu Qinggong Polytechnic University, Chengdu, China
e-mail: Zying1019@cdtc.edu.cn

M.-C. Lin
Department of Biomedical Engineering, College of Biomedical Engineering, China Medical University, Taichung City, Taiwan
e-mail: lmc9363@mail.cmu.edu.tw

C.-W. Lou
Department of Bioinformatics and Medical Engineering, Asia University, Taichung City, Taiwan
e-mail: cwlou@asia.edu.tw

T.-T. Li
Innovation Platform of Intelligent and Energy-Saving Textiles, School of Textile Science and Engineering, Tiangong University, Tianjin, China
e-mail: tingtingli@tiangong.edu.cn

J.-H. Lin (✉)
Advanced Medical Care and Protection Technology Research Center, Department of Fiber and Composite Materials, Feng Chia University, Taichung City, Taiwan
e-mail: jhlin@fcu.edu.tw

S.-L. Wang, M. Anpo (eds.), *Chitin and Chitosan*, Springer Series on Polymer and Composite Materials, https://doi.org/10.1007/978-981-95-7457-5_13

chemical modification, functional coating, wet spinning, and electrospinning. It is proposed that the journey of chitin fiber from laboratory research to large-scale application represents a development path marked by both opportunities and challenges under the background of increasing global consensus on sustainable development. In the future, its application will not only be limited to the field of traditional clothing but also expand to the depth of high value-added and interdisciplinary multi-scenario.

Keywords Chitin · Applications of chitin textiles · Fiber preparation method · Textile processing technology for chitosan

1 Introduction

Chitin is a combination of chitin and chitin polysaccharides, also known as chitin and chitosan. It is widely found in the wings or shells of arthropods and the cell walls of fungi and algae. About 10 billion tons of chitin are biosynthesized annually in nature, making it the second largest organic resource on Earth after cellulose. Chitin is a non-toxic, odorless, white or grayish-white translucent solid. It is insoluble in water, dilute acids, dilute alkalis, and most organic solvents. It dissolves in concentrated sulfuric acid, concentrated hydrochloric acid, and 85% phosphoric acid, undergoing degradation simultaneously.

Chitin is a linear amino polysaccharide composed of 2-acetylglucosamine linked by β-1,4 glycosidic bonds. Its chemical name is (1,4) -2-acetylamino-2-deoxy-β-D-glucose. Chitosan (CS) is obtained by deacetylation of chitin through treatment with concentrated alkali. It is also known as deacetylated chitin, soluble chitin, or chitinamine, with the chemical name (1,4)-2-amino-2-deoxy-β-D-glucose. The structural formulas of chitin and chitosan are shown in Figs. 1 and 2.

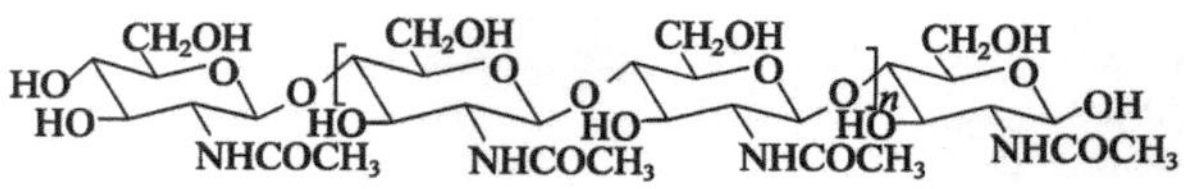

Fig. 1 The chemical structure of chitin

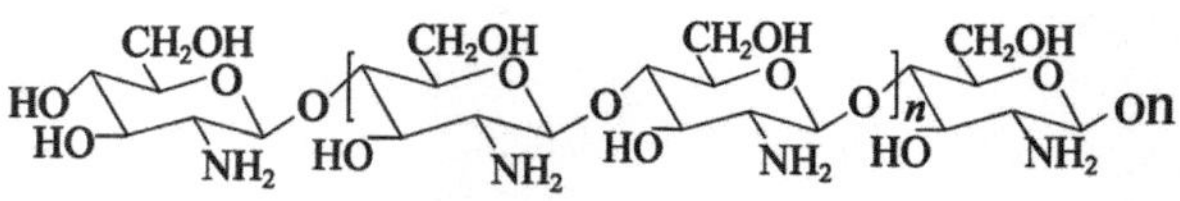

Fig. 2 The chemical structure of chitosan

2 Properties of Chitin and Its Derivatives

2.1 Biocompatibility

Chitin and its derivatives are non-toxic, side-effect-free natural polymeric materials. Their chemical properties are similar to human tissues, so there is no rejection issue for using chitosan in textile products. Moreover, chitin can be enzymatically degraded and absorbed by living organisms. It is generally non-toxic to humans, does not accumulate in the body, and its degradation products do not react with bodily fluids. It also does not cause tissue rejection, thus exhibiting excellent biocompatibility. Due to these characteristics, it serves as an ideal biomaterial for manufacturing various medical products.

2.2 Antibacterial Properties

Chitin exhibits outstanding antibacterial activity. The cationic amino groups in its molecules form ionic bonds with the anionic sialic acids or phospholipids that constitute microbial cell walls, thereby inhibiting microbial life processes. In addition, chitosan also has inhibitory effect on plant pathogens. As a result, fabrics made from chitin and its derivatives exhibit excellent antibacterial and deodorizing properties even without antimicrobial finishing, and can also effectively prevent skin diseases.

2.3 Biodegradability

Chitin can be decomposed under the action of enzymatic catalysts such as chitinase, deacetylchitinase, lysozyme, and snail enzyme. Furthermore, the final product of enzymatic hydrolysis is low-molecular-weight glucosamine, which poses no environmental pollution and serves as an excellent green textile raw material.

2.4 Bioactivity

Chitin and chitosan have good biological activity, which can stimulate immune cells to release growth factors, inhibit the growth of fibroblasts and promote the growth of epithelial cells, thus promoting wound healing. In addition, chitosan has a positive charge, which can separate platelet factor 4 from serum, increase the level of pH 6 in serum, or promote platelet aggregation, and has hemostatic function. Chitosan can enhance the phagocytosis of macrophages and the activity of hydrolases, stimulate macrophages to produce lymph factors, initiate the immune system, and do not increase the production of antibodies.

3 Advantages of Chitin and its Derivatives in Textile Science and Engineering

3.1 Morphology of Chitin Fiber

The molecular structure of chitin is a linear phthalamide chain, and the cross--sectional morphology of the fiber depends on the shape of the spinneret during the spinning process. Figs. 3 and 4 show the morphology of chitosan fibers prepared by wet spinning. It could be observed that chitin fibers prepared via wet spinning using a circular spinneret exhibit a circular or near-circular cross-sectional shape, with a deep yellow coloration. Meanwhile, the wet spinning process involves extruding the spinning solution through circular spinnerets, followed by solidification in a sodium hydroxide bath to form continuous filaments. Therefore, the longitudinal section of chitin fiber is smooth, and the macromolecules of the fiber are arranged neatly after stretching, and the orientation degree is high, which can withstand relatively high tensile strength.

3.2 Mechanical Properties

The mechanical properties of chitin fiber are given in Table 1. As given in the table, the breaking strength of chitin fiber is not significantly different from that of viscose fiber. However, the fiber strength decreased significantly after moisture absorption,

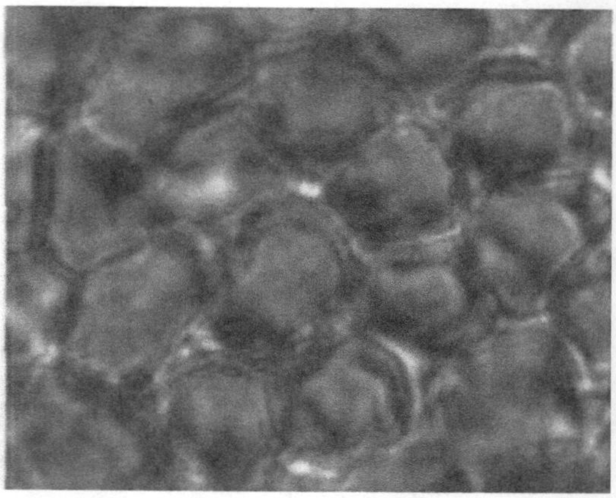

Fig. 3 The cross-section SEM image of chitin fiber

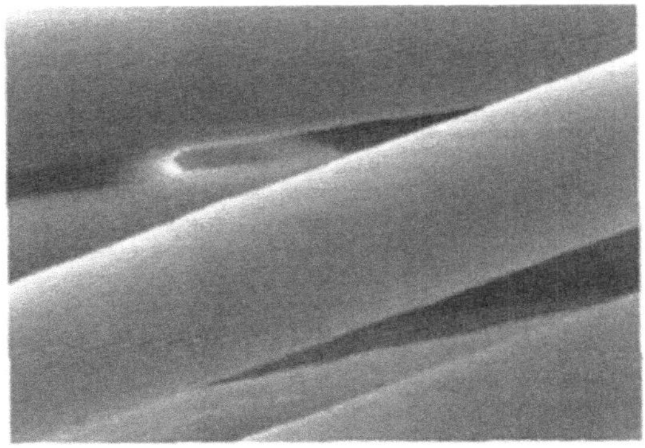

Fig. 4 The longitudinal cross-section SEM image of chitin fiber

Table 1 The comparison of mechanical properties between chitin fiber and several common fibers

Index fiber types	Breaking strength (cN/dtex)	Breaking elongation (%)	Initial modulus (cN/dtex)	Constant elongation rebound rate (%, 3% elongation)
Chitin fiber	2.27	4.51	95.20	56.00
Cotton	3.40	7.90	65.00	64.00
Viscose	2.20	19.84	52.10	58.67
Silk	2.80	20.00	44.10	54.00
Tencel	4.52	15.80	64.50	55.60
Bamboo	2.12	21.19	54.70	58.52

and the wet strength was about 1.42 cN/dtex. Therefore, the moisture content of the fibers should be appropriately controlled during the spinning process. Also, due to the high crystallinity of chitin fibers, their initial modulus is higher than that of cotton fibers, and their elongation at break is lower.

3.3 *Moisture Absorption and Water Retention*

The chitin fibers exhibit a relatively high moisture absorption rate, approximately twice that of cellulose fibers. This property is attributed to the unique microstructure of chitosan fibers. On the one hand, there are many fine pores within the core layer. On the other hand, the longitudinal interface of the fiber is uneven and there are many grooves, an uneven longitudinal interface characterized by abundant grooves. In turn, these microstructural features contribute to the fibers' moisture absorption and vapor permeability.

3.4 Thermal Resistance

Chitin fiber has no melting point and exhibits a relatively high thermal decomposition temperature, so the fiber could resist high temperature. This facilitates the thermal treatment of fibers during processing, which broadens their range of applications.

3.5 Spinnability

The volume resistivity of chitin fiber is approximately 4.22×10^9 Ω·cm. It showed that the resistivity of chitin fiber is higher than that of natural fibers, but lower than that of chemical fibers. Although this property is advantageous for spinning, chitin fiber has low strength and a low coefficient of friction. Additionally, it has fewer crimps and a low crimp density, resulting in poor inter-fiber cohesion, which makes it difficult to textile process.

3.6 Dyeing Properties

The chitin molecule in turn could be regarded as the product obtained by replacing the hydroxyl group (-OH) at the second carbon atom of the cellulose macromolecule with an acetylamino or amino group. Therefore, chitin fiber has basic chemical activity, good dyeing properties, and high dye uptake. In summary, chitin fibers exhibit a high affinity for reactive dyes and direct dyes.

4 Chitin and Its Derivatives: Fibers and Preparation Methods

Based on the exceptional properties of chitin and its derivatives, chitin and its derivatives could be made into fibers or films by textile-related processes. This fiber or film could be used as surgical absorbable surgical suture, artificial skin, hemostatic cotton, medical nonwovens, gauze, etc., with anti-inflammatory, hemostatic, antibacterial, promote wound healing and other effects.

Currently, the most widely used method for preparing chitin fibers is wet spinning technology. Specifically, chitin is first dissolved in a suitable solvent to form a spinning solution of a specific concentration. This solution is then filtered and degassed before being extruded under pressure through a spinneret into a coagulation bath. The resulting fibers undergo post-treatment processes such as drawing, washing, and drying to produce chitin fibers.

There are various production processes for chitin fiber, but their fundamental principles and operational procedures are similar. The slight difference is that the selection of solvents and coagulants, as well as slight variations in dissolution, spinning, and post-treatment processes. The process routes can generally be categorized into two types:

1. Chitin → Dissolved → Spinning solution →Filtration → Defoaming →Metering →First solidification→ Extension → Second solidification → Shaping → Washing→Drying →Fiber
2. Chitin → Modification → Spinning solution →Filtration → Defoaming →Metering→Solidification→Extension→Shaping→Washing→Drying→Fiber

Compared with the above two processes, the strength of the second fiber is twice as high as that of the former, while other properties do not change much. In addition to chitin and chitosan being capable of producing fibers, their derivatives could also produce fibers for different applications.

Chitin and chitosan fibers can be spun into filaments or short fibers. Among these, the filaments are mainly made into absorbable medical sutures by untwisting or weaving. The filaments are cut into short fibers of a certain length. Subsequently, these short fibers undergo loosening, carding, spinning, and weaving to produce medical gauze in various specifications. Furthermore, the loosened chitin or chitosan fibers could be combed into a net, and then layered, sized and bonded, dried, or needle-punching to form medical nonwoven fabrics. The gauze or nonwovens made by the above method not only have porous structure but also have good air permeability and water absorption, and their air permeability could also reach 1500 L/(m·s), and the water absorption capacity is 15%. These gauze or nonwoven fabrics are then cut to the required specifications as needed, packaged, and sterilized to become ideal medical dressings. Additionally, chitin and chitosan short fibers can be manufactured into fiber papers or fiber felts of various specifications and applications as needed, serving in water and air purification. This could be attributed to the fact that chitin fiber and chitosan fibers possess greater adsorption capacity than activated carbon materials, featuring safety, cleanliness, selective adsorption, diverse product forms, and low cost.

5 Applications in Textile Science and Engineering

5.1 Application of Fabric in the Field

Chitin and chitosan fibers have good biocompatibility, increasing lysozyme levels on the skin by 1 to 1.5 times. At the same time, they also have the advantages of broad-spectrum antibacterial properties, moisture absorption and odor elimination, mold resistance, moisture retention, and breathability. These fibers are suitable for health-oriented underwear, socks, infant apparel, and bedding.

1. Children's clothing: Babies and kids often lick and suck on their clothes, and children have weaker immune systems, making them more susceptible to external pathogens. Chitin and chitosan fibers, with their excellent antibacterial properties, are particularly suitable for the production of children's clothing.
2. Uniform: Chitin and chitosan fibers have an inhibitory effect on methicillin-resistant Staphylococcus aureus. A certain amount of chitin fiber/chitosan fiber incorporated into workwear could effectively suppress cross-infections caused by drug-resistant bacteria commonly found in hospitals and food processing plants, thereby protecting the health and safety of workers [1].
3. Jeans: The exceptional moisture absorption, moisture retention, and softness of chitin and chitosan fibers enhance the wearability of denim garments. Furthermore, the antibacterial properties of chitin and chitosan fibers ensure the clothing remains clean and odor-free.
4. Socks: The moisture produced by feet exceeds that of other body parts by several to dozens of times. Combined with the confinement of shoes, bacteria thrive easily, causing odors and athlete's foot. Blending chitin and chitosan fibers with cotton creates comfortable, antibacterial health socks.
5. Bedding: The soft, comfortable, highly absorbent, and strongly adsorbent properties of chitin and chitosan fibers make them ideal materials for towel and blanket fabrics. Using chitin and chitosan fibers blended with cotton to produce towels and bedding could inhibit the reproduction of bacteria on the towel quilt and keep clean.
6. Interior decoration: The amino groups on chitin and chitosan could adsorb formaldehyde under acidic conditions. If chitin and chitosan fibers are processed into home textile fabrics for use in carpets, curtains, sofa covers, and similar applications, this characteristic of the fibers could be leveraged to suppress formaldehyde pollution caused by indoor decoration.

5.2 Medical and Health Field

Fibers made from chitin and its derivatives possess natural antibacterial, anti-mite, moisturizing, and skin care properties. When these fiber products are in contact with the skin, they could activate lysozyme in human body fluids and revitalize skin tissue, making them ideal materials for health care, beauty, and skin care. Secondly, fibers made from chitin and its derivatives are soft, non-irritating, and hypoallergenic, making them suitable for the production of sanitary materials, such as protective mask, facial mask, diapers, adult incontinence, and other sanitary products. Additionally, chitin and its derivative fibers exhibit excellent biocompatibility and antibacterial properties, showing inhibitory effects against *Staphylococcus aureus*, *Escherichia coli*, *Pseudomonas aeruginosa*, and *Candida albicans*. The fiber made from chitin and its derivatives possesses hemostatic and wound-healing properties, preventing wound infections. It allows for the production of various medical dressings for treating burns, scalds, and traumatic injuries.

5.2.1 Medical Dressing

Medical dressings made from chitin and its derivative fibers possess the following features: ① It could reduce pain at the wound site by providing a cooling sensation. ② The dressing has excellent oxygen permeability and prevents wound hypoxia. ③ The dressing can absorb moisture and undergo degradation through enzymes within the body. ④ The dressing degrades to produce glucosamine, which accelerates wound healing and thereby improves the rate of wound healing. In addition, by controlling the degree of deacetylation and relative molecular weight of chitin, water-soluble chitin could be made to have better wound-healing ability [2].

5.2.2 Surgical Suture

The absorbable surgical suture made from chitin and its derivatives has an elongation of 17–20% in a wet state, similar to that of natural muscle tissue. Compared with the catgut, it has the characteristics of softness, easy knotting, high mechanical strength, easy absorption, no change of skin collagen hydroxyproline content, no inflammatory reaction, and so on. It could be sterilized using conventional methods and enhances wound tensile strength, thereby accelerating wound healing. In the late 1980s, DuPont had a patent for high-strength chitin fiber. Japan United Pharmaceutical has planned to mass-produce medical sutures made from chitin and its derivatives, and is preparing to sell them worldwide. The chitin suture developed by Donghua University has been used in clinical practice for surgical sutures such as digestive tract surgery and plastic surgery. Shanghai National Engineering Research Center for Nanoscale Applications Co., Ltd. has developed a non-removable antimicrobial surgical suture using chitosan. To address the current issue of weak mechanical properties (wet strength) in chitosan fibers, researchers at Wuhan University have invented a wet spinning method for producing high-strength chitosan fibers.

5.2.3 Artificial Skin

The chitin fiber was first treated with serum protein to improve its adsorption, and then water was used as a dispersant and polyvinyl alcohol was used as an adhesive to make nonwovens. Afterwards, the woven fabric could be cut and sterilized for use as artificial skin. The artificial skin prepared with chitin has good adhesion, which is convenient for epithelial cells to grow in, and has the effects of analgesia, hemostasis and promoting wound healing. The artificial skin could also be served as a substrate to culture epidermal cells in large quantities. After the cell loading, they are attached to the surface of deep burn. With the degradation of chitin fibers, a complete regenerated dermis is formed.

5.2.4 Hard Tissue Repair Materials

Absorbable materials used on fracture wounds must come into direct contact with human tissue, blood, and bodily fluids. Therefore, they must not only be biocompatible but also possess certain mechanical properties, formability, and physical properties that match the elastic modulus of bone tissue. Chitin and chitosan can serve as hard tissue stimulants, promoting the recovery and regeneration of hard tissues, particularly bone tissue [3]. Modified chitosan fibers serve as reinforcing fibers, while high molecular weight polylactic acid (PLA) functions as the resin matrix. Through winding and molding processes, modified chitosan fiber-reinforced PLA composites can be fabricated for use in fracture repair materials. The initial flexural strength of acylated modified chitosan fiber-reinforced polylactic acid composite panels is 114.72 MPa, with an initial flexural modulus of 3980.05 MPa. After 16 weeks of immersion in sodium lactate Ringer's solution at 37 °C, the flexural strength decreased to 31.42 MPa with the obvious in vitro hydrolysis resistance and strength retention properties significantly exceeding those of PGA/PLA (polyhydroxyacetic acid/ polylactic acid) and self-reinforced PGA composites.

5.2.5 Artificial Kidney Membrane

The artificial kidney membrane purifies blood by removing solutes from it, thereby sustaining the lives of patients with chronic kidney disease. Chitosan is a natural cationic polymer. The artificial kidney dialysis membrane made of chitosan has sufficient mechanical strength. It could penetrate small molecular organic matter such as urea and inosine, but not inorganic ions such as Na^+, K^+, and serum proteins. With excellent permeability, it serves as an ideal membrane material for artificial kidneys.

5.2.6 Drug-Delivery Materials

Drug controlled-release technology is becoming an increasing interest. As chitin and its derivatives possess excellent biodegradability and biocompatibility, they could be considered ideal materials for controlled and sustained drug release. In addition, the free amino groups of chitosan exhibit high affinity for various proteins, making it suitable as a carrier for immobilized enzymes, antigens, antibodies, and other substances.

6 Textile Processing Technologies and Their Applications

Chitin and its derivatives can be used in the field of textile and garment by chemical modification, coating method, wet spinning method, blended yarn method, and other methods.

6.1 Chemical Modification

The chemical modification method combines chitin and its derivatives with cotton or other fibers through chemical bonds using crosslinking agents and other chemical substances. However, the chemical structure of chitin or cellulose has been changed during this process due to the introduction of exogenous, non-natural chemical substances and chemical bonds. Consequently, some of the inherent properties of chitin and its derivatives, such as biodegradability, biocompatibility, and moisture absorption—are lost or diminished. Md. Mofakkharul Islam et al. [4] synthesized chitosan, N-octyl chitosan (NOCh) and carboxymethyl chitosan (CMCh) derivative from prawn shell wastes and identify their applications as modifiers on cellulosic fibers, jute and cotton, to develop quality textile fibers. The modified fibers exhibit excellent dyeability and colorfastness and other properties. Moreover, synthesizing chitosan from shrimp and crab waste for use as a modifier not only avoids the use of petroleum-based chemical modifiers but also contributes to the reuse of shrimp and crab waste. Blessy Joseph et al. [5] have explored the extraction of nanochitin from marine resources and the modification of its nanocomposites (Fig. 5). Shanthana Lakshmi et al. [6] concluded based on this approach that chemically modified nanochitin serves as a functional material and reinforcing agent, with applications in water purification and treatment. Wang et al. [7] developed a dual-crosslinked

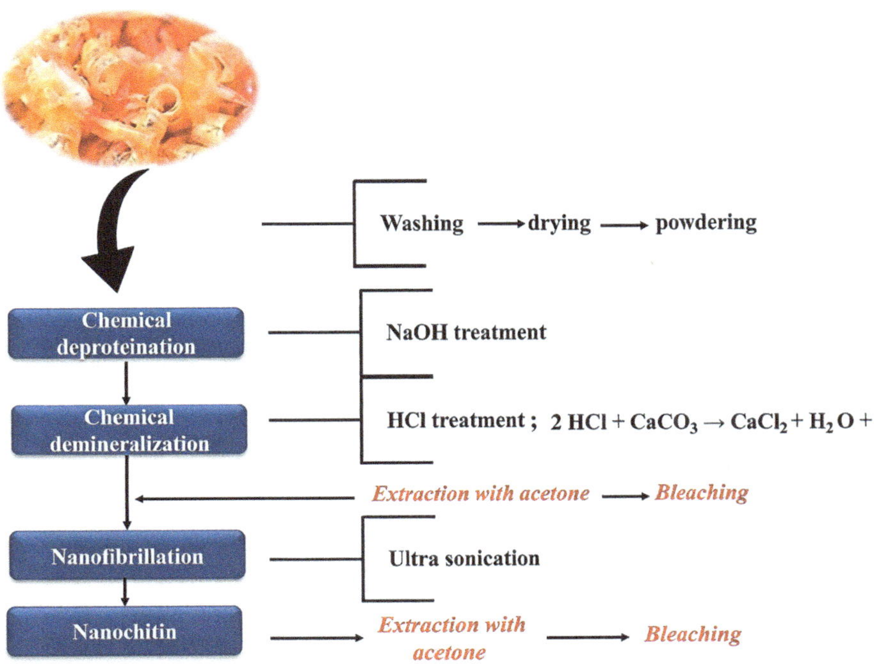

Fig. 5 Extraction of nanochitins

networked chitosan hydrogel with high strength and toughness using crosslinking agents. Meanwhile, the double-crosslinked chitosan hydrogels were fabricated for manufacturing eco-friendly sensors and flexible energy storage devices, based on the biodegradability and biocompatibility of chitin (Fig. 6). Zhang et al. [8] developed high-performance gas-sensitive sensors by combining chitin fibers (CFs) with molybdenum disulfide (MoS_2) via a one-step hydrothermal process. The developed CFs/MoS_2 composite exhibited better surface properties and fire resistance compared to CFs, while also demonstrating higher gas sensitivity than MoS_2 monomer. The CFs/MoS_2 composite could realize real-time signal conversion during tensile and motion performance testing, thereby assisting inspectors in making analytical judgments based on the results (Fig. 7).

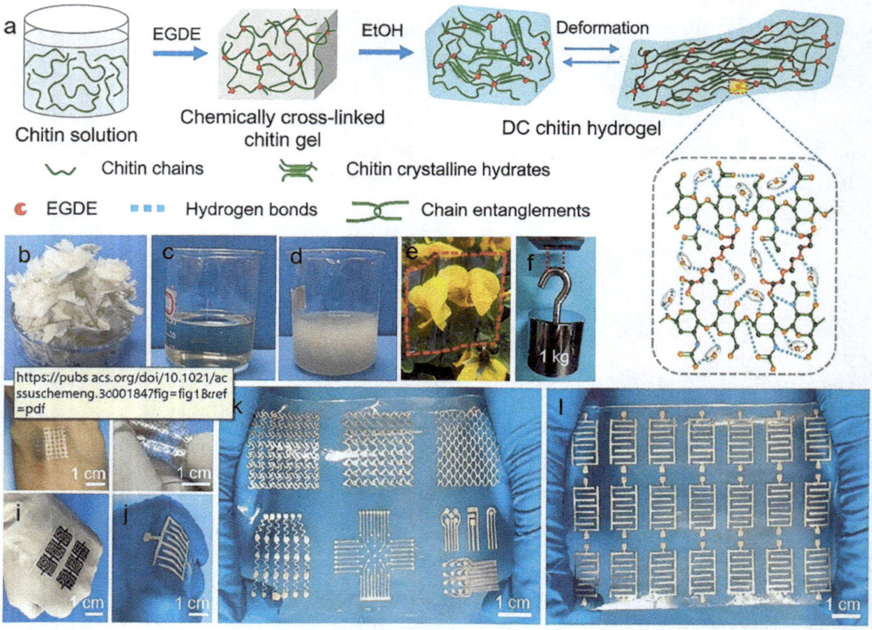

Fig. 6 (**a**) Schematic of the synthesis of DC chitin hydrogels by sequential chemical and physical crosslinking. Photographs of (**b**) purified shrimp shells (chitin patches), (**c**) chitin solution, (**d**) chitin-EGDE solution, (**e**) transparent dual cross-linked (DC) chitin hydrogels, and (**f**) a DC chitin hydrogel (width, 10 mm and thickness, 0.3 mm) strip used for hanging a 1 kg weight. Photographs of (**g**) hydrogel-printed electronics attached to the skin of the subject and (**h**) stretchable flexible electronics. Photograph of a supercapacitor printed with (**i**) graphene ink and (**j**) conductive silver ink that is tightly attached to the rubber-gloved fist of the subject. (**k** and **l**) Digital photographic images showing the large surface areas and mass preparation used for the hydrogel-printed electronics

Fig. 7 Schematic diagram of the process of synthesizing MoS_2/CFs composites via a hydrothermal reaction and dry

6.2 *Functional Coating*

With the development of research in fluorochemicals, textiles, and other industries, and the increasing realization of environmental protection, the bio-based polymer finishing agents have become an effective means to improve environmental pollution and energy substitution issues inherent in the synthesis of traditional polymer finishing agents. This type of finishing agent not only ensures effective finishing results but also reduces environmental pollution, thereby lowering costs and promoting energy substitution. Among them, chitin and its derivatives-based polymer fabric finishing agents could be mainly used for antibacterial finishing, anti-wrinkle finishing, antistatic finishing, hydrophilic finishing and washing resistance finishing [9, 10]. A layer of chitin or its derivative finishing agent can be uniformly covered on the surface of the fiber fabric though padding dyeing process. This treatment could give the fabric new properties without destroying the original properties of the fiber. This process possesses the advantages of simple operation, environmental friendliness, remarkable effectiveness, and low cost. Therefore, it is still a hot research topic in the synthesis, study, and application of functional coatings for textile finishing agents. Xiao et al. [11] have developed antibacterial cotton fabrics with persistent antimicrobial properties by grafting two natural antimicrobial materials, carboxymethyl chitosan (CMC) and betaine (Bet), on cotton fibers surface, through a simple pad-dry-cure modification process. The results indicate that the modification process had no effect on the crystal structure of the cotton fabric. Instead, the fabric's water vapor permeability and water absorption properties were enhanced. Furthermore, the modified fabric demonstrated persistent antibacterial activity against both *Escherichia coli* and *Staphylococcus aureus*. In addition, the antibacterial efficacy of the fabric remained above 95% even after 20 times of washing (Fig. 8). R.R. Gadkari et al. [12] employed poly(styrenesulfonate) (PSS) and synthesized silver loaded chitosan (CS-Ag) nanoparticles as anionic and cationic agents, respectively, and employed a layer-by-layer technique to impart antibacterial properties to cotton fabrics. The results demonstrated that the alternate L-B-L deposition of PSS and CS-Ag nanoparticles on fabric was done up to 15 bi-layers. Besides, no adverse effect on the physical and mechanical properties of the fabric, such as air permeability, tensile strength, and bending (flexural) rigidity, was observed after L-B-L coating of nanoparticles (Fig. 9). Based on the synthesis of polyaniline (PANI) with a nitrogen-containing aromatic structure, Gao et al. [13] developed PANI/chitosan coating to improve the sustainable flame retardant (FR) and UV protection performance of silk fabric. The addition of bio-based chitosan was applied to improve the binding capability and dispersion uniformity of PANI on the silk surface. The results indicated that the PANI/chitosan coating with 3 wt% PANI endowed the silk fabric with effective flame retardancy. The coated silk fabric self-extinguished during vertical flammability test and possessed a decreased char length of 9.5 cm compared with the untreated fabric. The coated silk fabric also exhibited excellent UV protection with the UPF of 50+. Additionally, the coated silk fabrics also demonstrate excellent good laundering resistance (Fig. 10).

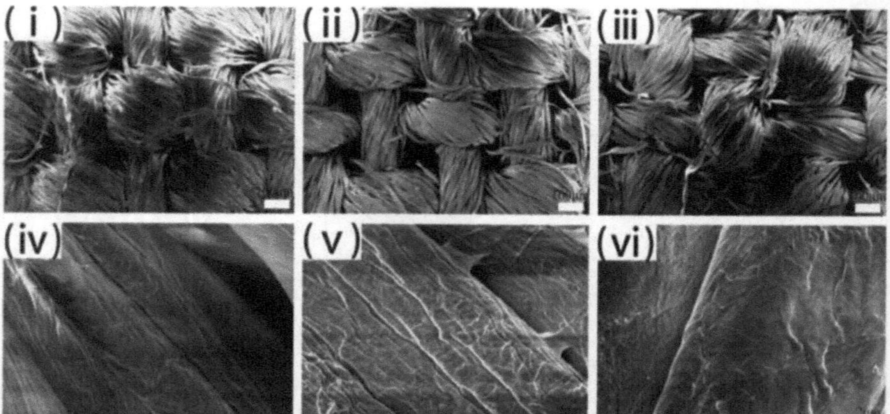

Fig. 8 SEM images of original cotton, CMC-Co and CMC-Bet-Co fabrics

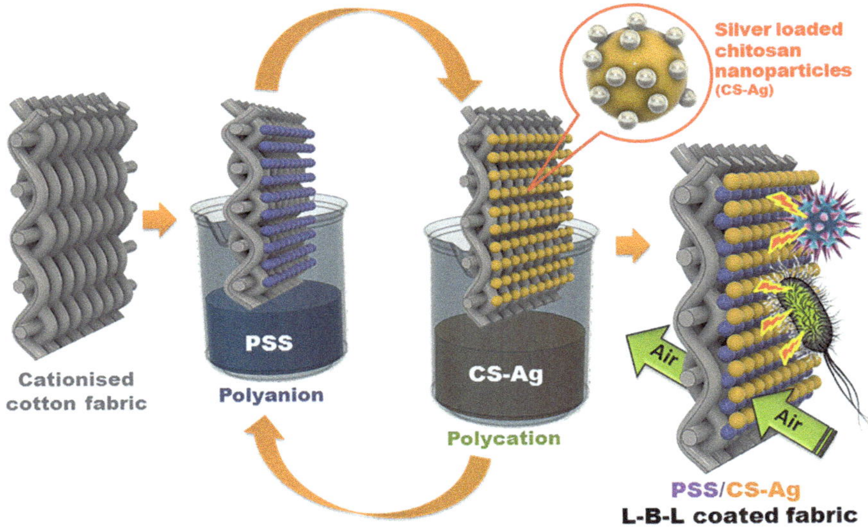

Fig. 9 Schematic diagram of layer-by-layer self-assembled coated cotton fabric

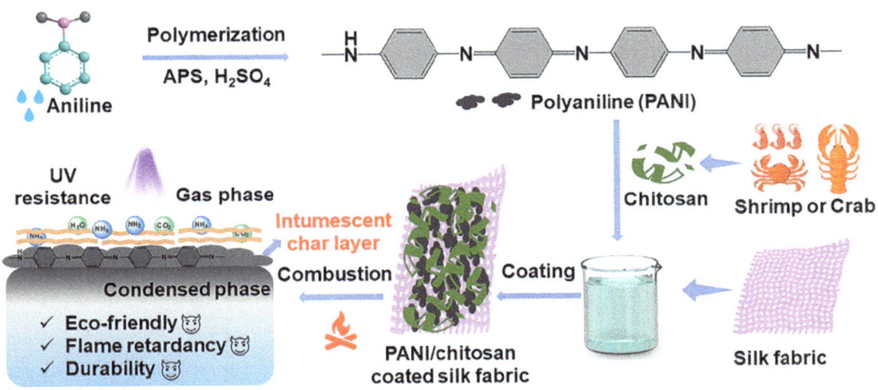

Fig. 10 Schematic diagram of the development of PANI and coated silk fabrics

6.3 *Wet Spinning*

Wet spinning is currently the most widely used continuous spinning strategy [14, 15]. It is a process involving the extrusion of a solution containing chitin and its derivative precursors into a coagulation bath through specific equipment. Subsequently, a phase separation occurs among the polymers or molecules in the solution to form a gel fiber. After that, further treatment steps such as drawing,

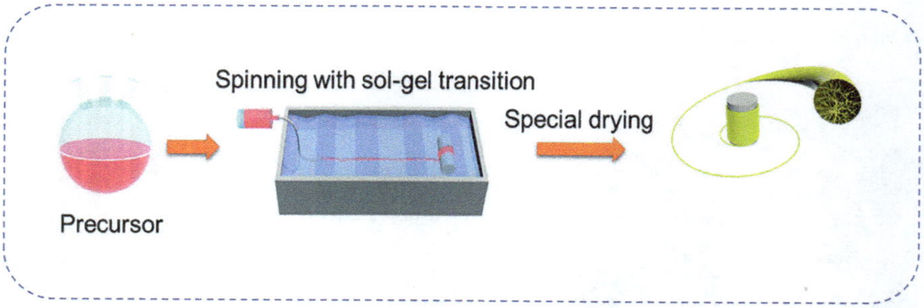

Fig. 11 The wet spinning processes

washing, and drying are performed to achieve the finished fibers (Fig. 11). Currently, chitosan with a deacetylation degree of 80–90% is predominantly used as raw material for wet spinning. The obtained fibers could be utilized in the production of woven, knitted, and nonwoven fabrics. Additionally, the fibers produced are widely used in medical and healthcare products, smart textiles, biosensors, and structural reinforcement composites. Khoa Dang Nguyen [16] investigated the effect of coagulation bath temperature on the properties of fibers produced by the wet spinning process. The results show that the prepared fiber structure becomes denser with the increase of coagulation bath temperature. Moreover, the fiber strength increases with the increase of temperature when using the same concentration of precursor solutions. Chen et al. [17] employed wet spinning technology to develop high-strength all-chitin filaments suitable for smart textiles, biosensors, and structural reinforcement materials. Partially deacetylated chitin nanofibers were isolated from shell residues derived from crab biomass by Wang and colleagues [18] and used for hydrogel preparation. Subsequently, the prepared hydrogel was converted into continuous microfibers with a layered structure and transversely aligned fibrils using the wet spinning. The cell proliferation experiments with rat cardiac myoblasts H9c2 and mouse osteoblasts K7M2 indicate that cells exhibit significant proliferative activity once the fibers come into contact with them.

6.4 Electrostatic Spinning

Electrospinning is a new spinning technique that has emerged in recent years. It is a technology to obtain nanofibers by overcoming the surface tension of the spinning solution and the viscous force inside the spinning solution through electrostatic force. It is a low-cost, simple, and highly reproducible manufacturing technique that

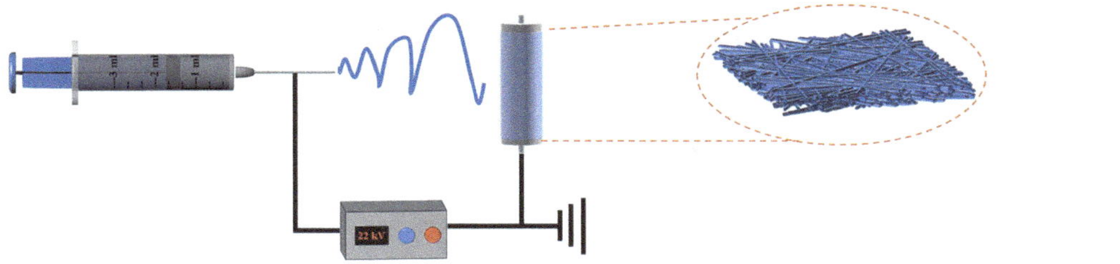

Fig. 12 The Schematic diagram of the electrostatic spinning process

enables efficient access to a wide variety of ultrafine fibers with diameters extending from nanometers to micrometers [19]. Therefore, it is widely recognized as one of the most important methods with the potential for mass production of nanofiber materials. The electrostatic spinning principal apparatus is shown in Fig. 12.

When a certain high-voltage electrostatic charge is applied to a polymer melt or solution, an electrostatic field forms between the spinneret and the receiving plate during the electrospinning process. If the electrostatic force of the spinning solution is equal to the sum of the surface tension and the internal viscous force of the spinning solution, a small hemispherical droplet will be formed at the spinneret. Once the electric field strength exceeds a certain threshold, the surface tension of the polymer solution is overcome by the electrostatic force, yielding a fine liquid jet that is attracted toward the counter electrode. During this process, the solvent evaporates or the melt solidifies, leading to the deposition of elongated solid fibers that aggregate on the collector [20–22]. Different types of fiber structures could be achieved by adjusting the electrospinning apparatus and optimizing process parameters. Qi et al. [23] employed chitosan as the raw material to fabricate chitin helical scaffolds through mold casting and acetylation, followed by electrospinning a fiber membrane onto the outer layer of the chitin scaffold. Moreover, the lumen of this structure was filled with human bone mesenchymal stem cell-derived extracellular matrix and fibers, which provide nutrition and topographical guidance, respectively. Furthermore, animal experiments validated its efficacy in repairing sciatic nerve defects in rats, offering a novel approach for restoring clinical peripheral nerve defects (Fig. 13). Sofia et al. [24] have developed poly(ε-caprolactone) (PCL) fiber membranes doped with EL-2 orthorhombic crystals via electrospinning. Furthermore, the electrospun membranes were modified using spray coating technology with chitin nanocrystals to enhance the biocompatibility and bioactivity of the material for localized cancer photodynamic therapy. Both of these researches will provide new insights into the potential for in vivo application and eventual clinical translation of the PCL membrane.

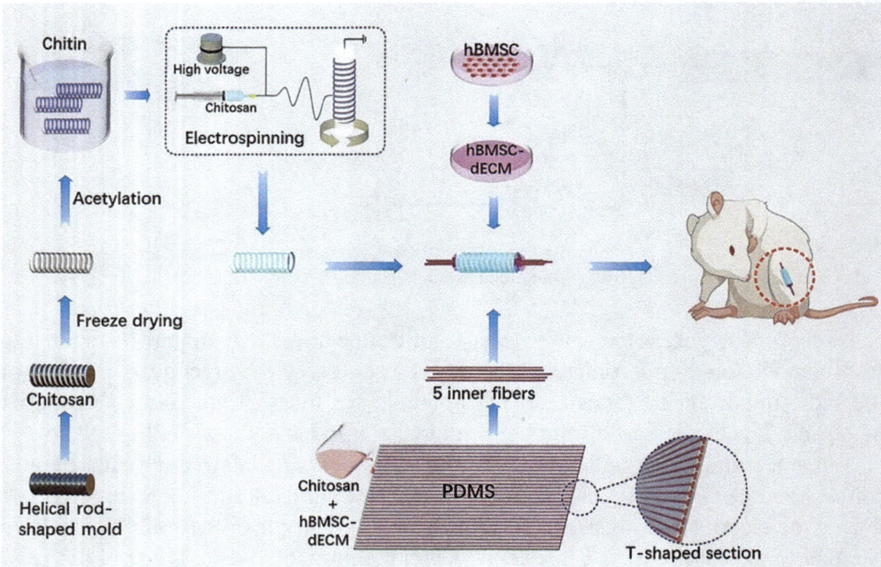

Fig. 13 Schematic representation of the tissue-engineered nerve graft production

7 Challenges and Development Trends

Chitin fiber, especially its deacetylated product chitosan fiber, is known as "the green fiber of the twenty-first century" due to its unique properties such as biocompatibility, antibacterial activity, and biodegradability. However, its application from laboratory to scale production still faces many challenges. (1) Raw material sources are constrained by seasonal variations, geographical limitations, and fishery yields. (2) The extraction process involves multiple steps, featuring lengthy procedures, high energy consumption, and the generation of substantial acidic and alkaline waste liquids, making it complex. (3) Production costs are prohibitively high: Chitin and chitosan dissolve only in harsh chemicals, which not only corrode production equipment but also increase costs and environmental burdens. (4) Industry standards remain inadequate, with lack of unified, authoritative international or industry standards established for evaluating chitin fiber content, antimicrobial performance, or biodegradability certification.

While facing challenges, under the dual drive of sustainable development and technological innovation, the journey of chitin fiber in the textile field is still full of opportunities, and the application of chitin fiber will also penetrate deeply into high-value and interdisciplinary fields.

References

1. Shi P, Zhu M, Chen W et al (2025) Biodegradable dual-network chitosan composite aerogel by in situ mineralization for substituting petroleum-based counterparts. Small:e06668
2. Lou C-W, Tu M-G, Lu C-T et al (2014) Manufacturing technique and mechanical properties of polylactide acid net/chitosan composite membrane. Fibers Polym 15(8):1739–1744
3. Li TT, Zhang Y, Ren HT et al (2021) Two-step strategy for constructing hierarchical pore structured chitosan-hydroxyapatite composite scaffolds for bone tissue engineering. Carbohydr Polym 260:117765
4. Islam MM, Mondal MIH, Ahmed F (2017) Study on prawn shell waste into chitosan and its derivatives as value added products for cellulosic Fibres. Res J Text Apparel 21(2):134–145
5. Joseph B, Mavelil Sam R, Balakrishnan P et al (2020) Extraction of Nanochitin from marine resources and fabrication of polymer nanocomposites: Recent Advances. Polymers 12(8)
6. Lakshmi DS, Ravindran L, Mary PJM et al (2024) Valorization of crustacean shells: preparation, characterization, and chemical modification of Nano-chitin biopolymer membrane for application in water purification. Waste Biomass Valoriz 15(11):6067–6090
7. Wang Y, Wang L, Lu Y et al (2023) Stretchable, biodegradable dual cross-linked chitin hydrogels with high strength and toughness and their potential applications in flexible electronics. ACS Sustain Chem Eng 11(18):7083–7093
8. Zhang Y, Wu Z, Sun J et al (2023) Synthesis and sensing performance of chitin fiber/Mos(2) Composites. Nanomater (Basel) 13(9)
9. Fernandes M, Padrao J, Ribeiro I A et al (2022) Polysaccharides and metal nanoparticles for functional textiles: A Review. Nanomaterials (Basel) 12(6)
10. Abate MT, Ferri A, Guan J et al (2019) Single-step disperse dyeing and antimicrobial functionalization of polyester fabric with chitosan and derivative in supercritical carbon dioxide. J Supercrit Fluids 147:231–240
11. Xiao Y, Duan P, Shen G et al (2022) Durably antibacterial cotton fabric prepared by a combination of betaine and Carboxymethyl chitosan. Fibers Polym 23(3):617–625
12. Gadkari RR, Ali SW, Joshi M et al (2020) Leveraging antibacterial efficacy of silver loaded chitosan nanoparticles on layer-by-layer self-assembled coated cotton fabric. Int J Biol Macromol 162:548–560
13. Gao Y-J, Li M-X, Cui M-L et al (2024) Polyaniline/chitosan coating as the novel and sustainable flame retardant and Uv protection route for silk fabric. Prog Org Coat 186
14. Ren N, Qiao A, Cui M et al (2023) Design and fabrication of nanocellulose-based microfibers by wet spinning. Chem Eng Sci 282
15. Zhang Y, Li T-T, Shiu B-C et al (2022) Multifunctional sodium alginate@ Urushiol fiber with targeted antibacterial, acid corrosion resistance and flame retardant properties for personal protection based on wet spinning. Appl Surf Sci 584
16. Nguyen KD (2021) Temperature effect of water coagulation Bath on chitin fiber prepared through wet-spinning process. Polymers 13(12)
17. Chen C, Wu Q, Zao Y et al (2022) Mechanically strong all-chitin filaments: wet-spinning of beta-chitin nanofibers in aqueous Naoh. Int J Biol Macromol 222(Pt B):3243–3249
18. Wang L, Ezazi NZ, Liu L et al (2020) Microfibers synthesized by wet-spinning of chitin nanomaterials: mechanical, structural and cell proliferation properties. RSC Adv 10(49):29450–29459
19. Sun F, Zheng Y, Li M, et al. Construction of self-cleaning membrane integrating underwater superoleophobicity, photocatalysis, and antibacterial activity for water purification. Langmuir, 2025, 41(4): 2562–2571

20. Zheng Y, Sun F, Zeng P et al (2025) Recent development of special wettability filtration membrane for selective oil/water separation applications: a review. Prog Org Coat 198
21. Li T-T, Sun L, Zhang Y et al (2023) Chitosan-based antibacterial microspheres loaded multifunctional Janus composite membranes with unidirectional biofluid transport, antioxidant and Ph-responsive monitoring. Chem Eng J 472
22. Dzolkifle N A N, Wan Nawawi W M F. A review on chitin dissolution as preparation for electrospinning application. Int J Biol Macromol, 2024, 265(Pt 1): 130858
23. Qi T, Zhang X, Gu X et al (2023) Experimental study on repairing peripheral nerve defects with novel bionic tissue engineering. Adv Healthc Mater 12(17):e2203199
24. Costa SM, Mattos BD, Calhelha RC et al (2025) Electrospun polycaprolactone membranes functionalized with nanochitin for enhanced bioactivity in localized cancer photodynamic therapy. Carbohydr Polym Technol Appl 11

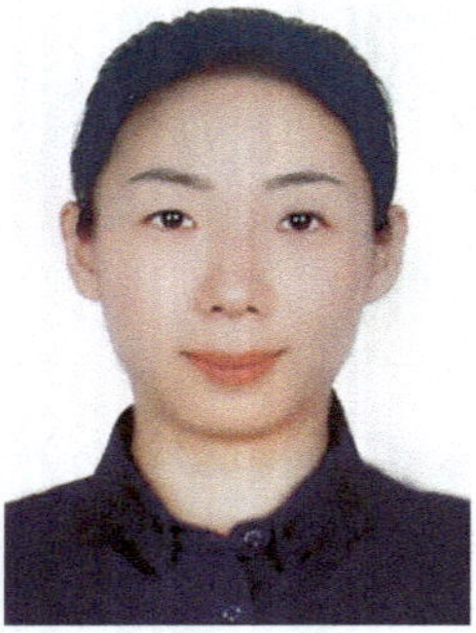

Ying Zhang

Mei-Chen Lin

Ching-Wen Lou

Ting-Ting Li

Jia-Horng Lin

Application of Chitosan in Membranes

Wei-Song Hung

Abstract Membrane separation technology plays a crucial role in advancing human life, through its solutions for the challenges associated with water pollution, environmental protection, and sustainable management of natural resources. Among various polymers, chitosan has attracted growing attention as a membrane material owing to its natural abundance, renewability, biodegradability, and outstanding film-forming properties. Derived from chitin found in marine waste, chitosan is an eco-friendly biopolymer suitable for many membrane applications. In this chapter, we provide an overview of chitosan and its extraction method, followed by its utilization in various membrane fabrication methods. Various classes of chitosan-based membranes, such as neat chitosan membranes, chitosan composite membranes, and chitosan blend membranes, are discussed in detail. Special emphasis is placed on their applications in wastewater treatment, specifically for the removal of dyes and heavy metals from wastewater. This chapter demonstrates the potential of chitosan as a sustainable membrane material for developing an effective and eco-friendly advanced membrane systems.

1 Introduction

Membrane science and technology have gone through a long historical development in laboratory study before realizing its first notable industrial application in the 1960s. With nearly 50 years of rapid advancement, in present days, membrane-based processes enjoy numerous industrial applications and have brought great benefits to improve quality of human life [1]. Membranes make the separation process less energy intensive, since they allow a continuous separation process without the requirement of sorbent regeneration or desorption by varying the temperature of the

W.-S. Hung (✉)
Advanced Membrane Materials Research Center, Graduate Institute of Applied Science and Technology, National Taiwan University of Science and Technology, Taipei, Taiwan
e-mail: wshung@mail.ntust.edu.tw

S.-L. Wang, M. Anpo (eds.), *Chitin and Chitosan*, Springer Series on Polymer and Composite Materials, https://doi.org/10.1007/978-981-95-7457-5_14

pressure. Moreover, the membrane modules are compact, static, easy to control operating parameters, and to scale up [2]. From the past two decades, the effort on research and technology development activities in the common area of membrane technology has been rapidly increasing. As a result, membrane technology has been used in diverse applications such as wastewater treatment and desalination, pollution control and gas separation, and pharmaceutical industry, food industries. The membrane performs the crucial role in deciding the separation process efficiency in membrane technology [3].

An ideal membrane exhibits various morphological and performance characteristics such as high permeation flux, high selectivity, long-term performance stability, antifouling, and physicochemical and mechanical robustness under operating conditions [2]. Due to the growing concern on public health and environmental protection accompanied by the strict environmental regulations on waste management, focus has been given on the application of biopolymers from conventional resources as alternatives to nondegradable synthetic polymers. Natural biopolymers obtained from plants and living organisms, involve in the biocycle, and are ultimately get decomposed and reabsorbed in ecosystem [4]. Chitin or poly (b-(1,4)-N-acetyl-D-glucosamine), one of the most abundant biopolymers in nature [5]. It is mainly obtained from the shells of crustaceans, crabs, and shrimps, as the by-product of seafood processing plants and also from the cell walls of some fungi. Chitosan (CS) is obtained from the deacetylation of chitin, which consists of linear b-1,4-linked D-glucosamine (GlcN) and N-acetyl glucosamine (GlcNAc) units. Figure 1 depicts the chemical structures of chitin and CS. The degree of deacetylation (%DD) generally decides the physicochemical and biological properties of CS. As a renewable naturally available biopolymer, CS offers various inherent advantages like biocompatible as well as biodegradable, non-toxic, inexpensive, and excellent sorption properties. Also, in particular, the free $-NH_2$ and $-OH$ groups present in the structure enable antimicrobial

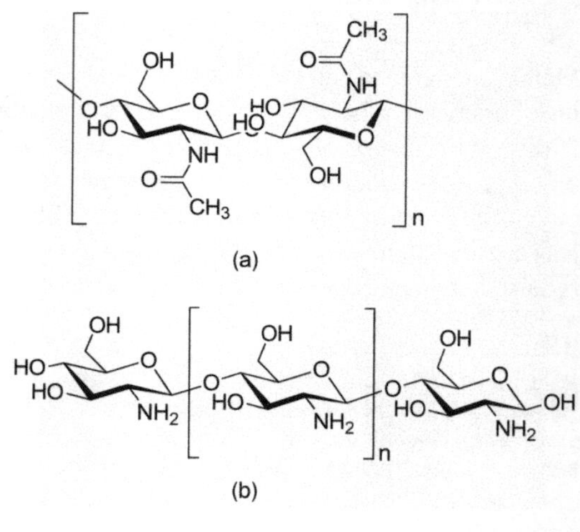

Fig. 1 Chemical structures of (**a**) Chitin and (**b**) Chitosan (CS) biopolymers [6]

properties, heavy metal ions chelation ability, gel-forming ability, great affinity toward proteins, solubility in water, and enables facile chemical modification and processing [6].

So far, many CS derivatives have been prepared for numerous applications such as drug delivery, biocompatible scaffold material for tissue engineering and bone substitution, sterile material for wound healing. CS is used as a feasible and effective adsorbent for the removal of dyes, heavy metal ions and other contaminants in water purification applications. In addition, CS plays an important role in agriculture, cosmetics and food industries [6]. CS is used to prepare microcapsules, microspheres, fibers, gel beads, and membranes for various applications. As a membrane, CS offers the advantage of energy and cost reduction and ease of operation, which allow the selective transport of specific components, solving the challenges in health, energy, and environmental issues [4]. This chapter provides an overview of CS and its extraction method, followed by its utilization in various membrane fabrication methods. Various classes of CS-based membranes, such as neat CS membranes, CS composite membranes, and CS blend membranes, are discussed in detail. Special emphasis is placed on their applications in wastewater treatment, specifically for the removal of dyes and heavy metals from wastewater.

2 Chitosan in Membrane Preparation

CS is extracted from seafood waste as shown in the schematic Fig. 2. Crustacean shells are crushed and ground into small particles, and the shell particles are then treated with dilute hydrochloric acid extraction process to remove calcium

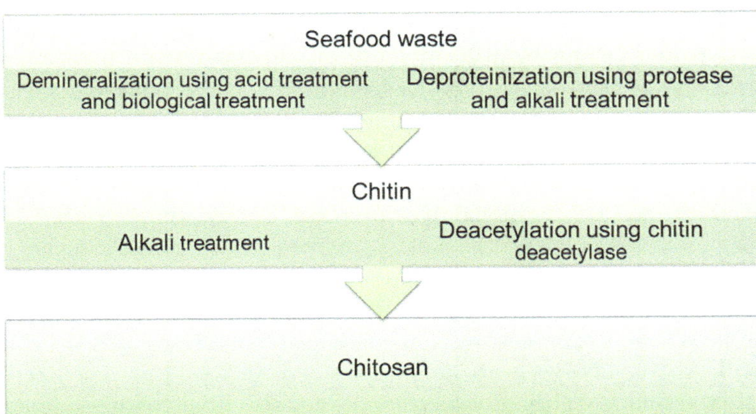

Fig. 2 Schematic for the steps involved in the CS extraction from seafood waste and shells [6]

carbonate and other minerals, followed by solubilization to remove proteins. The decolorization process was performed to remove the color. The resultant colorless chitin is deacetylated by treating with concentrated sodium hydroxide (NaOH, 40–45 wt%) at 120 °C for 1–3 h, with the variables controlling the degree of deacetylation [7]. The methods for the preparation of CS membranes are as follows.

2.1 Neat Chitosan Membranes

Facile and effective solution casted film nanofiltration membrane for dye and salt rejection [8], dry-wet spinning method fabricated hollow fiber membranes with CS nanoparticles loaded with silver for dyes removal [9], electrospun CS membranes for the removal of dyes are some examples for preparation methods. Electrospinning is the famous technology for nanofiber membranes fabrication due to diverse opportunities for membrane property tuning and ease of upscaling for large-scale production. Several electrospinning techniques have been used such as multilayer electrospinning, needleless electrospinning, pore forming agent assisted electrospinning, template-assisted electrospinning and three-fluid electrospinning through suitable approach in water purification [10].

2.2 Chitosan Composite Membranes

Generally, CS composite membranes are prepared for the applications such as dyes removal, desalination and oil/water separation. For example, CS nanofiltration membrane was fabricated with graft copolymerization modification for dyes and salts removal [11], CS and multiwalled carbon nanotube composite layer was constructed on polyethersulfone membrane to prepare thin film nanocomposite (TFN) membrane to remove malachite green from water [12]. CS-ZIF-8 nanofiltration membrane was prepared for dyes removal [13]. CS composite membranes retain their performance stability, even under high pressure filtration conditions, for instance, Cai Zhijiang et al. have reported a combined approach of grafting, electrospinning and surface coating to fabricate nanofiber composite nanofiltration membrane to remove oil and dyes from wastewater with enhanced flux and rejection performance. Facile free radical polymerization method to obtain molecular imprinted membranes, complex forming and ion pairing, ionic cross-linking, facile cross-linking, interfacial polymerization (IP) technique [14], nonsolvent-induced phase conversion method to prepare mixed matrix membranes (MMMs) [15], hydrogel composite with antibacterial properties [16], precipitation by immersion technique [17] are specifically applied to prepare CS composite membranes.

2.3 Chitosan Blend Membranes

The development of CS blend membranes for wastewater treatment plays a crucial role in addressing water scarcity issues and also in mitigating the negative impacts of water pollution on the environment. CS blend membranes are familiar for their superior selectivity and durability in water treatment, offering potential solutions to overcome the challenge associated with water contamination. Recently, researchers have developed dense membranes, flat sheet, and hollow fiber CS membranes for water treatment, to enhance the application potential of CS compared to its bead, flake, or powder form adsorbents.

The solution casting method is a common and versatile method employed to synthesize CS blend membranes for wastewater treatment applications, in which CS is dissolved in a suitable volatile organic solvent to form a viscous and homogeneous CS solution and then it is cast onto a clean glass or Teflon substrate, so that the solvent is allowed to evaporate slowly, forming a thin CS membrane. Different fillers or additives can also be dispersed the CS solution to improve the membrane performance and properties. For instance, Zarghami et al. fabricated CS- polyvinyl alcohol (PVA)/ carbon nanotubes (CNTs) membranes via the solution casting method. First, CS and PVA polymer solutions were prepared and blended to form homogeneous solution, then the functionalized CNTs dispersed in water was subsequently mixed with the polymer blend solution, then the cross-linking agent glutaraldehyde (GA) was added. The resultant solution mixture was cast on a substrate, dried at room temperature, finally, the resultant membrane was treated with NaOH, the fabrication method is as depicted in Fig. 3. The membranes were employed for the Zn^{2+} and Pb^{2+} heavy metal ions adsorption from the wastewater [18].

In another study by Caprarescu et al., a chitosan—cellulose acetate biopolymer infused with silver ions was developed for the adsorption of iron from the simulated effluent. The addition of CS—silver ions into the cellulose acetate matrix resulted in complexation in the membrane. The membrane network, interactions, and the mechanisms underpinning the separation performance are presented in Fig. 4 [19].

The phase inversion method is a commonly used method for the fabrication of CS blend membranes. In this method, a homogeneous polymer solution will get

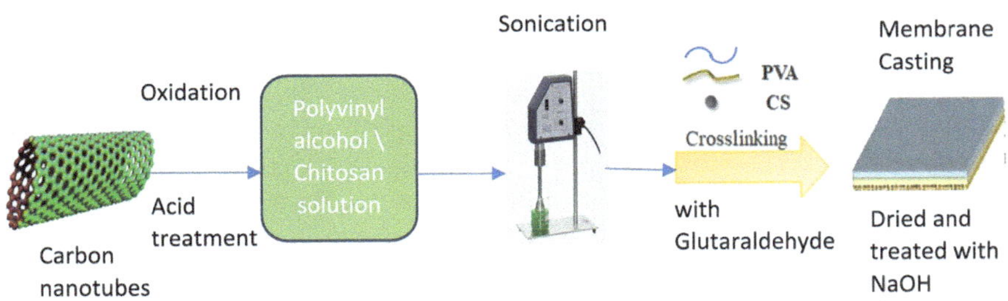

Fig. 3 Schematic diagram for the fabrication of CS-PVA/CNTs membrane [18]

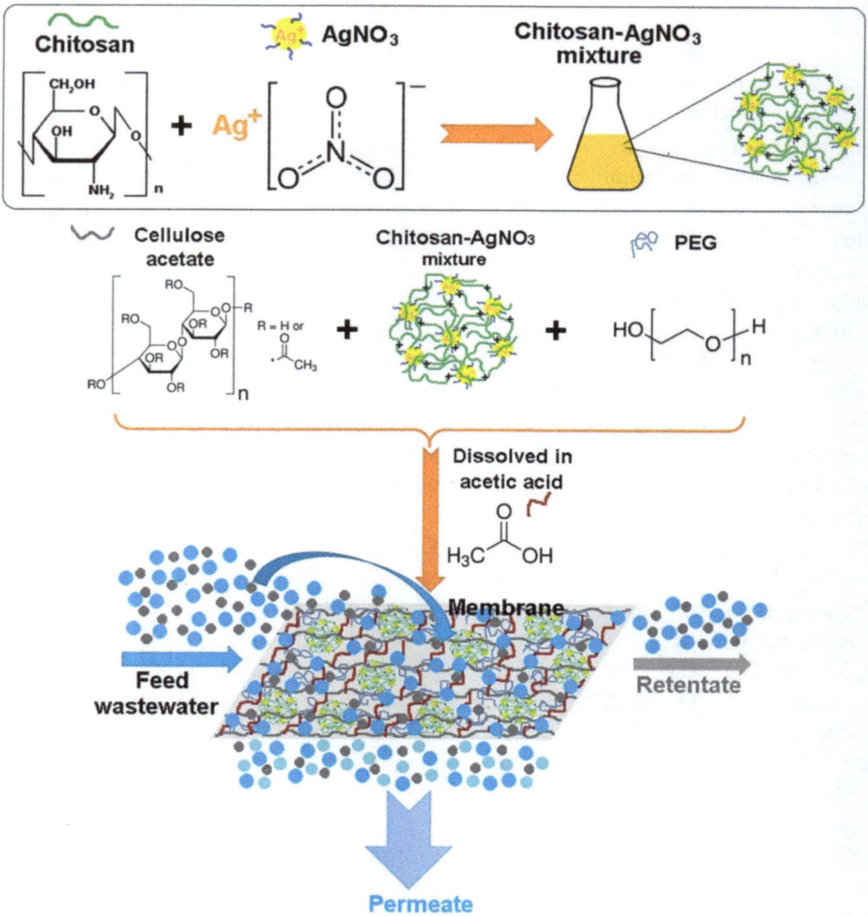

Fig. 4 The schematic overview of the synthesis of cellulose acetate—CS—silver ion biopolymeric membrane and its application for the treatment of iron [19]

transformed into a membrane structure via nonsolvent-induced phase separation. During the phase separation, porous structure is formed in the membrane matrix. First, CS is dissolved in a solvent, generally acetic acid, to obtain a homogeneous polymer solution for casting onto a substrate, through different methods such as casting using doctor blade, dip coating, or spray coating. As the phase separation occurs, the membrane will be formed. The temperature and the nonsolvent play crucial roles in deciding the structural characteristics of the membrane. Syeda Samia Nayab et al. employed phase inversion method to fabricate polyether sulfone membrane blended with functionalized thiolated CS and activated carbon (AC) [20]. The membrane exhibited improved water flux performance owing to its hydrophylic property. The schematic representation of the membrane development is shown in Fig. 5.

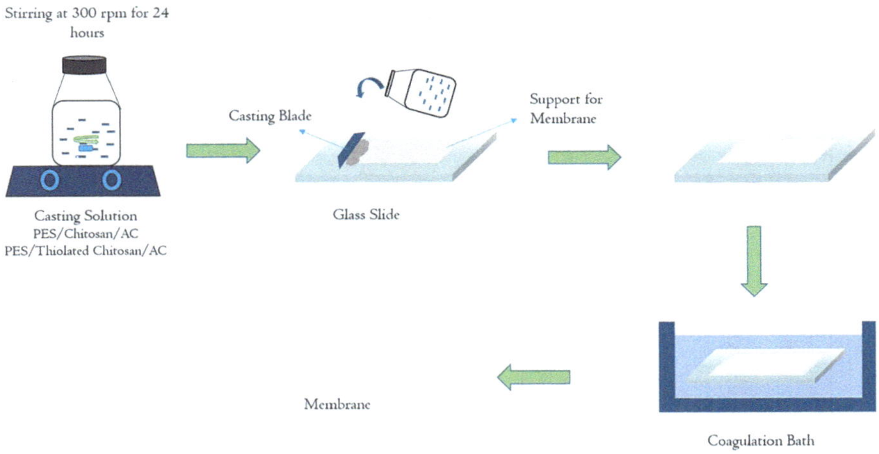

Fig. 5 The schematic depiction of PES membrane blended with modified thiolated CS and activated carbon (AC) [20]

From the last 2–3 decades, extensive research has been conducted on electrospun membranes focusing on wastewater treatment. CS membranes are used among electrospun membranes to separate heavy metals and dyes from wastewater. This method is famous due to its ability to develop CS membranes with nanoscale roughness and porosity. In this method, CS is electrospun into nanofibers utilizing an electric field. The as obtained membrane offers high surface area and porosity for adsorbing heavy metal ions and contaminants. In a study conducted by Yang et al. [21], CS Electrospun nanofiber membranes functionalized with -NH_2 groups were prepared, which were then sequentially grafted with poly (glycidyl methacrylate) (PGMA) and polyethyleneimine (PEI) polymers. The developed membrane exhibited excellent adsorption properties for Cr^{6+}, Cu^{2+}, and Co^{2+}, with the highest removal of Cr^{6+}. The membranes exhibited reproducible and stable performance. The membrane synthesis is presented in Fig. 6a, and the heavy metal ions mechanism is illustrated in Fig. 6b.

3 Applications of Chitosan in Other Membranes

Other than the above-mentioned water treatment applications, CS membranes have been used extensively in other areas such as biomedical, drug delivery, fuel cells, gas separation, tissue engineering, wound dressing, and catalysis. Owing to the cell compatibility, and antibacterial properties, CS films are used for various biomedical applications as summarized in Fig. 7.

In fuel cell technology, proton exchange membranes pose lot of challenges such as CO poisoning, poor durability, fuel crossover, high cost, and so on. Some studies

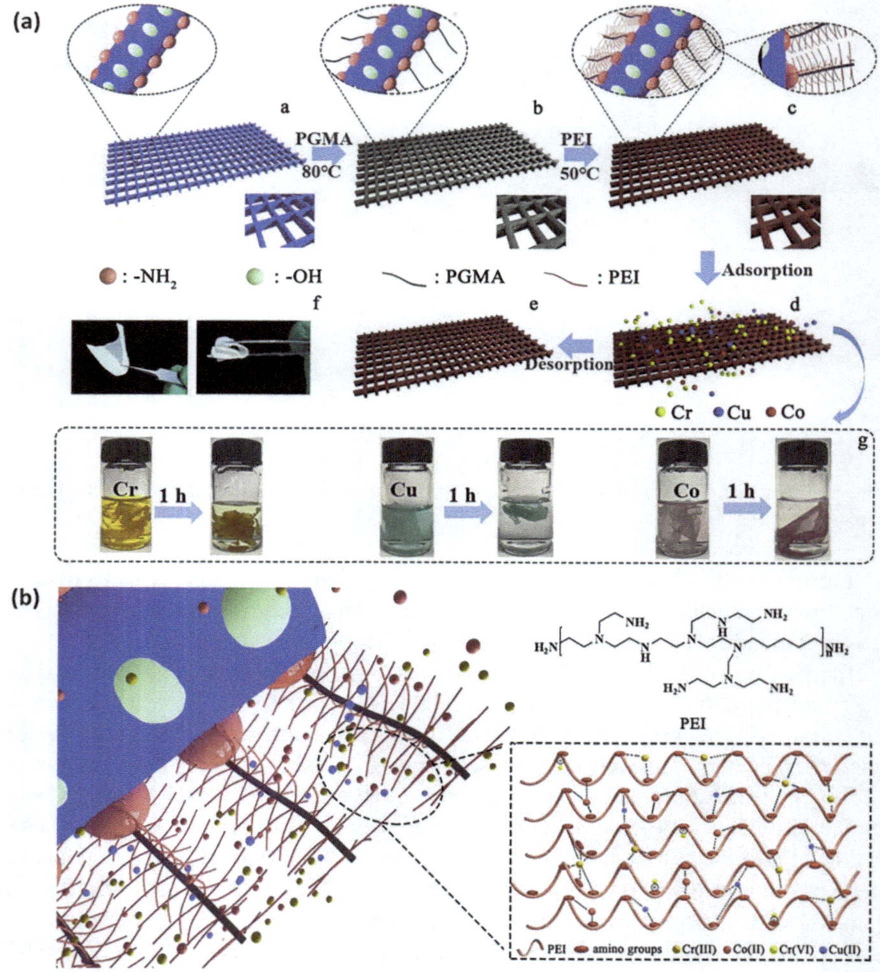

Fig. 6 (**a**) The preparation of CS-PGMA-PEI nanofiber membrane and demonstration as adsorbent, (**b**) the adsorption mechanism for metal ions on the electrospun CS-PGMA-PEI membrane [22]

have shown that the CS membranes could replace polyelectrolyte membranes. Metal nanoparticles incorporated CS hybrid membranes have shown enhanced proton conductivity. Inorganic/organic electrolyte CS composite membranes can substitute HPAs and polybenzimidazole (PBI) for hydrogen fuel cells and polymer membrane fuel cells. Some of the important applications of CS membranes is shown in Fig. 8 [6].

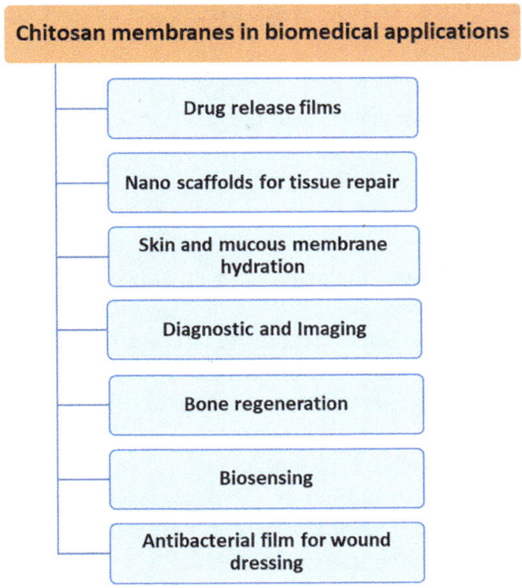

Fig. 7 CS membranes in various biomedical applications

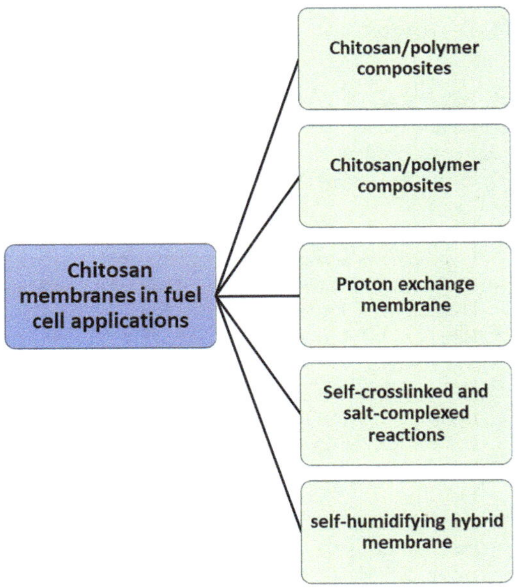

Fig. 8 Applications of CS membranes in fuel cell technology

4 Conclusion and Future Perspective

As an abundant biopolymer, CS has demonstrated to be a suitable and sustainable material for membrane preparation applications, owing to the advantages like biocompatibility, renewability, and excellent film-forming properties. In wastewater treatment, mainly in the removal of dyes and heavy metals, it has proven its effectiveness for solving environmental challenges through membrane separation technology. CS-based membranes whether neat, blended, or composite exhibit promising performance; however, further improvements in mechanical properties, chemical stability, and durability for long-term real-world applications are still needed for large-scale implementations. Future research needs to focus on h chemical functionalization of CS, cross-linking, and use of advanced nanomaterials as additives to enhance performance and antifouling properties. The use of CS membranes in advanced separation technologies, such as hybrid or multifunctional membrane systems, can offer new pathways for cost effective and sustainable filteration technologies. With continuous developments in membrane design and processing, CS membranes are anticipated to play a crucial role in the progress of next-generation sustainable separation systems.

References

1. Baker RWJM (2004) Overview of membrane science and technology and applications. In: Membrane technology and applications, vol 3, pp 1–14
2. Yeo ZY, Chew TL, Zhu PW, Mohamed AR, Chai S-P (2012) Conventional processes and membrane technology for carbon dioxide removal from natural gas: a review. J Nat Gas Chem 21:282–298
3. Kang G-d, Cao Y-m (2014) Application and modification of poly(vinylidene fluoride) (PVDF) membranes—a review. J Membr Sci 463:145–165
4. Krajewska B (2005) Membrane-based processes performed with use of chitin/chitosan materials. Sep Purif Technol 41:305–312
5. Rinaudo MJ (2006) Chitin and chitosan: Properties and applications. Progr Polym Sci 31:603–632
6. Vedula SS, Yadav GD (2021) Chitosan-based membranes preparation and applications: challenges and opportunities. J Indian Chem Soc 98:100017
7. Bakshi PS, Selvakumar D, Kadirvelu K, Kumar NS (2020) Chitosan as an environment friendly biomaterial—a review on recent modifications and applications. Int J Biol Macromol 150:1072–1083
8. Long Q, Zhang Z, Qi G, Wang Z, Chen Y, Liu Z-Q (2020) Fabrication of chitosan Nanofiltration membranes by the film casting strategy for effective removal of dyes/salts in textile wastewater. ACS Sustain Chem Eng 8:2512–2522
9. Kolangare IM, Isloor AM, Karim ZA, Kulal A, Ismail AF, Inamuddin et al (2019) Antibiofouling hollow-fiber membranes for dye rejection by embedding chitosan and silver-loaded chitosan nanoparticles. 17:581–587
10. T. M S AAB, Lin PT, Widakdo J, H. K M AHFM et al (2021) A review of recent progress in polymeric electrospun nanofiber membranes in addressing safe water global issues. RSC Adv 11:9638–9663

11. Liu S, Wang Z, Song P (2018) Free radical graft copolymerization strategy to prepare Catechin-modified chitosan loose Nanofiltration (NF) membrane for dye desalination. ACS Sustain Chem Eng 6:4253–4263
12. Mousavi SR, Asghari M, Mahmoodi NM (2020) Chitosan-wrapped multiwalled carbon nanotube as filler within PEBA thin film nanocomposite (TFN) membrane to improve dye removal. Carbohydr Polym 237:116128
13. Vatanpour V, Salehi E, Sahebjamee N, Ashrafi M (2020) Novel chitosan/polyvinyl alcohol thin membrane adsorbents modified with detonation nanodiamonds: preparation, characterization, and adsorption performance. Arab J Chem 13:1731–1740
14. Weng R, Huang X, Liao D, Xu S, Peng L, Liu X (2020) A novel cellulose/chitosan composite nanofiltration membrane prepared with piperazine and trimesoyl chloride by interfacial polymerization. RSC Adv 10:1309–1318
15. Chaudhary M, Maiti A (2020) Fe–Al–Mn@chitosan based metal oxides blended cellulose acetate mixed matrix membrane for fluoride decontamination from water: removal mechanisms and antibacterial behavior. J Membr Sci 611:118372
16. Ulu A, Birhanlı E, Köytepe S, Ateş B (2020) Chitosan/polypropylene glycol hydrogel composite film designed with TiO2 nanoparticles: a promising scaffold of biomedical applications. Int J Biol Macromol 163:529–540
17. Remanan S, Ghosh S, Das TK, Sharma M, Bose M, Bose S et al (2020) Gradient crystallinity and its influence on the poly(vinylidene fluoride)/poly(methyl methacrylate) membrane-derived by immersion precipitation method. J Appl Polym Sci 137:48677
18. Zarghami S, Tofighy MA, Mohammadi T (2015) Adsorption of zinc and Lead ions from aqueous solutions using chitosan/polyvinyl alcohol membrane incorporated via acid-functionalized carbon nanotubes. J Dispers Sci Technol 36:1793–1798
19. Căprărescu S, Zgârian RG, Tihan GT, Purcar V, Eftimie Totu E, Modrogan C et al (2020) Biopolymeric membrane enriched with chitosan and silver for metallic ions removal. Polymers 12:1792
20. Nayab SS, Abbas MA, Mushtaq S, Khan Niazi B, Batool M, Shehnaz G et al (2021) Antifoulant ultrafiltration polymer composite membranes incorporated with composite activated carbon/chitosan and activated carbon/thiolated chitosan with enhanced hydrophilicity. Membranes (Basel) 11:827
21. Xu S, Liu Y, Yu Y, Zhang X, Zhang J, Li Y (2020) PAN/PVDF chelating membrane for simultaneous removal of heavy metal and organic pollutants from mimic industrial wastewater. Sep Purif Technol 235:116185
22. Yang D, Li L, Chen B, Shi S, Nie J, Ma GJP (2019) Functionalized chitosan electrospun nanofiber membranes for heavy-metal removal. Polymer 163:74–85

Wei-Song Hung

Treatment of Chitin Production Wastewater Containing High Salinity from Shrimp Scraps with By-Product Chitosan, Chemical Coagulants and Biological Processes

Tan Phong Nguyen and Nguyen Van Nhi Tran

Abstract Chitin production wastewater is characterized by high organic load, high salinity, and the presence of nutrients and minerals, making it challenging to treat by conventional methods. In this chapter, a combined approach of preliminary sedimentation, coagulation, and biofilm-based biological treatment is described. Among the coagulants tested, by-product chitosan emerged as an effective and eco-friendly alternative to chemical agents such as PAC and PAA, offering both pollutant removal and sludge valorization. Sedimentation was highly efficient in removing suspended solids, while coagulation enhanced the reduction of organic matter, nutrients, and calcium ions. The subsequent biological process, employing anaerobic–anoxic–aerobic biofilm reactors with Stick-bed Biofix and Swim-bed Biofringe carriers, sustained high organic removal under saline conditions, although nitrogen and phosphorus removal declined at higher salinities. Notably, the system consistently reduced calcium, contributing to overall salinity control. This integrated approach highlights a sustainable pathway for managing saline wastewater in the chitin industry while promoting resource recovery.

Keywords Chitin production wastewater · Chitosan · Coagulation · Stick-bed Biofix · Swim-bed Biofringe · High salinity

T. P. Nguyen (✉)
Ho Chi Minh City University of Industry and Trade, Ho Chi Minh City, Vietnam
e-mail: phongnt@huit.edu.vn

N. V. N. Tran
Institute of Biotechnology and Environment, Nha Trang University, Nha Trang, Khanh Hoa, Vietnam
e-mail: vannhi@ntu.edu.vn

S.-L. Wang, M. Anpo (eds.), *Chitin and Chitosan*, Springer Series on Polymer and Composite Materials, https://doi.org/10.1007/978-981-95-7457-5_15

1 Introduction

The rapid growth of the seafood processing industry, particularly shrimp aquaculture, has generated substantial volumes of by-products such as heads, shells, and tails, which represent approximately 40–60% of total shrimp biomass [1]. These residues are rich in proteins, minerals, carotenoids, fatty acids, and bioactive compounds, with shrimp heads containing higher levels of fat, protein, and amino acids, while shells provide more chitin and calcium [2]. Valorizing these materials reduces environmental burdens while yielding high-value products such as astaxanthin, protein, and chitin, which have applications in pharmaceuticals, agriculture, food, and biodegradable packaging [3]. In doing so, shrimp waste management aligns with circular economy practices and sustainable resource utilization.

Chitin extraction, however, traditionally relies on concentrated hydrochloric acid and sodium hydroxide for demineralization and deproteinization. This chemical-intensive approach generates wastewater with highly variable and extreme pH (<1 or > 12), elevated COD (>4000 mg/L), excessive turbidity, suspended solids, and high salinity dominated by sodium and calcium ions [4]. In developing countries, small- and medium-sized facilities often lack decentralized wastewater treatment systems, leading to ecological degradation, soil salinization, and associated socio-economic impacts [5]. Moreover, the protein-rich particulates and mineral residues in these effluents hinder conventional treatment processes and complicate biological treatment operations [6].

Among pre-treatment options, coagulation–flocculation is a widely applied and practical method to remove colloids, turbidity, and particulate organic matter from industrial wastewater [7]. Chemical coagulants such as aluminum and iron salts, polyaluminum chloride (PAC), and polyacrylamide (PAA) have been extensively used due to their low cost and operational simplicity [8]. However, their application raises concerns about toxic residuals, non-biodegradable sludge, and long-term health risks such as neurological disorders linked to aluminum exposure [9].

In response, natural coagulants have attracted increasing attention, with chitosan standing out for its biodegradability, non-toxicity, broad pH effectiveness, and ability to yield nutrient-rich sludge suitable for reuse as feed or fertilizer [4]. Furthermore, utilizing industrial by-product chitosan as a coagulant not only improves wastewater treatment performance but also contributes to resource circularity by reintegrating processing residues into the treatment process.

This section examines the role of coagulation–flocculation in treating nutrient-rich, saline chitin wastewater, comparing the effectiveness of conventional coagulants (PAC, PAA) with by-product chitosan. Beyond pollutant removal, emphasis is placed on sludge valorization and sustainability, positioning coagulation–flocculation as a key pre-treatment step within integrated wastewater management strategies for the chitin industry.

2 Material and Methods

2.1 Wastewater Characteristics and Preparation

Real chitin production wastewater (CPW) was collected from Vietnam Food Joint Stock Company (VNF), a large-scale seafood waste processing facility located in Vietnam. The effluent originated from the demineralization and deproteinization of shrimp by-products and was characterized by high turbidity, elevated total suspended solids (TSS), abundant organic matter, substantial calcium ion content, and intense coloration (Table 1).

The use of real wastewater ensured that coagulation performance was assessed under authentic and variable industrial conditions, reflecting the complexity and challenges of full-scale effluents. This approach provides practical insights into coagulant efficiency and sludge characteristics, making the results directly applicable to industrial chitin wastewater treatment.

2.2 Coagulant Preparation

By-product chitosan obtained from VNF was used as the primary natural coagulant. This industrial by-product, considered substandard relative to commercial-grade chitosan, exhibited a viscosity of 583.8 cP (25 °C) and a deacetylation degree of 79.55%. A stock solution of 10 g/L was prepared by dissolving chitosan in 1% (v/v) acetic acid under continuous stirring at 100 rpm for 4 h at room temperature (RT).

For comparative purposes, two conventional coagulants were employed. Polyaluminum chloride (PAC 31) was supplied by Weifang Tenor Chemical Company, Qingdao, China, and polyacrylamide (PAA, cationic polymer C1492) was obtained from Specfloc KMR, England. PAC and PAA stock solutions were prepared at concentrations of 50 g/L and 5 g/L, respectively, under constant stirring at 100 rpm at RT. Complete dissolution was achieved after approximately 5 mins for PAC and 2 h for PAA.

Table 1 Characteristics of chitin production wastewater

Parameters	Values
pH	3.55–7.41
COD (mg/L)	4245–23,600
TKN (mg/L)	639–1395
NH_4^+-N (mg/L)	145–842
TP (mg/L)	53–366
TSS (mg/L)	1880—10,400
Ca^{2+} (mg/L)	1200—3163
Crude protein (mg/L)	3994—8719
Turbidity (NTU)	363–1254

2.3 Experimental Design for Coagulation

Preliminary sedimentation tests were conducted at 30 °C using a two-factor design with pH (4–11) and settling time (5–45 min). 1-L wastewater samples were adjusted to the target pH values, allowed to settle, and the turbidity of the supernatant was measured. Each condition was tested in triplicate to ensure reproducibility.

Subsequently, coagulation–flocculation experiments were carried out using chitosan, PAC, PAA, and PAC–PAA combinations. The experimental design adopted the Box–Wilson Central Composite Design (CCD), which provides a statistically rigorous approach to optimize treatment performance. Unlike conventional one-factor-at-a-time experiments, CCD enables simultaneous evaluation of multiple variables and their interactions, thereby reducing the number of experiments while enhancing predictive accuracy. By incorporating both factorial and axial points around the central values, CCD allows the construction of second-order response surfaces that capture nonlinear behaviors often observed in coagulation processes. This ensures that the derived operating conditions are not only statistically robust but also directly applicable to variable industrial effluents.

For two-factor designs (k = 2), 11 experimental runs were conducted, while three-factor designs required 17 runs. The axial distances (α) were set at 1.414 for k = 2 and 1.353 for k = 3. Each 500 mL sample was adjusted for pH (4–10) and treated with coagulants: chitosan or PAA (40–140 mg/L), PAC (30–90 mg/L), or PAC—PAA combined (PAA: 20–60 mg/L). Coagulant addition was followed by rapid mixing at 150 rpm for 2 min, flocculation at 20 rpm for 15 min, and settling for 30 min. Turbidity removal efficiency (TRE) was measured, and the data were analyzed using Minitab 16.0 to determine optimal operating conditions. Under these conditions, additional water quality parameters including COD, TKN, NH_4^+–N, TP, TSS, Ca^{2+}, and crude protein were measured to provide a comprehensive evaluation of treatment performance.

Further optimization with chitosan was performed using both conventional experiments and CCD, testing pH values between 5 and 9 and dosages from 50 to 100 mg/L. The experimental design matrix and data processing, including optimal response surface based on TRE data (Table 2), were carried out by the Minitab 16.0.

The variations in initial turbidity values given in Table 2 reflect day-to-day differences in fresh wastewater batches, characteristic of real chitin effluents. To ensure comparability, TRE (%) was used as a normalized indicator. Coagulant dosages and pH ranges were tailored to each agent's effective range and optimized using the Box–Wilson design. Since the aim was to identify optimal conditions, uniform dosing would not have been appropriate. For cross-comparison, Table 3 presents the removal efficiencies of tCOD, sCOD, TKN, NH_4^+–N, TP, Ca^{2+}, and crude protein under each coagulant's optimal conditions using the same wastewater source and consistent methodology.

Table 2 Coagulation experimental conditions

No	Experiment 1 (Chitosan)				Experiment 2 (PAC)				Experiment 3 (PAA)				Experiment 4 (PAC and PAA)					Experiment 5 (chitosan optimization)			
	Initial turbidity: 523 NTU Turbidity after settling: 79 NTU				Initial turbidity: 493 NTU Turbidity after settling: 81 NTU				Initial turbidity: 511 NTU Turbidity after settling: 97 NTU				Initial turbidity: 511 NTU Turbidity after settling: 100 NTU					Initial turbidity: 963 NTU Turbidity after settling: 184 NTU			
	pH/ chitosan (mg/L)/ turbidity (NTU)			*TRE (%)	pH/ PAC (mg/L)/ turbidity (NTU)			*TRE (%)	pH/ PAA (mg/L)/ turbidity (NTU)			*TRE (%)	pH/ PAC (mg/L)/ PAA (mg/L)/ turbidity (NTU)				*TRE (%)	pH/ chitosan (mg/L)/ turbidity (NTU)			*TRE (%)
1	10.0	140.0	6	92.4	10.0	90.0	10	87.7	10.0	140.0	13	86.6	10.0	90.0	60.0	8	92.0	5.0	50.0	148	19.6
2	4.0	140.0	67	15.2	4.0	90.0	29	64.2	4.0	140.0	92	5.2	4.0	90.0	60.0	37	63.0	9.0	50.0	23	87.5
3	10.0	40.0	12	84.8	10.0	30.0	9	88.9	10.0	40.0	8	91.8	10.0	30.0	60.0	8	92.0	5.0	100.0	150	18.5
4	4.0	40.0	70	11.4	4.0	30.0	40	50.6	4.0	40.0	86	11.3	4.0	30.0	60.0	49	51.0	9.0	100.0	20	89.1
5	11.2	90.0	8	89.9	11.2	60.0	5	93.8	11.2	90.0	2	97.9	10.0	90.0	20.0	6	94.0	9.8	75.0	19	89.7
6	2.8	90.0	58	26.6	2.8	60.0	43	46.9	2.8	90.0	87	10.3	4.0	90.0	20.0	32	68.0	4.2	75.0	153	16.8
7	7.0	160.7	49	38.0	7.0	102.4	13	84.0	7.0	160.7	18	81.4	10.0	30.0	20.0	6	94.0	7.0	110.4	12	93.48
8	7.0	19.3	27	65.8	7.0	17.6	16	80.2	7.0	19.3	14	85.6	4.0	30.0	20.0	44	56.0	7.0	39.7	13	92.93
9	7.0	90.0	12	84.8	7.0	60.0	13	84.0	7.0	90.0	15	84.5	11.1	60.0	40.0	2	98.0	7.0	75.0	21	88.6
10	7.0	90.0	13	83.5	7.0	60.0	13	84.0	7.0	90.0	18	81.4	3.0	60.0	40.0	38	62.0	7.0	75.0	19	89.7
11	7.0	90.0	19	75.9	7.0	60.0	17	79.0	7.0	90.0	17	82.5	7.0	100.6	40.0	13	87.0	7.0	75.0	18	90.2
12													7.0	19.4	40.0	13	87.0				
13													7.0	60.0	67.1	13	87.0				
14													7.0	60.0	12.9	13	87.0				
15													7.0	60.0	40.0	14	86.0				
16													7.0	60.0	40.0	13	87.0				
17													7.0	60.0	40.0	13	87.0				
Optimal results based on the calculation of Minitab 16.0 software																					
18	10.6	86.4	–	99.4	11.2	17.6	–	93.5	9.8	79.3	–	99.5	11.0	48.1	39.2	–	98.9	8.3	77.5	–	99.3

All measured turbidity values represent the mean of triplicate experiments (n = 3)

TRE: Turbidity Removal Efficiency

Table 3 Removal efficiencies of preliminary sedimentation and coagulation experiments using different coagulants under controlled pH conditions

Coagulants	Samples	pH	tCOD (mg/L)	sCOD (mg/L)	TKN (mg/L)	NH_4^+-N (mg/L)	TP (mg/L)	TSS (mg/L)	Ca^{2+} (mg/L)	Crude protein (mg/L)
Preliminary sedimentation	Inlet wastewater	5.6	4245 ± 158	1621 ± 91	639 ± 11	145 ± 8	53 ± 5	1880 ± 53	1200 ± 20	3996 ± 67
	After settling	4.0	2612 ± 157	1584 ± 111	579 ± 8	133 ± 7	40 ± 4	139 ± 10	1233 ± 12	3617 ± 51
	*RE (%)		38 ± 3	2 ± 2	9 ± 1	8 ± 2	25 ± 2	93 ± 0.5	−3 ± 1	9 ± 1
Chitosan 86.4 mg/L	After coagulation	10.6	1741 ± 33	1463 ± 151	453 ± 4	103 ± 1	5.2 ± 0.3	39 ± 2	587 ± 12	2829 ± 25
	**REC (%)		33 ± 4	8 ± 3	22 ± 2	23 ± 3	87 ± 2	72 ± 2	52 ± 1	22 ± 2
	***ToRE (%)		59 ± 2	10 ± 5	29 ± 2	29 ± 3	90 ± 1	98 ± 0	51 ± 0.5	29 ± 2
PAC 17.6 mg/L	After coagulation	11.2	2116 ± 148	1403 ± 11	448 ± 2	95 ± 5	4.9 ± 0.5	51 ± 6	673 ± 31	2800 ± 10
	**REC (%)		19 ± 5	11 ± 6	23 ± 1	29 ± 1	88 ± 1	63 ± 4	45 ± 3	23 ± 1
	***ToRE (%)		50 ± 2	13 ± 5	30 ± 1	35 ± 3	91 ± 1	97 ± 0.4	44 ± 3	30 ± 1
PAA 79.3 mg/L	After coagulation	9.8	2298 ± 52	1246 ± 54	520 ± 8	146 ± 7	4.2 ± 0.3	38 ± 1	587 ± 12	3252 ± 141
	**REC (%)		12 ± 5	21 ± 2	10 ± 1	−10 ± 4	90 ± 1	73 ± 2	52 ± 1	10 ± 5
	***ToRE (%)		46 ± 3	23 ± 2	19 ± 2	−1 ± 1	92 ± 1	98 ± 0.1	51 ± 1	19 ± 5
Mixture of PAC and PAA 87.3 mg/L	After coagulation	11.0	1875 ± 21	1645 ± 80	462 ± 8	98 ± 3	4.3 ± 0.3	44 ± 1	660 ± 20	2888 ± 44
	**REC (%)		28 ± 5	−4 ± 2	20 ± 1	26 ± 6	89 ± 1	68 ± 2	46 ± 2	20 ± 1
	***ToRE (%)		56 ± 2	−2 ± 1	28 ± 0.2	32 ± 5	92 ± 0.2	98 ± 0.1	45 ± 3	28 ± 0.2

The reported pH values represent the preset experimental conditions prior to coagulant addition

Data represent mean ± SD from three replicates (n = 3)

**RE: Removal Efficiency*

***REC: Removal Efficiency of Coagulation*

****ToRE: Total Removal Efficiency*

2.4 Analytical Methods

Chemical oxygen demand (COD), total Kjeldahl nitrogen (TKN), ammonium nitrogen (NH_4^+–N), total phosphorus (TP), calcium ion (Ca^{2+}), total suspended solids (TSS), and pH were analyzed following the Standard Methods for the Examination of Water and Wastewater [10]. COD (5220 C), TKN (4500-Norg C), NH_4^+–N (4500-NH_3 C), TP (4500-PE), Ca^{2+} (2340 A), TSS (2540 D), and pH (4500-H^+B) were determined according to the respective methods. Crude protein content was estimated indirectly by quantifying total Kjeldahl nitrogen (TKN) and multiplying by a nitrogen-to-protein conversion factor of 6.25, consistent with standard protocols for organic biomass and wastewater sludge [11]. This factor assumes that nitrogen originates primarily from proteins and provides an approximate measure of total protein content. Analytical instruments included a COD reactor (HI839800–02, HANNA Instruments, Romania), a UV-Vis spectrophotometer (DR5000–03, HACH Company, USA), a Kjeldahl digestion system (Behr Labor Technik, Germany), and a pH meter (SevenEasy™ S20, Mettler Toledo, Switzerland).

All experimental data were collected in triplicate ($n = 3$) and are reported as mean ± standard deviation (SD). While no ANOVA was performed, the consistency of replicates and the use of response surface modeling support the robustness of the results.

3 Results and Discussion

3.1 Preliminary Sedimentation

Preliminary sedimentation at pH 4 markedly improved solid removal from chitin wastewater, with turbidity and TSS reductions of 80% and 93%, respectively (Table 3). This efficiency is attributed to protonation of amino groups ($-NH_2 \rightarrow -NH_3^+$), which neutralizes surface charges and promotes floc aggregation by reducing electrostatic repulsion and disrupting hydrogen bonding and hydrophobic interactions [12]. At pH > 6, deprotonation weakens these forces, leading to lower removal. However, the effect on dissolved pollutants was minimal, as sCOD decreased by only 2% and Ca^{2+} slightly increased (3%), confirming that sedimentation mainly targets suspended solids.

3.2 Comparison of Chitosan and Conventional Coagulants

Chitosan exhibited the highest overall coagulation efficiency (Table 3). At an optimal dose of 86.4 mg/L and pH 10.6, determined by the Box–Wilson model, TRE reached 99.4%, TSS 98%, and tCOD 59%. Nitrogen compounds (TKN and NH_4^+–N) were reduced by approximately 29%, TP removal was 90%, Ca^{2+} decreased by 51%, and crude protein recovery reached 29%. TRE was strongly influenced by both pH and dosage, with stable performance across pH 7–12 and limited improvement above 80 mg/L. However, soluble COD removal remained modest (~10%), confirming that coagulation primarily removed particulates.

PAC achieved a maximum TRE of 93.5% at 17.6 mg/L and pH 11.2, with 97% TSS, 50% tCOD, ~30% nitrogen, and 91% TP removal. Protein recovery was about 30%, though aluminum residues raise sustainability concerns. The high pH reported reflects experimental adjustment in CCD design rather than PAC's inherent behavior. PAA, optimized at 79.3 mg/L and pH 9.8, achieved 99.5% TRE and 98% TSS removal but lower tCOD (46%), nitrogen (19%), and protein recovery (19%), despite higher sCOD removal (23%). Combined PAC–PAA yielded 98.9% TRE, 56% tCOD, ~30% nitrogen, and 28% protein recovery, but did not substantially improve sCOD removal.

Mechanistically, chitosan acts via protonated amino groups under acidic conditions and through polymer bridging and hydrogen bonding at neutral–alkaline pH, offering broader applicability than PAC or PAA, which rely mainly on charge neutralization and sweep flocculation [13]. Chitosan outperforms conventional coagulants in turbidity and COD removal, while offering biodegradability and sludge valorization for sustainable wastewater treatment [4].

From an economic perspective, the by-product chitosan used in this study was derived directly from the same facility, consisting of off-specification residues unsuitable for sale. With minimal processing, its cost was estimated at \$4–5 per kg, resulting in treatment costs of \$0.31–0.43/m^3, comparable to PAA (\$0.24–0.48/m^3) and higher than PAC (\$0.01–0.04/m^3). However, sludge rich in protein and nutrients can be reused as feed or fertilizer, offsetting disposal costs and aligning with circular economy principles.

In summary, while all coagulants effectively removed turbidity and TSS, chitosan provided superior nutrient and organic removal, produced reusable sludge, and offered environmental and economic advantages. These findings position by-product chitosan as a sustainable alternative to conventional coagulants for treating CPW.

3.3 Comparative Evaluation of Optimization Approaches

In traditional experiments, TRE increased with chitosan dosage at pH 7, stabilizing above 96% at approximately 80 mg/L, while efficiency also improved with pH, reaching ~90% at pH 7 and remaining stable between pH 7 and 11 (Figs. 1 and 2). Complementary optimization using the Box–Wilson Central Composite Design (CCD) identified optimal conditions at pH 8.3 and 77.5 mg/L chitosan, achieving a maximum TRE of 99.3% (Table 2). This design outperformed conventional approaches by capturing interactive effects between variables and enabling precise optimization.

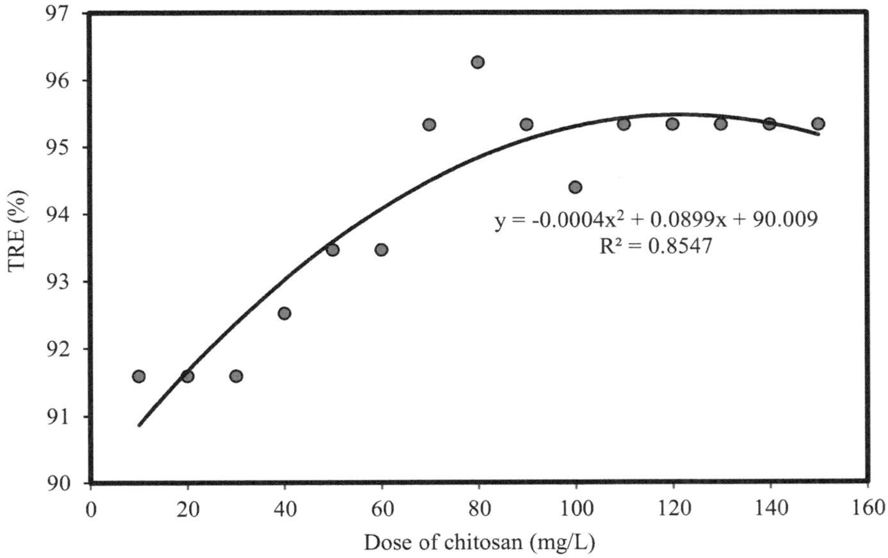

Fig. 1 Correlation between turbidity removal efficiency (TRE) and chitosan dose at pH 7

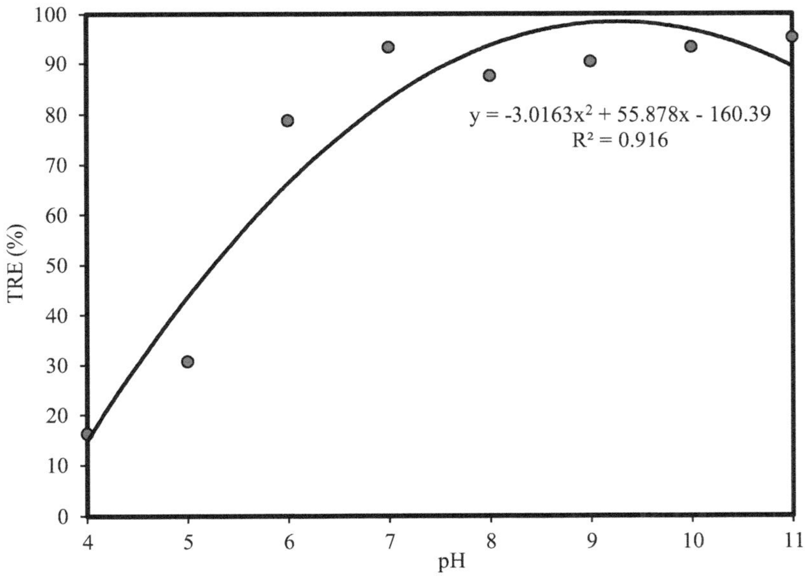

Fig. 2 Correlation between turbidity removal efficiency (TRE) and pH at chitosan dose of 80 mg/L

The high coagulation performance of chitosan at pH 8.3 reflects a dual mechanism: partial charge neutralization through residual protonated amine groups (–NH_3^+) and polymer bridging via hydrogen bonding and Van der Waals interactions. The resulting flocs were compact and crystalline, consistent with previous reports that electrostatic neutralization dominates under acidic conditions, whereas bridging mechanisms prevail under neutral to alkaline pH [13].

CCD-based optimization outperformed the traditional experimental approach, achieving higher removal efficiencies for both organic and inorganic pollutants (Fig. 3). This design also produced sludge with higher protein (up to 55 mg/g) and mineral content, but lower moisture, enhancing its valorization potential as feed or fertilizer (Fig. 4). Turbidity removal further correlated with the crude protein/Ca^{2+} ratio, with efficiencies maximized at ratios above 4.5 (Fig. 5), underscoring the critical role of proteinaceous compounds in floc formation [13].

From an environmental perspective, chitosan flocs are non-toxic, biodegradable, and safer than metal-based sludges, minimizing risks of heavy metal accumulation [9]. Nonetheless, as chitin wastewater may contain residual antibiotics or heavy metals, further risk assessment is required before recommending reuse, which should initially exclude direct human-related applications and be limited to safer, indirect uses.

Overall, the optimized conditions (pH 8.3, 77.5 mg/L) achieved superior pollutant removal while producing reusable, nutrient-rich sludge, confirming by-product chitosan as a cost-effective and eco-friendly alternative for sustainable chitin wastewater treatment.

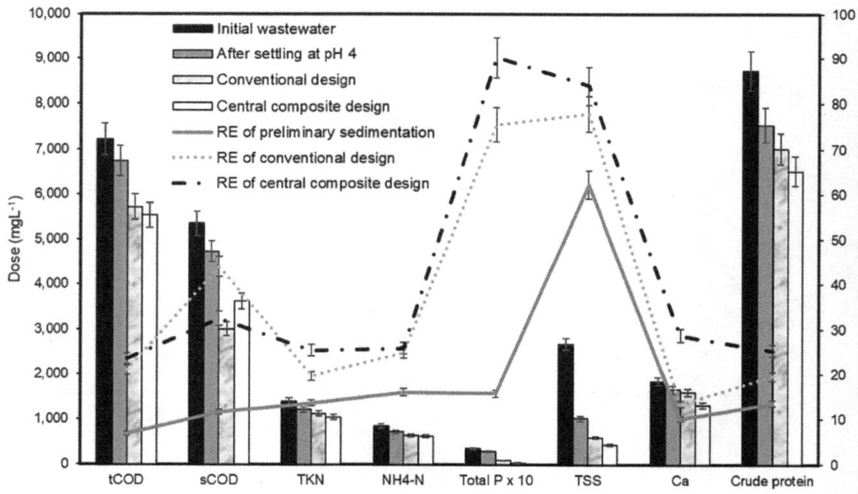

Fig. 3 Changes of characteristics of wastewater after preliminary sedimentation and coagulation at optimum experimental conditions (left axis) and removal efficiencies (right axis)

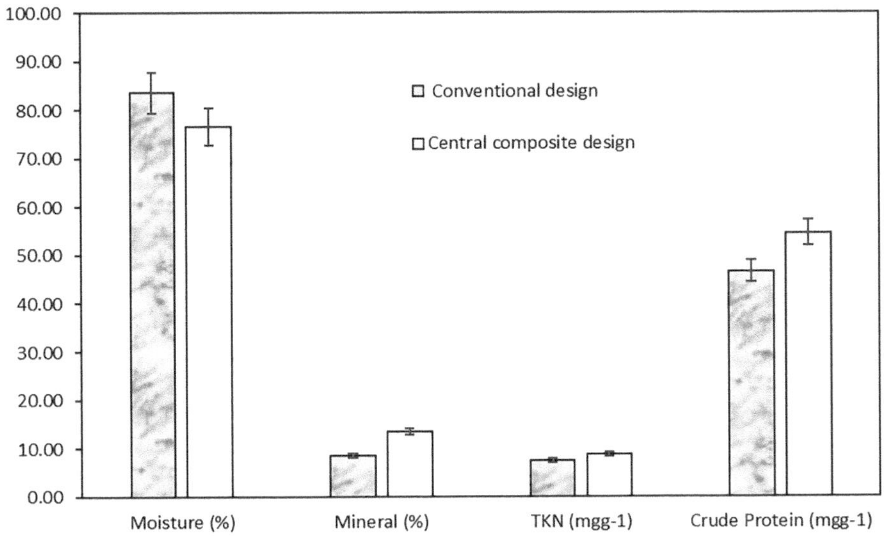

Fig. 4 Basic components of post-treatment sediment with two experimental designs

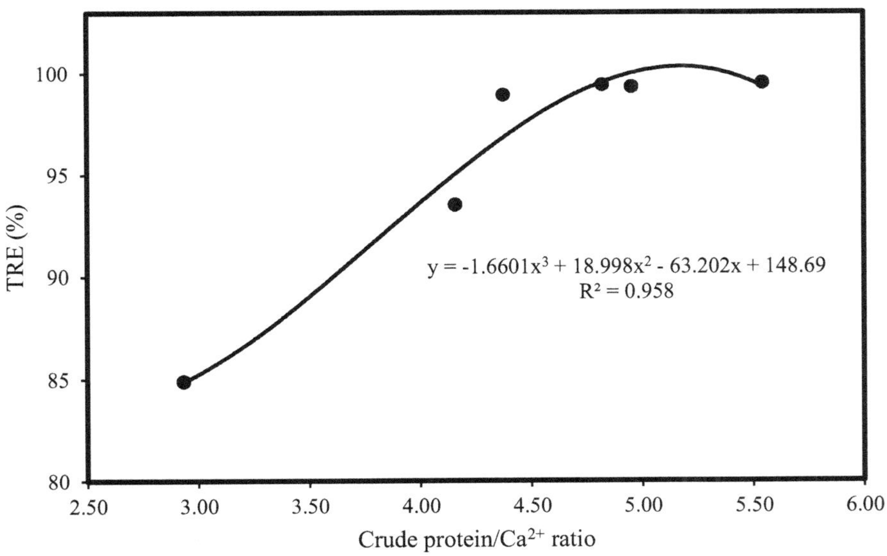

Fig. 5 Effects of ratio of crude protein and calcium ion on turbidity removal efficiencies

4 Conclusion

This study demonstrated that coagulation–flocculation is an effective pre-treatment for CPW, with by-product chitosan showing superior performance compared to PAC, PAA, and their combinations. Under optimized conditions (pH 8.3, 77.5 mg/L), chitosan achieved nearly complete turbidity and TSS removal, alongside significant reductions in COD, nutrients, and calcium. Beyond treatment efficiency, chitosan generated nutrient-rich, biodegradable sludge with potential for valorization, unlike conventional metal-based coagulants that raise sustainability concerns. The application of Box–Wilson design ensured robust optimization, capturing interactive effects and providing reliable operational guidance. Overall, these findings highlight chitosan's dual role as both a coagulant and a pathway toward resource recovery in sustainable wastewater management for the chitin industry.

5 Biofilm-Based Biological Treatment Under Saline Conditions

5.1 Introduction

While coagulation–flocculation is effective for reducing suspended solids and turbidity, it cannot adequately address the extreme pH, high COD, nutrient loads, and salinity characteristic of chitin production wastewater (CPW). Such conditions impair conventional activated sludge due to osmotic stress, enzyme inhibition, and floc destabilization [14].

To overcome these limitations, biofilm-based biological systems have emerged as resilient alternatives, as attached microbial consortia improve biomass retention, nutrient uptake, and tolerance to saline stress [15]. Carriers such as Stick-bed BioFix and Swim-bed BioFringe further enhance performance through strong microbial adhesion, efficient oxygen transfer, and improved mass diffusion, enabling COD removals of 90–95% even under salinity up to 18,000 mg/L [16]. However, nitrogen and phosphorus removal efficiencies typically decline as salinity exceeds 10,000 mg/L due to inhibited nitrification and phosphorus uptake. Importantly, biofilm systems also facilitate partial calcium removal, alleviating salinity stress and further improving effluent quality. These findings highlight biofilm reactors as resilient and adaptable systems capable of treating the multifactorial challenges of chitin wastewater, though optimization remains necessary to enhance nutrient removal under high-saline conditions.

Given the increasing environmental pressures on the seafood processing industry, developing sustainable treatment strategies for chitin wastewater is imperative. By combining pre-treatment via coagulation–flocculation with biofilm-based biological processes, a holistic and synergistic treatment train can be established. Coagulation efficiently reduces particulate matter and turbidity, thereby mitigating fouling and salinity stress on subsequent biological stages. Biofilm reactors then provide the robustness required to degrade residual organics and nutrients under saline conditions. Together, these technologies form a complementary framework capable of achieving high pollutant removal efficiencies, sludge valorization, and resource recovery, while also supporting the circular economy goals of the chitin industry.

This section therefore emphasizes the role of biofilm-based biological treatment in addressing the unique challenges of saline, nutrient-rich chitin wastewater. It highlights both the potential of innovative carrier systems in enhancing treatment performance and the limitations that remain in achieving complete nutrient removal at high salinity. Ultimately, the integration of coagulation–flocculation with biofilm reactors represents a promising path toward scalable, cost-effective, and environmentally sustainable wastewater management solutions tailored to the chitin sector.

5.2 *Material and Methods*

5.2.1 Wastewater Characteristics and Preparation

Synthetic CPW was prepared in a controlled laboratory setting to ensure reproducibility and precise control of composition, while closely mirroring industrial processes. Shrimp head and shell residues were first demineralized with 4% (v/v) hydrochloric acid for 24 h at ambient temperature and rinsed thoroughly with deionized water. This was followed by deproteinization using 5% (w/v) sodium hydroxide under reflux for 24 h. The treated solids were then washed and dried at 105 °C to a constant weight. Wastewater generated from both acid and alkaline stages was pooled, resulting in wide pH variation (2.7–12.3), which was subsequently adjusted to 7.5 ± 0.5 with 1 N NaOH or H_2SO_4 prior to biological treatment. The synthetic wastewater exhibited average characteristics of COD 3138 ± 786 mg/L, TN 501 ± 101 mg/L, TP 41 ± 13 mg/L, hardness 565 ± 189 mg/L, and salinity 4000–6000 mg/L, reflecting the complex and saline nature of real chitin effluents.

This controlled composition provided consistent experimental conditions while retaining industrial relevance, thereby enabling systematic evaluation of biofilm performance under defined stressors and supporting the development of scalable, sustainable treatment strategies.

5.2.2 System Configuration

The laboratory system consisted of three sequential tanks (anaerobic, anoxic, aerobic), each 160 × 115 × 640 mm (10 L), connected in series with up-flow configuration (Fig. 6). Anaerobic and anoxic tanks were packed with Stick-bed BioFix (480 cm^2 surface; 4.8 m^2/m^3), while the aerobic tank contained Swim-bed BioFringe media (450 cm^2; 4.5 m^2/m^3).

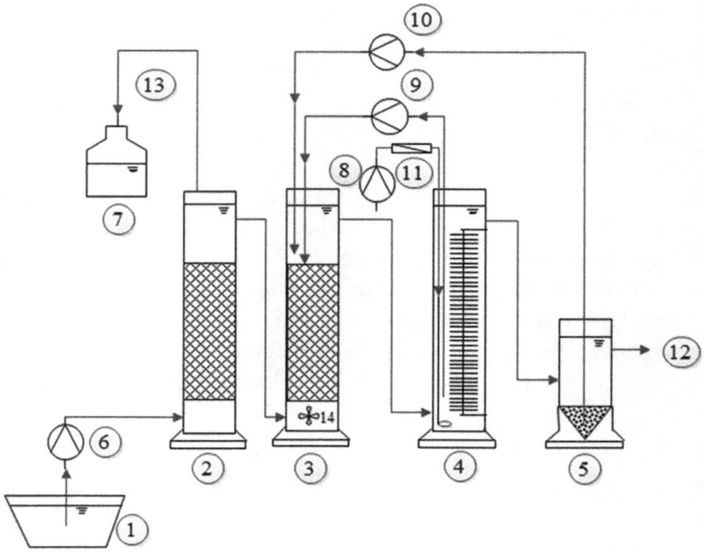

Fig. 6 Diagram of the experimental apparatus. 1—Wastewater feeding tank, 2—Biofix anaerobic compartment, 3—Biofix anoxic compartment, 4—Biofringe aerobic compartment, 5—Settling compartment, 6—Feeding pump, 7—Glass water trap, 8—Air compressor, 9—Recirculation pump, 10—Sediment circulation pump, 11—Flow control valve, 12—Effluent pipe, 13—Gas collecting pipe, 14—Stirrer

The anoxic tank was gently stirred (15–25 rpm), and the aerobic tank was aerated (1.0–1.5 L/min) to support nitrification and COD removal. An internal recycle (300%) from aerobic to anoxic tank provided nitrate for denitrification. DO concentrations were maintained at 0.10 ± 0.00 (anaerobic), 0.35 ± 0.11 (anoxic), and 3.15 ± 0.21 mg/L (aerobic). Temperature was ambient (26 ± 3 °C). Effluent passed through a sedimentation tank; settled sludge was returned to the aerobic tank.

5.2.3 Seed Sludge and Acclimatization

Seed sludge used in this study was sourced from the Vissan food processing plant in Ho Chi Minh City, Vietnam. The inoculum comprised two distinct fractions: black granular anaerobic sludge with particle sizes ranging from 2 to 3 mm, and fine brown sludge collected from anoxic and aerobic treatment tanks. Prior to use, the biomass was subjected to a 60-day acclimation process designed to enhance its tolerance to saline conditions characteristic of chitin production wastewater. During this period, the organic loading rate (OLR) was gradually increased from 1.5 to 2.5 kg $COD/m^3/d$ in a stepwise manner, allowing the microbial community to progressively adapt to both elevated salinity and organic stress. This acclimation strategy not only stabilized microbial activity but also promoted the enrichment of salt-tolerant and functionally diverse microbial populations, thereby ensuring reliable performance of the biofilm-based treatment system under saline and nutrient-rich wastewater conditions.

5.2.4 Operational Conditions

Following the acclimation phase, the biofilm system was subjected to five sequential operational conditions designed to evaluate both treatment performance and the influence of salinity stress. Hydraulic retention time (HRT) and influent flow rate were systematically varied to simulate different organic loading scenarios, thereby providing insights into the stability of the system under changing hydraulic regimes. In parallel, salinity was progressively increased through controlled NaCl supplementation, establishing stepwise gradients that reflect the fluctuating salt concentrations commonly encountered in real chitin production wastewater. Such an approach allowed a systematic assessment of the biofilm reactor's capacity to maintain pollutant removal efficiency under variable and challenging conditions. The detailed parameters defining these experimental periods are summarized in Table 4.

5.2.5 Analytical Methods

Chemical oxygen demand (COD), total Kjeldahl nitrogen (TKN), ammonium nitrogen (NH_4^+–N), total phosphorus (TP), calcium ion (Ca^{2+}), total suspended solids (TSS), and pH were analyzed following the Standard Methods for the Examination of Water and Wastewater [10]. COD (5220 C), TKN (4500-Norg C), NH_4^+–N (4500-NH_3 C), TP (4500-PE), Ca^{2+} (2340 A), TSS (2540 D), and pH (4500-H^+B) were determined according to the respective methods. Analytical instruments included a COD reactor (HI839800–02, HANNA Instruments, Romania), a UV-Vis spectrophotometer (DR5000–03, HACH Company, USA), a Kjeldahl digestion system

Table 4 Various operational conditions for the lab-scale system

	Period 1	Period 2	Period 3	Period 4	Period 5
Pre-treatment	Without	Without	Without	With	With
Flow rate (L/d)	12.96	14.4	16.4	16.4	16.4
HRT (h)	18.4	16.7	14.6	14.6	14.6
Organic loading rate (kg COD/m^3/d)	2.78 ± 0.36	4.26 ± 0.94	6.07 ± 1.09	5.92 ± 0.43	5.84 ± 0.79
Nitrogen loading rate (kg N/m^3/d)	0.57 ± 0.03	0.58 ± 0.13	0.87 ± 0.13	0.83 ± 0.08	1.03 ± 0.10
Phosphorus loading rate (kg P/m^3/d)	0.05 ± 0.00	0.07 ± 0.01	0.06 ± 0.03	0.03 ± 0.01	0.04 ± 0.01
Salinity (mg/L)	6000	10,000	12,000	14,000	18,000

Data represent mean ± SD from three replicates (n = 3)

(Behr Labor Technik, Germany), and a pH meter (SevenEasy™ S20, Mettler Toledo, Switzerland).

All experimental data were collected in triplicate (n = 3) and are reported as mean ± standard deviation (SD). While no ANOVA was performed, the consistency of replicates and the use of response surface modeling support the robustness of the results.

5.3 Results and Discussion

5.3.1 Effect of Operational Conditions on Pollutant Removal

Subsequent to acclimatization, the biofilm system was systematically operated under varying OLRs and salinity conditions to assess treatment effectiveness. Influent COD increased from 2137 ± 278 mg/L in Period 1 to 3691 ± 661 mg/L in Period 3, while HRT was reduced from 18.4 to 14.6 h. Salinity was gradually raised from 6000 to 12,000 mg/L to minimize osmotic shock. After approximately 100 days, influent COD was intentionally elevated to simulate high-strength CPW and test system stability under intensified stress (Fig. 7).

Effluent COD remained below the discharge limit of 150 mg/L in Periods 1 (121 mg/L) and 2 (72 mg/L), but exceeded the threshold in Periods 3 (175 mg/L) and 5 (235 mg/L), demonstrating the inhibitory effects of OLRs >5 kg COD/m^3/d combined with salinity ≥12,000 mg/L. In contrast, Period 4 (147 mg/L) remained compliant due to chitosan pre-treatment, which reduced organic load entering the

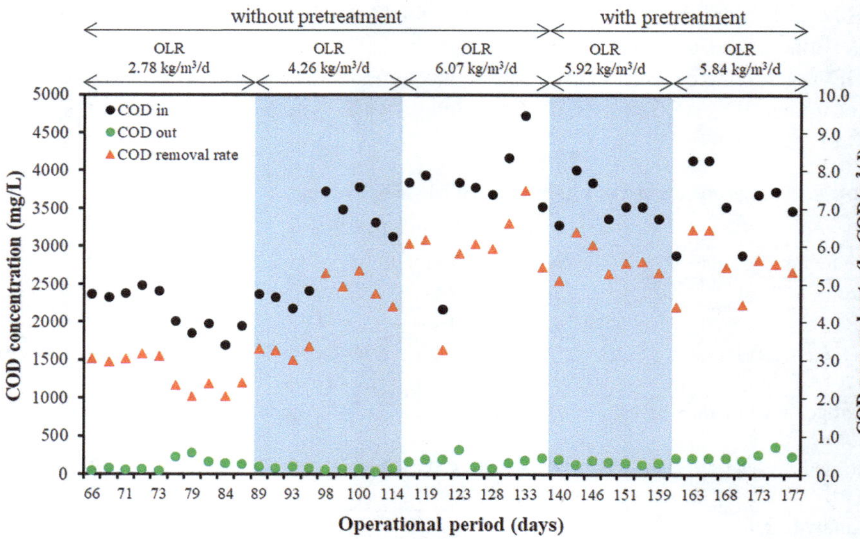

Fig. 7 Variation of COD concentration of the combined system at different operational conditions

Table 5 Comparison of effluent quality with QCVN 11-MT:2015/BTNMT (Column B—Seafood Processing Wastewater)

Parameter	Period 1	Period 2	Period 3	Period 4	Period 5	QCVN 11-MT:2015/ BTNMT (Column B)
COD (mg/L)	121 ± 80	72 ± 21	175 ± 68	147 ± 19	235 ± 54	150
Total N (mg/L)	89 ± 6	86 ± 21	259 ± 47	192 ± 26	479 ± 63	60
Total P (mg/L)	14 ± 4	10 ± 3	10 ± 6	5 ± 0	7 ± 3	20
Ca^{2+} (mg/L)	183 ± 28	258 ± 52	320 ± 96	271 ± 88	490 ± 125	Not regulated

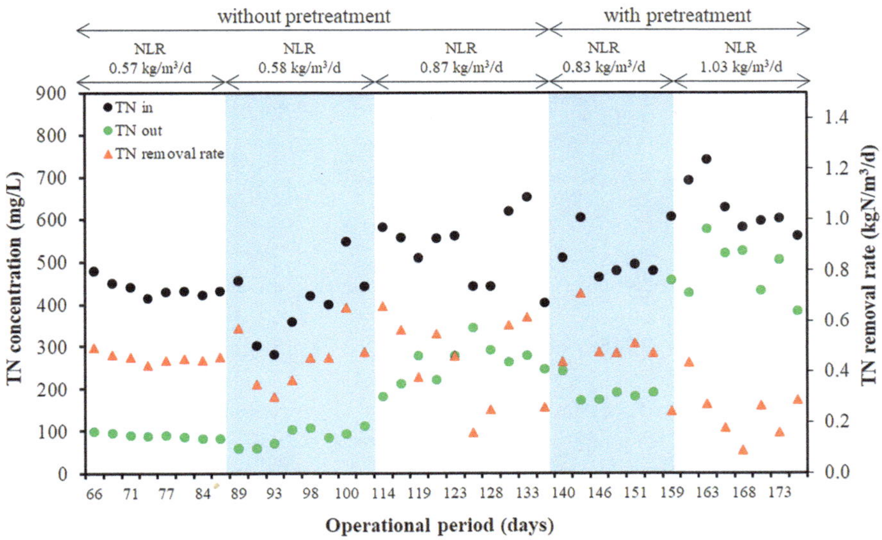

Fig. 8 Variation of TN concentration of the combined system at different operational conditions

biofilm reactor (Table 5). Overall COD removal was consistently high (95 ± 2%), with removal rates of 2.63 ± 0.45, 4.16 ± 0.96, and 5.78 ± 1.10 kg COD/m^3/d for Periods 1–3, respectively. Contributions by process units were 82 ± 4% (anaerobic), 47 ± 5% (anoxic), and 34 ± 3% (aerobic), highlighting the dominant role of the anaerobic stage in organic degradation (Fig. 7).

Nitrogen removal proved more sensitive to stress [17]. At moderate salinities (6000–12,000 mg/L), effluent TN decreased to 89 mg/L (Period 1) and 86 mg/L (Period 2), yielding stable removal rates of 0.45 ± 0.08 kg N/m^3/d with around 79% efficiency (Fig. 8). However, TN removal sharply declined under elevated salinity (>12,000 mg/L), with effluent values reaching 259 mg/L in Period 3 and 479 mg/L in Period 5, four to eight times above the discharge limit. Even with chitosan pre-treatment stabilizing OLRs, nitrogen removal rates fell to 0.24 ± 0.10 kg N/m^3/d at 18,000 mg/L. Compared to reported MBBR-based anammox enrichment (0.20–0.30 kg N/m^3/d) [18], the biofilm system in this study

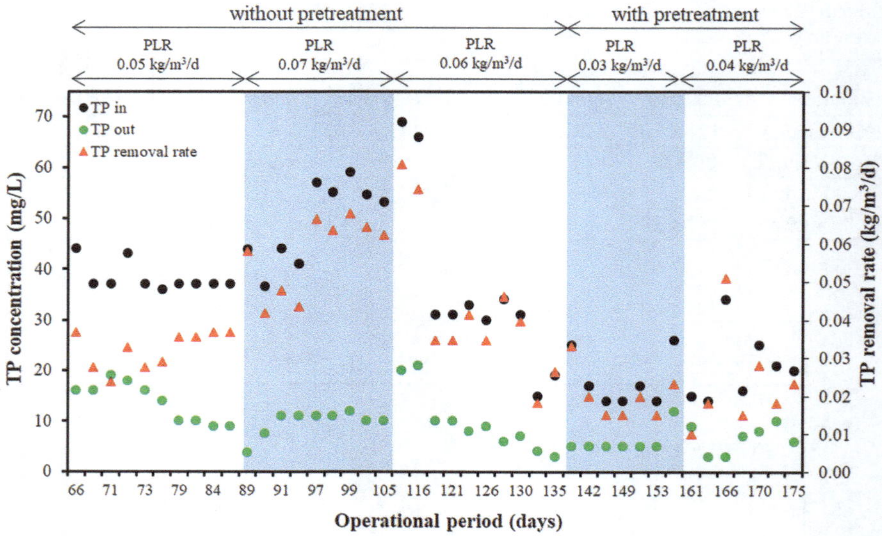

Fig. 9 Variation of TP concentration of the combined system at different operational conditions

demonstrated slightly higher performance, though clearly constrained under high salinity stress.

Phosphorus removal was more robust. Effluent TP consistently met the discharge limit (20 mg/L), reaching as low as 5 mg/L in Period 4 due to the combined effects of pre-treatment and biological uptake. TP removal rates increased moderately from 0.032 ± 0.005 to 0.057 ± 0.010 kg P/m^3/d as loading intensified, but declined by approximately 1.5-fold at salinity >14,000 mg/L (Fig. 9). This reduction suggests that both microbial assimilation and chemical precipitation (e.g., $Ca_3(PO_4)_2$ formation) were impaired under saline stress [19].

5.3.2 Effect of Salinity on Treatment Efficiency

Escalation of salinity to 18,000 mg/L markedly impaired nitrogen and phosphorus removal, although COD removal remained comparatively stable, decreasing by only ~6% relative to Period 3. In Period 5, the nitrogen loading rate rose to 1.03 ± 0.10 kg N/m^3/d, yet TN removal efficiency dropped by approximately 2.5-fold compared to Period 3. This sharp decline contrasts with earlier studies reporting tolerance up to 37,700 mg/L [20], suggesting that the combined stress of elevated nitrogen loading and high salinity critically constrained performance. Indeed, the nitrogen loading rate applied by She et al. (2016) was nearly one-tenth of that in the present study, highlighting the decisive role of loading intensity [21]. The suppression of microbial activity under these conditions aligns with Zhang et al. (2019), who attributed salinity-induced inhibition to severe osmotic pressure imbalances

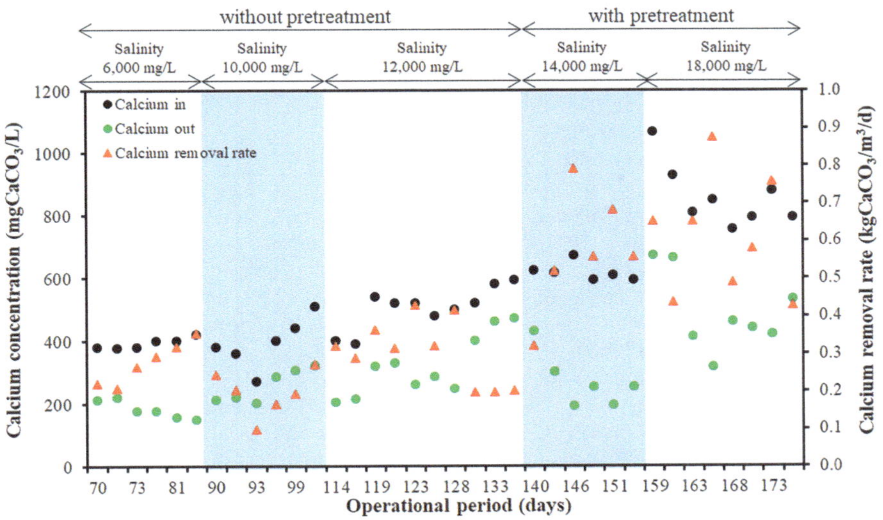

Fig. 10 Variation of Calcium concentration of the combined system at different operational conditions

that compromise bacterial cell viability and disrupt nitrogen transformation pathways [22].

Calcium accumulation further exacerbated microbial inhibition. As salinity increased, $CaCO_3$ concentrations rose from 393 ± 17 to 859 ± 101 mg/L (Fig. 10), with loading rates escalating from 0.20 to 0.61 kg $CaCO_3/m^3/d$. Elevated calcium is known to interfere with cell membrane integrity and enzymatic functions, thereby aggravating the reduction in nitrogen and phosphorus removal [23]. Nonetheless, the biofilm system consistently removed around 50% of influent calcium, underscoring its capacity to partially mitigate salinity-related stress through both physicochemical and biological mechanisms.

Overall, these findings demonstrate that the anaerobic–anoxic–aerobic biofilm configuration, employing structured carriers such as Stick-bed BioFix and Swim-bed BioFringe, effectively enhanced biomass retention and stability. This structural advantage contributed to superior resilience and pollutant removal when compared with conventional suspended-growth systems (Table 6).

5.4 Conclusion

The biofilm-based treatment system demonstrated robust performance in removing organic matter and nutrients from saline chitin production wastewater. COD removal was consistently high (95 ± 2%), with effluent concentrations meeting discharge standards under moderate OLR and salinity. However, combined stress from high

Table 6 Comparison of treatment methods for wastewater containing high salinity

Treatment methods	Operational conditions	Removal rate	References
Sequencing batch reactor (SBR)	HRT: 14.9 h COD: 200 mg/L TN: 63 mg/L Salinity: 5000–37,700 mg/L	COD: 0.01 kg/m^3/d TN: 0.004 kg/m^3/d	[14]
Fixed bed reactors (FBRs)	HRT: 9.6 h NH_4^+-N: 400 mg/L Salinity: 4000–16,000 mg/L	NH_4^+-N: 0.97 kg/m^3/d at 4000 mg/L, and 0.70 kg/m^3/d at 16,000 mg/L	[23]
Bioreactor using purple phototrophic bacteria	HRT: 25 h COD: 868 mg/L TN: 71 mg/L TP: 12 mg/L Salinity: 28,000 mg/L	COD: 0.71 kg/m^3/d TN: 0.04 kg/m^3/d TP: 0.006 kg/m^3/d	[17]
Airlift SBR	HRT: 8 h COD: 800 mg/L TN: 700 mg/L Salinity: 30,000 mg/L	COD: 2.18 kg/m^3/d TN: 1.53 kg/m^3/d	[24]
Biofilm reactor system	HRT: 14.6–18.4 h COD: 1692–4720 mg/L TN: 280–741 mg/L TP: 15–69 mg/L Salinity: 6000–18,000 mg/L	COD: 5.78 kg/m^3/d TN: 0.51 kg/m^3/d TP: 0.057 kg/m^3/d	This study

organic loading (>5 kg COD/m^3/d) and salinity above 12,000 mg/L led to reduced COD and nitrogen removal efficiencies. Nitrogen removal was particularly sensitive, declining to 0.24 ± 0.10 kg N/m^3/d at 18,000 mg/L, whereas phosphorus removal remained largely stable, supported by both biological uptake and chemical precipitation. Pre-treatment with chitosan effectively alleviated organic load, improving subsequent biofilm performance. Compared to conventional single-stage biofilm systems, the hybrid configuration provided superior resilience under saline stress. These findings highlight the potential of integrated biofilm-based processes, combined with natural coagulant pre-treatment, as a sustainable and scalable solution for managing high-strength, saline industrial wastewaters.

Glossary

Coagulation–Flocculation A physicochemical treatment process in which coagulants (such as metal salts or natural polymers) are added to wastewater to destabilize colloidal and suspended particles. This destabilization reduces electrostatic repulsion, allowing particles to aggregate. The subsequent flocculation step involves gentle mixing that promotes the collision and binding of destabilized particles into larger flocs, which can then be removed by sedimentation or filtration. It is widely applied as a pre-treatment to reduce turbidity, suspended solids, and organic matter.

AAO (Anaerobic–Anoxic–Aerobic process) A biological nutrient removal process that integrates three zones in sequence: anaerobic, anoxic, and aerobic. In the anaerobic stage, microorganisms degrade organic matter and release biogas while phosphorus is released. The anoxic stage supports denitrification, where nitrate and nitrite are reduced to nitrogen gas, and phosphorus compounds are transformed into more degradable forms. In the aerobic stage, organic matter is oxidized, ammonium is nitrified to nitrate, and phosphorus-accumulating organisms take up phosphate. Together, these steps enable simultaneous removal of organic matter, nitrogen, and phosphorus from wastewater.

OLR (Organic Loading Rate) A critical operational parameter that represents the amount of organic matter, typically expressed as chemical oxygen demand (COD), applied per unit reactor volume per day (kg COD/m^3·d). OLR directly influences microbial activity and system stability. Excessive OLR can lead to biomass washout, oxygen depletion, or incomplete degradation, whereas maintaining an optimal OLR ensures effective pollutant removal and sustainable operation of biological treatment systems.

NLR (Nitrogen Loading Rate) An operational parameter that quantifies the amount of nitrogen, usually expressed as total Kjeldahl nitrogen, ammonium, or total nitrogen, applied per unit reactor volume per day (kg N/m^3·d). NLR is used to assess the nitrogen removal capacity of biological systems, particularly in nitrification–denitrification and AAO configurations. High NLR may overload microbial communities and reduce treatment efficiency, while optimized NLR supports stable and effective nutrient removal.

References

1. Abuzar HR, Sharif MK, Sharif R, Arshad A, Rehman W, Ashraf A, Karim KA, Awan HR, Khalid W (2023) Potential industrial and nutritional applications of shrimp by-products: a review. Int J Food Prop 26(2):3407–3432
2. XiangGang Z, AiMei Z, XiaoXia L, YongQuan C, GongMing Y (2009) Comparative study of chemical compositions of white shrimp head and shell. Mod Food Sci Technol
3. Stephen J, Lekshmi M, Nayak BB, Kumar S (2022) Advances in fisheries biotechnology. Springer, pp 495–521
4. Tran NVN, Yu QJ, Nguyen TP, Wang S-L (2020) Coagulation of chitin production wastewater from shrimp scraps with by-product chitosan and chemical coagulants. Polymers 12(3):607
5. Lee E, Jung K, Jayakumar R (2024) The water, energy, and food security nexus in Asia and the Pacific: East and Southeast Asia. Springer Nature, Cham, pp 29–61
6. Trang ST (2009) Evaluation of product quality and environmental efficiency of modified chitin production process combining with enzyme treatment. J Fish Sci Technol 04:17–23
7. Aragaw TA, Bogale FM (2023) Role of coagulation/flocculation as a pretreatment option to reduce colloidal/bio-colloidal fouling in tertiary filtration of textile wastewater: a review and future outlooks. Front Environ Sci 11:1142227
8. Zeng D, Wu J, Kennedy JF (2008) Application of a chitosan flocculant to water treatment. Carbohydr Polym 71(1):135–139
9. Koumentakou I, Meretoudi A, Emmanouil C, Kyzas GZ (2025) Environmental toxicity and biodegradation of chitosan derivatives: a comprehensive review. J Ind Eng Chem 146:70–86

10. APHA (2025) Standard methods for the examination of water and wastewater. American Public Health Association, Washington, DC
11. FAO (2003) Agriculture organization of the United Nations. Food energy: methods of analysis and conversion factors. Report of a technical workshop. Food Nutr. Pap. 77
12. Saïed N, Aïder M (2014) Zeta potential and turbidimetry analyzes for the evaluation of chitosan/phytic acid complex formation. J Food Res 3(2):71
13. Sillanpää M, Ncibi MC, Matilainen A, Vepsäläinen M (2018) Removal of natural organic matter in drinking water treatment by coagulation: a comprehensive review. Chemosphere 190:54–71
14. Zhao Y, Zhuang X, Ahmad S, Sung S, Ni S-Q (2020) Biotreatment of high-salinity wastewater: current methods and future directions. World J Microbiol Biotechnol 36:1–11
15. Zhao C, Zhang W, Guo Y, Zhang M, Han F, Lei J, Zhou W (2025) Flocculent sludge outperforms filler biofilm for high salinity oilfield produced water treatment: performance, metabolic pathways, and microbial communities. J Hazard Mater 492:138217
16. Nguyen TP, Vo TKQ, Tran NVN, Tran NVN (2024) Biofilm reactors filled with Stick-bed Biofix and Swim-bed Biofringe biomass carriers in treating chitin production wastewater containing high salinity. Environ Technol 45(22):4457–4466
17. Ramaswami S, Uddin FMJ, Behrendt J, Otterpohl R (2019) High-rate nitrification of saline wastewaters using fixed-bed reactors. J Environ Manag 243:444–452
18. Zekker I, Rikmann E, Tenno T, Saluste A, Tomingas M, Menert A, Loorits L, Lemmiksoo V, Tenno T (2012) Achieving nitritation and anammox enrichment in a single moving-bed biofilm reactor treating reject water. Environ Technol 33(6):703–710
19. Metcalf L, Eddy HP, Tchobanoglous G, Burton FL, Tsuchihashi R, Stensel HD (2014) Wastewater engineering: treatment and resource recovery. McGraw-Hill Professional, New York
20. Giustinianovich EA, Aspé ER, Huiliñir CE, Roeckel MD (2014) Simultaneous C and N removal from saline salmon effluents in filter reactors comprising anoxic-anaerobic-aerobic processes: effect of recycle ratio. J Environ Sci Heal A 49(5):584–592
21. She Z, Zhao L, Zhang X, Jin C, Guo L, Yang S, Zhao Y, Gao M (2016) Partial nitrification and denitrification in a sequencing batch reactor treating high-salinity wastewater. Chem Eng J 288:207–215
22. Zhang Y, Kuroda M, Arai S, Kato F, Inoue D, Ike M (2019) Biological treatment of selenate-containing saline wastewater by activated sludge under oxygen-limiting conditions. Water Res 154:327–335
23. Li Y, Zou J, Zhang L, Sun J (2014) Aerobic granular sludge for simultaneous accumulation of mineral phosphorus and removal of nitrogen via nitrite in wastewater. Bioresour Technol 154:178–184
24. Hülsen T, Hsieh K, Batstone D (2019) Saline wastewater treatment with purple phototrophic bacteria. Water Res 160:259–267

Tan Phong Nguyen

Nguyen Van Nhi Tran

Kolibri 360 GmbH
Ferdinand-Jühlke-Straße 7
99095 Erfurt, Deutschland
produktsicherheit@kolibri360.de